Smart Applications of Artificial Intelligence and Big Data

Smart Applications of Artificial Intelligence and Big Data covers a wide range of topics related to AI and big data, including machine learning, deep learning, natural language processing, computer vision, data analytics, and data mining. It focuses on the integration of these technologies to create smart applications, such as intelligent transportation systems, smart healthcare, smart cities, and smart grids.

This book comprises 21 chapters, each providing technical details pertaining to research, practical examples, and case studies to help readers understand the real-world applications of AI and big data technologies. The book also highlights cutting-edge research on AI and big data, including novel algorithms, tools, and techniques. It discusses the challenges and opportunities of using AI and big data to develop smart applications and provides recommendations for the development of responsible and transparent AI-based systems. This book is a valuable resource for researchers and professionals looking to stay up-to-date with the latest advancements in AI and big data and how they can be applied to solve real-world challenges.

Smart Applications of Artificial Intelligence and Big Data

Edited by
Salwa Belaqziz, Salma El Hajjami,
Hicham Amellal, Redouan Lahmyed,
Lahcen Koutti, Boulouz Abdellah,
and Inam Ullah Khan

CRC Press
Taylor & Francis Group
Boca Raton London New York

CRC Press is an imprint of the
Taylor & Francis Group, an **informa** business

Designed cover image: Shutterstock image 2322632583

First edition published 2025
by CRC Press
2385 NW Executive Center Drive, Suite 320, Boca Raton FL 33431

and by CRC Press
4 Park Square, Milton Park, Abingdon, Oxon, OX14 4RN

CRC Press is an imprint of Taylor & Francis Group, LLC

ISBN: 9781032664279 (hbk)
ISBN: 9781032664286 (pbk)
ISBN: 9781032664293 (ebk)

DOI: 10.1201/9781032664293

Typeset in Sabon
by KnowledgeWorks Global Ltd.

Contents

Preface

In the ever-evolving landscape of technology, the emergence of advanced computing systems and smart applications marks a transformative era in our understanding of artificial intelligence (AI) and machine learning. This edited volume unfolds a tapestry of cutting-edge research and profound insights across diverse domains within the boundless realm of AI and smart applications. As we stand at the crossroads of technological evolution, the chapters compiled herein traverse the trajectory of AI trends and challenges, showcasing the profound impact on communication, healthcare, security, and more.

This edited book, titled *Smart Applications of Artificial Intelligence and Big Data*, presents machine learning-based technologies divided into three parts: Artificial Intelligence, Smart Applications, and Internet of Things (IoT) and Big Data. The first part introduces the vast potential of AI in various fields, setting the stage for innovative applications and breakthroughs. The second part shifts the focus to practical implementations of smart applications. The third part delves into the integration of AI with IoT, creating intelligent solutions for smart cities. It also addresses the transformative potential of AI in managing and interpreting big data.

This edited compilation endeavors to provide a comprehensive exploration of the intricate interplay between AI and the multifaceted domains it influences. The contributors to this volume offer insights, analyses, and solutions that contribute to the ongoing narrative of this transformative era. Each chapter represents a unique contribution, a stepping stone in the collective journey toward a future where intelligent systems enhance our lives and redefine the boundaries of what is possible. The chapters herein provide not only knowledge but also a compass for navigating the vast and dynamic landscape of AI and machine intelligence.

This book is divided into three parts. The first part introduces AI and its applications, addressing the challenges and trends of future developments. The second part spotlights smart applications of AI in various fields. The third part explores the integration of AI with IoT and also examines the transformative impact of AI on big data. Together, these parts present a holistic view of the current state and future potential of AI.

PART I: ARTIFICIAL INTELLIGENCE (AI)

- **Chapter 1** explores the efficiency and cost-effectiveness of container live migration within cloud environments, specifically focusing on implementation within AWS.
- **Chapter 2** introduces a deep learning framework designed for the classification and localization of gastrointestinal abnormalities in wireless capsule endoscopy (WCE) images.
- **Chapter 3** details the application of U-Net architecture and transfer learning techniques for the segmentation of breast cancer in medical imaging.
- **Chapter 4** discusses the transformation of GPU-accelerated machine learning environments through the use of Docker cloud containers for improved efficiency and scalability.
- **Chapter 5** reviews various image-based deep learning methodologies applied to estimate atmospheric visibility, highlighting the latest advancements and applications.
- **Chapter 6** evaluates the effectiveness of interpretable machine learning algorithms in detecting distributed denial of service (DDoS) attacks to enhance cloud security.
- **Chapter 7** explores a novel method for classifying gastrointestinal bleeding in WCE images by integrating a stabilizing block with the Xception model.
- **Chapter 8** presents an advanced approach combining convolutional neural networks (CNNs), transfer learning, and ensemble learning for the recognition and classification of lung tumors in CT scan images.
- **Chapter 9** introduces a new method for pedestrian detection in intelligent surveillance systems using motion information and GLBP-Color features with SVM classification, tested on the INO Video Analytics dataset.

PART II: SMART APPLICATIONS

- **Chapter 10** investigates the use of ADALINE neural networks to compensate for harmonic currents in shunt active power filters, enhancing power quality.
- **Chapter 11** discusses the implementation of maximum power point tracking (MPPT) and vector control techniques in grid-connected photovoltaic systems.
- **Chapter 12** introduces GeoArgania, a comprehensive geolocation mapping dataset of Argania trees, aimed at supporting ecological and agricultural studies in the Souss region.
- **Chapter 13** proposes a PSM (platform-specific model) for NoSQL key-value databases to improve data handling and efficiency through model programming.

- **Chapter 14** examines the implementation and effectiveness of body temperature screening systems used during the COVID-19 pandemic for public health safety.
- **Chapter 15** details a signal processing system designed for heart rate extraction using plethysmography analysis, developed on the LabVIEW platform.
- **Chapter 16** investigates the temporal impacts of the COVID-19 pandemic through sentiment analysis and topic modeling of public opinion and discourse.

PART III: INTERNET OF THINGS (IoT) AND BIG DATA

- **Chapter 17** explores smart electric vehicle (EV) routing to charging stations aimed at optimizing traffic flow in smart cities, with a case study in Agadir.
- **Chapter 18** presents innovative techniques for dimensionality reduction and spectral unmixing tailored for hyperspectral data analysis to improve data interpretability and accuracy.
- **Chapter 19** introduces a detection system designed to identify counterfeit medicines, leveraging advanced data analysis and machine learning techniques.
- **Chapter 20** discusses an intrusion detection system enabled by software-defined networking (SDN) that utilizes machine learning and neural network schemes for enhanced security.
- **Chapter 21** highlights the importance of customized information security awareness programs in higher education, advocating for tailored approaches to enhance security education and practices.

About the Editors

Dr. Salwa Belaqziz is a Full Professor in Computer Science at the Faculty of Sciences, Ibn Zohr University in Agadir, and an affiliate professor at the Center for Remote Sensing Applications of the Mohammed VI Polytechnic University. She received her Ph.D. degree in Computer Science from Cadi Ayyad University of Marrakech in 2014, which focused on the development of a decision support approach for gravity irrigation systems management, based on the combination of multi-agent modeling, remote sensing, and optimization by evolutionary algorithms. Salwa Belaqziz has worked at the National Institute of Agricultural Research in Toulouse as a postdoctoral researcher, where she participated in the development of an agent-based modeling and simulation platform to assess the environmental, economic, and social impacts of various management strategies and policies regarding the management and uses of water resources. Her main research interests include irrigation scheduling, smart farming, and decision support systems development, based on approaches combining artificial intelligence, big data, data science, agent-based modeling, physics-based modeling, remote sensing, IoT, and optimization algorithms for integrated water resources management. Salwa Belaqziz co-supervises four Ph.D. students and participates in national and international research projects (ASSIWAT, MorSnow, GeanTech, PRIMA, ...). She has published papers in international peer-reviewed journals and participated in several international conferences.

Dr. Salma El Hajjami has been an Assistant Professor and Researcher at the Faculty of Science, Ibn Zohr University, Agadir, Morocco since 2021. She received her Ph.D. in 2021 in Computer Science, at the Laboratory of Artificial Intelligence, Data Science and Emerging Systems from ENSA, Sidi Mohammed Ben Abdellah University, Fez, Morocco. She is a Computer Science Engineer, who graduated in 2015 from the National School of Applied Sciences Fez, Morocco. She has previous expertise acting in the Ministry of Interior in Morocco as a Research and Development Engineer from 2017 to 2021. She is a member of the International Association of Engineers (IAENG) and the International Association of Online

Engineering. Dr. Salma has made contributions in the fields of Social Big Data, Semantics Analytics, Anomaly Detection, and Imbalanced Big Data published at international conferences and journals. Her main research topics are machine learning, deep learning, imbalanced big data, data science, and blockchain. She has served and continues to serve on technical program and organizer committees of several conferences and also as a reviewer of numerous international journals.

Dr. Redouan Lahmyed received the Ph.D. degree in Computer Science from the Faculty of Science, University Ibn Zohr, Agadir, Morocco in 2019. He also received his Master's degree in Computer Science (M.Sc.) from the Department of Studies in Computer Science at the University of Mysore, Karnataka, India. His research interests include image processing and computer vision. Apart from his experience with teaching and working with researchers from different countries, he published and served as a reviewer at numerous international conferences and journals.

Dr. Hicham Amellal is Assistant Professor, Department of Computer Science Faculty of Sciences at Ibn Zohr University. He holds a doctorate in information security from Mohamed V University, Faculty of Sciences Rabat-Morocco, and computer engineering, laureate of Kazan State Technological University, Russian Federation. Dr. Amellal has different experiences in different regions and various higher educational institutions with diverse academic configurations and in different levels such as International Academy Mohammed VI of Civil Aviation, National School of Applied Sciences al Hoceima, Faculty of Sciences Agdal, and Faculty of Science and Technology Al Hoceima. He is a member of the organizing committee for the workshops "Cyber Crime and Quantum Information," held on October 18–19, 2016, in Rabat, organized by the VSST Association Chapter Morocco, and "Quantum Africa C: Advances in Quantum Sciences," held at the Faculty of Sciences, Rabat, from September 22–26, 2014. His research interests include IT security, quantum cryptography, post-quantum cryptography, and quantum algorithms.

Dr. Koutti Lahcen is a Professor at the Department of Computer Science within the Faculty of Science at Ibn Zohr University in Agadir, Morocco. He obtained his Ph.D. in computational physics from University Paul Verlaine in France in 1999 and his Habilitation degree from Ibn Zohr University in 2010. Prior to his academic career, Dr. Lahcen worked as a software engineer at the multinational company "ALDATA." His research interests revolve around artificial intelligence, computer vision, medical imaging, and feature extraction. Dr. Lahcen is a member of the Computer Systems and Vision Laboratory and has an h-index of 7, having coauthored 48 publications that have received 237 citations. Additionally, he currently manages the AL-KHAWARZIMI project (2020/20).

Dr. Boulouz Abdellah is a university professor in the IT department at the Faculty of Sciences, University Ibn Zohr in Agadir. He holds a Ph.D. in Microelectronics, Microsensors, and Systems, which he obtained from the University of Montpellier 2, IES (Institut d'Électronique et des Systèmes). He also has a Master's degree in Computer Engineering and Open Systems/ Computer Networks (MSIO) from Ecole Centrale Paris (ECP France) and a "Diplôme des Études Approfondies" in Instrumentation Sciences and Physics from the University of Reims Champagne-Ardenne. He worked as a researcher (postdoctoral position) at the École Polytechnique Fédérale EPFL in Lausanne, Switzerland, and the Leibniz-Institut für Festkörper und Werkstoffforschung, IFW Dresden, Germany. He is also the president and founder of the association Sciences Pour Tous and has several publications in peer-reviewed, specialized journals and international conferences.

Dr. Inam Ullah Khan is a visiting researcher at King's College London, UK. He has also been a faculty member at different universities in Pakistan which include Center for Emerging Sciences Engineering & Technology (CESET), Islamabad, Abdul Wali Khan University, Garden Campus, Timergara Campus, and the University of Swat & Shaheed Zulfikar Ali Bhutto Institute of Science and Technology (SZABIST), Islamabad Campus. He completed his Ph.D. in Electronics Engineering from the Department of Electronic Engineering, Isra University, Islamabad Campus, School of Engineering & Applied Sciences (SEAS). He previously completed his M.S. degree in Electronic Engineering at the Department of Electronic Engineering, Isra University, Islamabad Campus, School of Engineering & Applied Sciences (SEAS). His undergraduate degree is the Bachelor of Computer Science from Abdul Wali Khan University Mardan, Pakistan. His Master's thesis is published as a book, *Route Optimization with Ant Colony Optimization (ACO)*, which is available on Amazon. Additionally, he used to teach subjects like computer network security, artificial intelligence, evolutionary computing, professional practice, software engineering, data communication & networks, database, cyber security, visual programming, and introduction to programming. He has authored/coauthored more than 30 research articles in reputable journals, conferences, and book chapters.

List of Contributors

Sadoon Hussein Abdullah
Department of Physics, College of Science, Mosul University
Mosul, Iraq

Rachida Abounacer
Department of Mathematics, Ibn Zohr University
Agadir, Morocco

Abdallah Abrada
Lab of Mathematical Modeling and Economic Calculations, Hassan First University
Settat, Morocco

Karim Afdel
LabSIV, Department of Computer Science, Faculty of Sciences, Ibn Zohr University
Agadir, Morocco

Elhafed Agherrabi
LabIRF-SIC, Department of Mathematics, Faculty of Science, Ibn Zohr University
Agadir, Morocco

Essadki Ahmed
Electrical Engineering Department, ENSAM, Mohammed V University
Rabat, Morocco

Abdullah Akbar
Department of Electrical Engineering, National University of Computer and Emerging Sciences
Peshawar, Pakistan

Mohamed Akouhar
Lab Partial Differential Equations, Algebra and Spectral Geometry, University IBN Tofail
Kenitra, Morocco

Rachid Alaoui
Research Laboratory in Computer Science and Telecommunications (LRIT), Mohammed V University
Rabat, Morocco

Husam Alfergani
Department of Electrical and Computer Engineering, Rowan University
Glassboro, New Jersey, USA

Mubashar Ali
Department of Computer Science Sciences, Shaheed Zulfikar Ali Bhutto Institute of Science and Technology Islamabad
Islamabad, Pakistan

Manuel Arbelo
Universidad de La Laguna, San Cristobal de La Laguna
Tenerife, Spain

Reismary Armas
University Canada West
Vancouver, Canada

Rithika Badam
Stanley College of Engineering and Technology for Women
Hyderabad, India

Noura Bentaher
Department of Computer Science, Ibnou Zohr University
Agadir, Morocco

Amine Bouaouda
Department of Computer Science, Ibn Zohr University
Agadir, Morocco

Taha Bouhsine
IRF-SIC Laboratory, Ibn Zohr University
Agadir, Morocco

Addaali Bouthayna
Laboratory of Systems Engineering and Information Technology (LISTI), National School of Applied Sciences, Ibn Zohr University
Agadir, Morocco

Abderrahim Beni-Hssane
LAROSERI Laboratory, Computer Science Department, University of Chouaib Doukkali, Faculty of Sciences
El Jadida, Morocco

Nidhal Carla Bouaynaya
Department of Electrical and Computer Engineering, Rowan University
Rowan, New Jersey, USA

Enrique Casas-Mas
Universidad de La Laguna, San Cristobal de La Laguna
Tenerife, Spain

Said Charfi
Laboratory of Computer Systems and Vision, Faculty of Science, Ibn Zohr University
Agadir, Morocco

Noureddine Ech-Cherki
Univ Chouaib Doukkali
El Jadida, Morocco

Amine Aaz el Aarab
InterDisciplinary Applied Research Laboratory, International University of Agadir
Agadir, Morocco

Mohamed El Ansari
Informatics and Applications Laboratory, Department of Computer Science, Faculty of Sciences, Moulay Ismail University
Meknes, Morocco

Ibrahim El-Fedany
Higher School of Technology, Meknes, Moulay Ismail University
Meknes, Morocco

Zakaria El Khadiri
Laboratory of Systems Engineering and Information Technology (LISTI), National School of Applied Sciences, Ibn Zohr University
Agadir, Morocco

Ahmed Fouad El Ouafdi
LabSIV, Department of Computer Science, Faculty of Sciences, Ibn Zohr University
Agadir, Morocco

Azeddine Elhassouny
Mohammed V University
Rabat, Morocco

Y. Errami
Univ. Chouaib Doukkali
El Jadida, Morocco

Ahmed Essadki
Electrical engineering department ENSAM, Mohammed V University
Rabat, Morocco

Anass Garbaz
Laboratory of Computer Systems and Vision, Faculty of Science, Ibn Zohr University
Agadir, Morocco

Fatima Guerouate
LASTIMI Laboratory, Superior School of Technologies of Sale, Mohammadia School of Engineering, Mohammed V University
Rabat, Morocco

Mustapha Hankara
LAROSERI Laboratory, Computer Science Department, University of Chouaib Doukkali, Faculty of Sciences
El Jadida, Morocco

Waqar Hussain
Department of Computer Science, Shaheed Zulfikar Ali Bhutto Institute of Science and Technology Islamabad
Islamabad, Pakistan

Soufiane Idbraim
IRF-SIC Laboratory, Ibn Zohr University
Agadir, Morocco

Charles Cliff Johnson
William J. Hughes Technical Center, Federal Aviation Administration
Atlantic City, New Jersey, USA

Younes Kabbadj
Department of Computer Science, Ibn Zohr University
Agadir, Morocco

Younes Karmoude
Ibn Zohr University, Agadir
Souss, Morocco

Mohammed Kasrib
High National School of Artificial Intelligence and Data Science
Taroudant, Morocco

Vijaya Sindhoori Kaza
Stanley College of Engineering and Technology for Women
Hyderabad, India

Zakaria Kerkaou
LabSIV, Department of Computer Science, Faculty of Sciences, Ibn Zohr University
Agadir, Morocco

Driss Kiouach
Laboratory of Applied Physics, Informatics and Statistics (LPAIS), Sidi Mohamed Ben Abdellah University
Fez, Morocco

Lahcen Koutti
LabSIV, Department of Computer Science, Faculty of Sciences, Ibn Zohr University
Agadir, Morocco

Rachid Latif
Laboratory of Systems Engineering and Information Technology LISTI, National School of Applied Sciences, Ibn Zohr University
Agadir, Morocco

Redouan Lahmyed
Department of Computer Science, Faculty of Science, Ibn Zohr University
Agadir, Morocco

Antoine Masse
Collecte Localisation Satellites
Villeneuve, France

Nour-Eddine Moumni
Research Laboratory in Computer Science and Telecommunications (LRIT), Mohammed V University
Rabat, Morocco

Tali Mouna
Electrical Engineering Department, ENSAM, Mohammed V University
Rabat, Morocco

Toufik Mzilia
LAROSERI Laboratory, Computer Science Department, University of Chouaib Doukkali, Faculty of Sciences
El Jadida, Morocco

I. Nassar-Eddine
Univ Chouaib Doukkali
El Jadida, Morocco

Ahmed Sami Nori
Department of Cyber Security, College of Computer Science and Mathematics
Mosul University, Iraq

A. Obbadi
Univ Chouaib Doukkali
El Jadida, Morocco

Kabira Ait Ouadil
IRF-SIC Laboratory, Ibn Zohr University
Agadir, Morocco

Mohamed Ouhssini
LabSIV, Department of Computer Science, University Ibn Zohr
Agadir, Morocco

Yassine Oukdach
LabSIV, Department of Computer Science, Faculty of Sciences, Ibn Zohr University
Agadir, Morocco

Muhammad Allah Rakha
Department of Computer Science, FAST National University of Computer and Emerging Sciences
Peshawar, Pakistan

Amine Saddik
Laboratory of Systems Engineering and Information Technology (LISTI), National School of Applied Sciences, Ibn Zohr University
Agadir, Morocco
Faculty of Applied Sciences, Ibn Zohr University
Ait Melloul, Morocco

S. Sahnoun
Univ Chouaib Doukkali
El Jadida, Morocco

Souad Saidi
Ibn Zohr University, Agadir
Souss, Morocco

Muhammad Saleem
Coopers & McGill
Istanbul, Turkey

Mohamed Ben Salah
Department of Computer Science, Ibnou Zohr University
Agadir, Morocco

Oussama Smimite
InterDisciplinary Applied Research Laboratory, International University of Agadir
Agadir, Morocco

Aziz Srai
ERCI2A, FSTH, Abdelmalek Essaadi University
Tetouan, Morocco

Hamed Taherdoost
University Canada West, Hamta Business Corporation, Q Minded | Quark Minded Technology Inc.
Vancouver, Canada

Nasser Tamou
Electronic Communication Network Department, Mohammed V University
Rabat, Morocco

Abida Tahsin Tawfik
Department of Physics, College of Science, Mosul University
Mosul, Iraq

Part I

Artificial intelligence (AI)

Chapter 1

Resource utilization and cost implications of container live migration in clouds

An approach performed on Amazon Web Services (AWS)

Amine Bouaouda, Karim Afdel, and Rachida Abounacer

1.1 INTRODUCTION

Containerization has revolutionized the deployment of applications in cloud environments, offering lightweight and portable solutions that enhance resource utilization and simplify application deployment across diverse platforms [1–3]. As the adoption of containerized workloads continues to rise, the need for effective management techniques, including live migration, becomes increasingly critical to ensure seamless operation, workload balancing, and resource optimization [1, 4].

Container live migration, which involves transferring active containers from one host to another without interrupting their execution [5–7], is essential for scenarios such as load balancing, hardware maintenance, and system upgrades. This capability guarantees uninterrupted service availability, improved fault tolerance, and optimal resource utilization [8–10]. Furthermore, live migration supports dynamic resource allocation and workload consolidation, enhancing the performance and resilience of containerized applications [10–12].

In the dynamic realm of cloud computing, understanding the impact and cost of container live migration is crucial when deploying containerized applications. This knowledge is vital for informed decision-making and developing strategies to achieve efficient and cost-effective cloud deployments [13, 14]. Key performance metrics, such as CPU utilization, memory usage, network traffic, and latency, offer insights into how live migration affects application performance. Concurrently, monitoring costs allows for assessing financial implications and optimizing resource utilization during the migration process [13, 15].

This study aims to illuminate the complex relationship between container live migration, performance metrics, and associated costs. By exploring these dimensions, we provide valuable insights that promote efficient and cost-effective cloud deployments, contributing to the ongoing discourse on optimizing containerized workloads in production environments.

DOI: 10.1201/9781032664293-2

1.2 BACKGROUND AND LITERATURE REVIEW

Numerous studies have explored the impact and benefits of live migration in diverse cloud computing environments, particularly focusing on containers. These studies emphasize the critical role of live migration in optimizing resource allocation within multi-tenant cloud platforms [5, 12, 16–18]. Findings from various investigations have shown that live migration techniques significantly enhance consolidation ratios and reduce operational costs by dynamically reallocating resources based on workload demands [17–21].

Furthermore, the advancement of live migration techniques aims to bolster fault tolerance and system availability. Certain studies have highlighted that live migration enhances the fault tolerance of microservices by seamlessly transferring them to healthy hosts in the event of failures or performance degradation [18, 22]. However, despite the numerous benefits of container live migration, it also presents challenges and considerations, especially regarding performance and cost [17–19].

The migration process incurs overheads, such as network transfer and memory copying, which can impact the performance of containerized applications. Researchers have explored these performance impacts and proposed optimization techniques to mitigate performance degradation during the migration process [16–18].

In addition to performance considerations, the cost implications of container live migration are significant. Cloud providers often charge for network bandwidth, storage, and compute resources used during migration. Understanding these cost implications and optimizing resource utilization can lead to substantial cost savings in cloud deployments [17, 18, 20, 21, 23]. This study focuses on investigating the impact and cost aspects of container live migration in cloud computing environments through a practical approach, rather than relying solely on simulations. By leveraging cloud infrastructure, such as Amazon Web Services (AWS), we assess the performance impact and cost considerations associated with container live migration. Through experiments, performance metric collection, and cost data analysis, we aim to provide valuable insights into the effectiveness, efficiency, and cost-effectiveness of live migration techniques. These findings can guide cloud practitioners and researchers in making informed decisions when deploying containerized workloads in production environments.

The remainder of this chapter is organized as follows: Section 1.3 defines the problem and outlines our proposed methodology. Section 1.4 discusses the application and evaluation of our approach. The chapter concludes in Section 1.5.

1.3 PROBLEM DEFINITION AND METHODOLOGICAL APPROACH

In this section, we explore the problem definition and outline the methodology used in our study to assess the impact and cost of container live migration in cloud environments. We start by examining the fundamental concepts of container-based virtualization and live migration, followed by an overview of AWS as our chosen cloud computing platform.

1.3.1 Containerization-based virtualization

Containerization-based virtualization has emerged as a groundbreaking technology within cloud computing, enabling the encapsulation of applications and their dependencies within lightweight, isolated containers [2, 24, 25]. Leveraging technologies such as Docker, these containers create a standardized and portable environment for application deployment, as illustrated in Figure 1.1, ensuring consistent execution across various computing platforms [1, 24, 26].

The adoption of containerization offers numerous benefits, including improved resource utilization, rapid scalability, and streamlined application management [12, 24]. By abstracting the underlying infrastructure, containerization decouples applications from the host operating system, enhancing portability and enabling faster deployment, thereby reducing time-to-market [25].

This technology is particularly advantageous in cloud computing environments, where the need for flexible and efficient resource allocation is critical [8]. Containers allow developers to package their applications with all

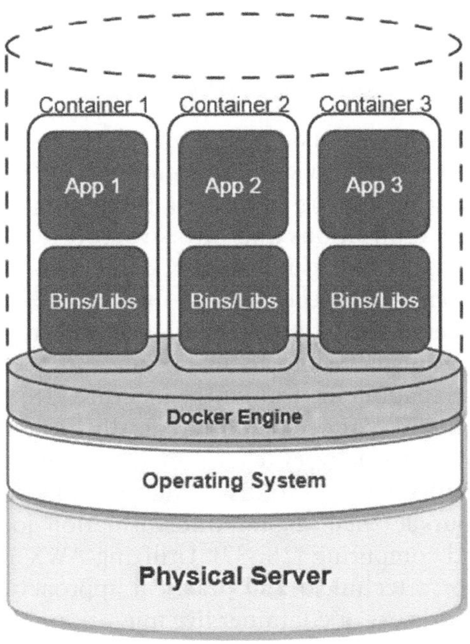

Figure 1.1 Docker-based cloud container operation.

necessary dependencies, ensuring uniform execution across different cloud instances or clusters [2, 26]. By utilizing containerization-based virtualization, cloud providers can optimize their infrastructure resources, leading to enhanced performance and efficiency of cloud-based applications.

1.3.2 Live migration

Live migration is an essential technique in cloud computing, allowing the seamless transfer of running virtual machines (VMs) or containers from one physical host to another without interrupting their operations [17, 22, 27]. In the cloud computing domain, live migration is critical for maintaining high availability, fault tolerance, load balancing, and resource optimization [8, 10–12]. For VMs, live migration ensures continuous service and application operations, even during maintenance activities, hardware failures, or when resource demands necessitate the relocation of virtualized workloads [17, 27].

Regarding cloud containers, live migration involves the real-time transfer of running containers between different host nodes within a container cluster or orchestration platform [5, 6]. This dynamic migration ensures uninterrupted application execution, enhancing the elasticity and scalability of containerized workloads in cloud environments [7, 10]. Container orchestration tools, such as Kubernetes and Docker Swarm, provide mechanisms for managing the live migration of containers. These capabilities facilitate automatic scaling, load balancing, and efficient resource allocation based on the real-time demands of the application [10, 19].

1.3.3 Amazon Web Services (AWS)

AWS is a highly popular cloud computing platform offered by Amazon. With an extensive array of services and solutions, AWS enables individuals, organizations, and governments to harness the power of the cloud to build scalable and flexible applications and infrastructures [28–30]. AWS services encompass a wide range, including storage, databases, artificial intelligence, the Internet of Things (IoT), security, and more [31, 32]. These services are meticulously engineered for high reliability, cost-effectiveness, and seamless scalability, allowing businesses to focus on innovation and growth without the complexity of managing intricate infrastructure [28]. AWS's global infrastructure, with numerous data centers strategically located in various regions, ensures high-performance delivery to customers worldwide. As a leading cloud service provider, AWS continues to drive innovation in the industry, enabling organizations to pursue their digital transformation journeys through the capabilities of cloud computing [28, 32]. Utilizing AWS allows us to leverage multiple services for a technical and practical approach, demonstrating the real-world impact and cost of container live migration on AWS. Before diving into the specifics of our approach, we will outline the key services integral to our proposed methodology.

1.3.3.1 Amazon Elastic Compute Cloud (EC2)

Amazon Elastic Compute Cloud (EC2) is a fundamental service for scalable cloud computing [28, 33]. This service enables users to quickly and easily provision virtual servers, known as instances, and dynamically scale their compute resources to meet specific needs [31, 34]. EC2 offers a wide variety of instance types, operating systems, and configurations, addressing the diverse requirements of applications and workloads. Known for its flexibility, reliability, and security features, EC2 allows cloud users to pay only for the compute capacity they actually use [34, 35].

1.3.3.2 Amazon Elastic Kubernetes Service (EKS)

Amazon Elastic Kubernetes Service (EKS) is a managed service that facilitates the deployment and management of containerized applications using Kubernetes [36, 37]. One of the primary benefits of EKS is that it relieves users from the complex tasks of managing the underlying Kubernetes infrastructure, including control plane installation, scaling, and patching. EKS integrates seamlessly with various AWS services, streamlining the creation of a secure and scalable containerized application stack [37].

1.3.3.3 AWS CloudWatch

AWS CloudWatch is an extensive monitoring and observability service that provides real-time insights into users' AWS resources and applications [33]. CloudWatch offers a comprehensive approach by collecting metrics, monitoring log files, and setting alarms. This service gives users a clear view of the operational health and performance of their applications, infrastructure, and services [33]. By leveraging CloudWatch, users can monitor resource utilization, configure automated actions based on predefined thresholds, and gain detailed insights through custom metrics and dashboards [33].

1.3.4 Methodology proposed

To demonstrate the impact and cost of container live migration in cloud environments, we utilized an AWS cloud account (free trial). Our approach involved the following steps:

- Creation of an Amazon EC2 instance: We launched an EC2 instance from the AWS Management Console, tailored to meet specific requirements and workloads.
- Setting up Docker and containers: Following the official Docker documentation, we installed Docker on our EC2 instance. Subsequently, we created and deployed containers on the EC2 instance using Docker, as shown in Figure 1.2.

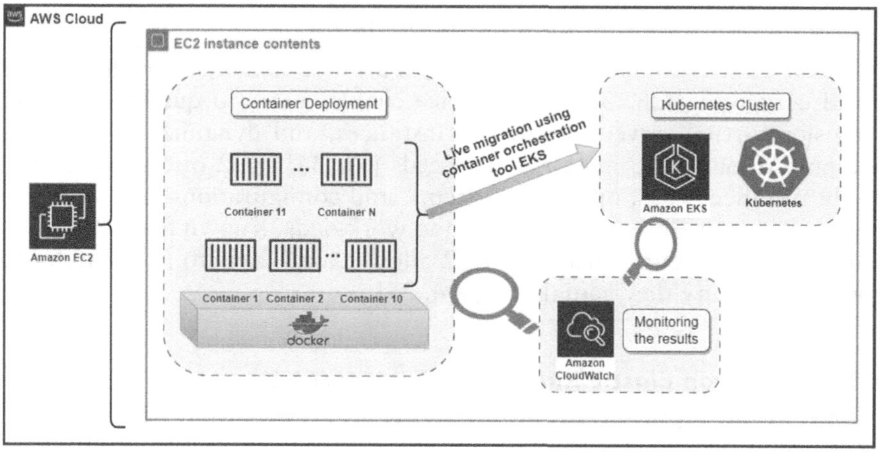

Figure 1.2 Overview of the proposed AWS approach.

- Performing container live migration: To showcase container live migration, we employed Amazon EKS, a container orchestration tool that supports live migration. We set up an EKS cluster, configured the necessary resources, and performed live migration of containers between nodes in the cluster.
- Monitoring and measuring the impact: Using AWS CloudWatch, we collected performance metrics such as CPU utilization, memory usage, network traffic, and latency. We monitored these metrics before, during, and after the live migration process to assess the impact on application performance.
- Cost analysis: We tracked and monitored the cost of running our containers during the live migration process. AWS provides a billing dashboard that allows us to view detailed cost breakdowns.

1.3.5 Methodological considerations

1.3.5.1 Platform-specific focus

Our study concentrates on AWS as the selected cloud computing platform. Although AWS is extensively utilized and provides a representative sample, the findings may contain platform-specific characteristics. To address this, we selected AWS due to its widespread use and extensive array of container-related services, including Amazon EKS and Amazon EC2, which facilitate a thorough investigation. Nonetheless, future research could apply similar methodologies to other cloud platforms to gain a more comprehensive understanding.

1.3.5.2 Real-world constraints

Our study's experiments are grounded in real-world scenarios, utilizing an AWS account (free trial) for practical implementation. While this approach enhances the relevance of our findings to actual cloud deployments, it is important to acknowledge the inherent variability and unpredictability of real-world environments. To mitigate this, we have meticulously documented our methodology, ensuring transparency in our approach and adhering to established best practices in the field.

1.3.5.3 Cost implications

The cost analysis component of our study is based on the pricing models and services provided by AWS. While we aim to deliver a thorough cost analysis, variations in pricing models or specific configurations may affect the generalizability of our cost-related findings. We have meticulously detailed the AWS services used for cost tracking, ensuring transparency in our reporting so readers can understand the cost implications within the AWS framework.

In summary, our methodology is designed to balance the complexities of real-world cloud environments with the need for controlled experimentation. We have accounted for potential limitations and strived for transparency in our approach to ensure the validity and reliability of our findings.

1.4 APPLICATION AND RESULTS

1.4.1 Setting up an Amazon EC2 instance

Upon accessing the AWS Management Console, we proceeded to the EC2 service and initiated the process of launching a new EC2 instance by adhering to the steps outlined in the console. We selected the Amazon Linux AMI for our instance and opted for the t2.micro instance type due to its eligibility for the free tier and its appropriateness for our needs. The specifications for our EC2 instance are detailed in Figure 1.3.

1.4.2 Docker setup and container initialization

After successfully launching our EC2 instance, as outlined in Figure 1.4, we established a connection via SSH (Secure Shell). To facilitate secure remote access, we first downloaded PuTTY, a dependable SSH client. Following the installation, we opened PuTTYgen, included with PuTTY, and clicked the "Load" button to select the AmineKP.pem key file previously downloaded. This action loaded the key file and displayed its relevant details. In SSH and other cryptographic applications, a PEM file typically contains a public key, a private key, or both, essential for secure communication and authentication between systems.

Figure 1.3 EC2 instance requirements.

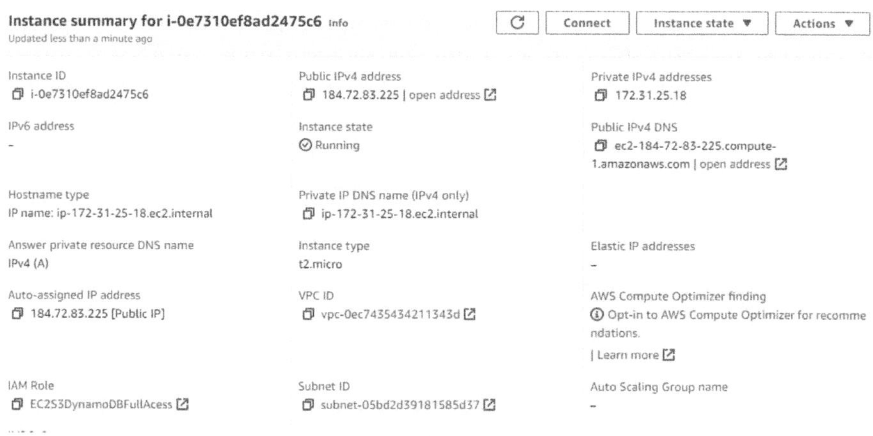

Figure 1.4 Detailed view of the EC2 instance.

Our EC2 instance uses the AmineKP.pem key file for SSH connections. For added security and ease of use, we clicked the "Save private key" button, acknowledging the warning about not setting a passphrase, and saved the private key with a "ppk" extension in a convenient location. This process effectively set up our SSH key, enabling us to securely access remote servers using PuTTY with the generated "ppk" key.

After that, we launched PuTTY (the SSH client) and accessed the PuTTY Configuration window. We entered the public DNS or IP address of our EC2 instance in the "Host Name (or IP address)" field and set the port to 22, the default SSH port. Navigating through the left pane, we went to "Connection" → "SSH" → "Auth" and clicked the "Browse" button next to the "Private key file for authentication" field to select the "ppk" private key file we generated with PuTTYgen. We then provided a name for our session in the "Saved

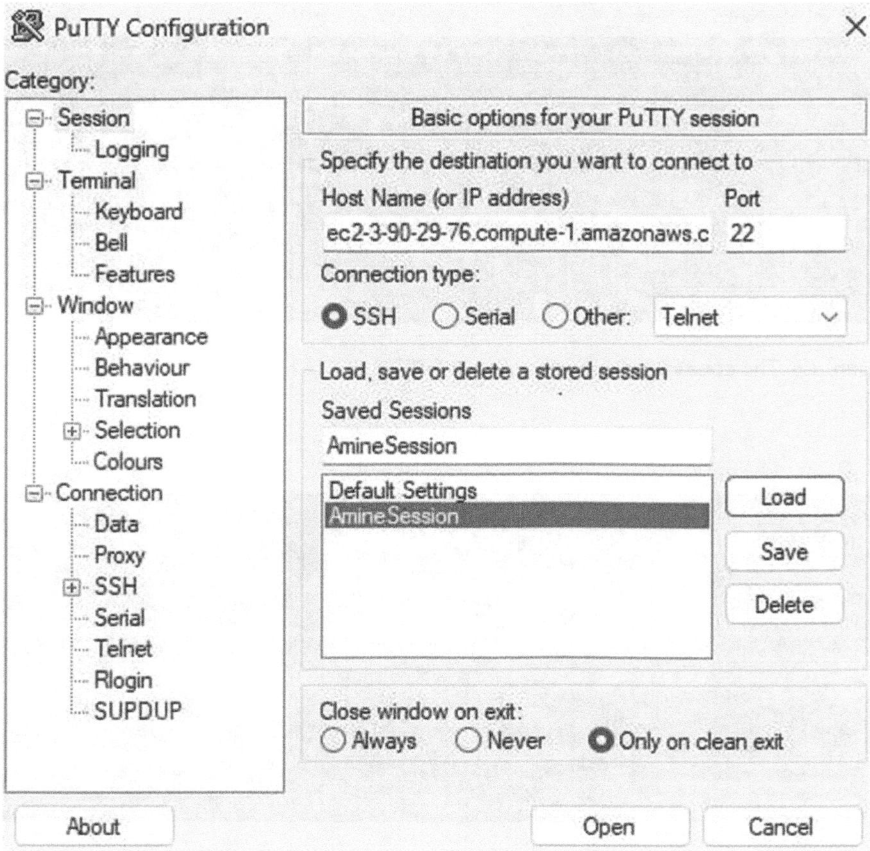

Figure 1.5 SSH connection via PuTTY.

Sessions" field and saved our settings. Finally, we clicked the "Open" button to start the SSH connection, as depicted in Figure 1.5.

Upon initiating the connection, a console appeared prompting us to enter the password "ec2-user" to establish the connection with our EC2 instance, as illustrated in Figure 1.6. Through this console, we proceeded to install Docker on the EC2 instance by following the official Docker documentation for Amazon Linux. The installation process involved several steps. First, we updated the package index and installed the necessary dependencies, as outlined in Figure 1.7, using the following commands:

- sudo yum update -y
- sudo yum install -y yum-utils device-mapper-persistent-data lvm2.

Next, we installed Docker itself by executing, as displayed in Figure 1.8. To ensure the installation was successful, we verified the Docker version with the

Figure 1.6 The connection console through PuTTY.

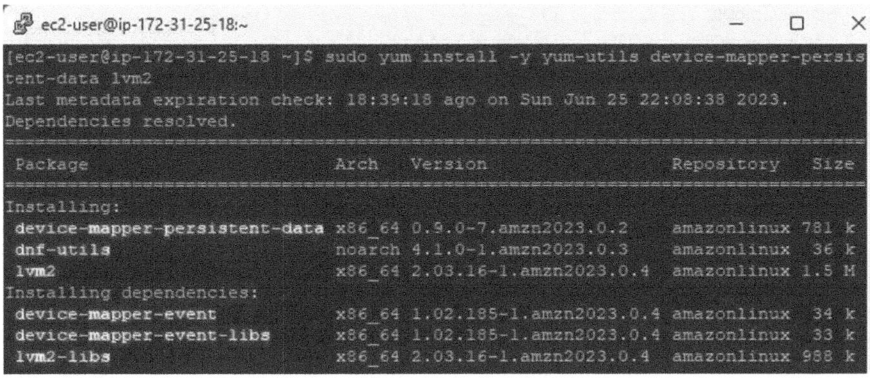

Figure 1.7 Installation of dependencies.

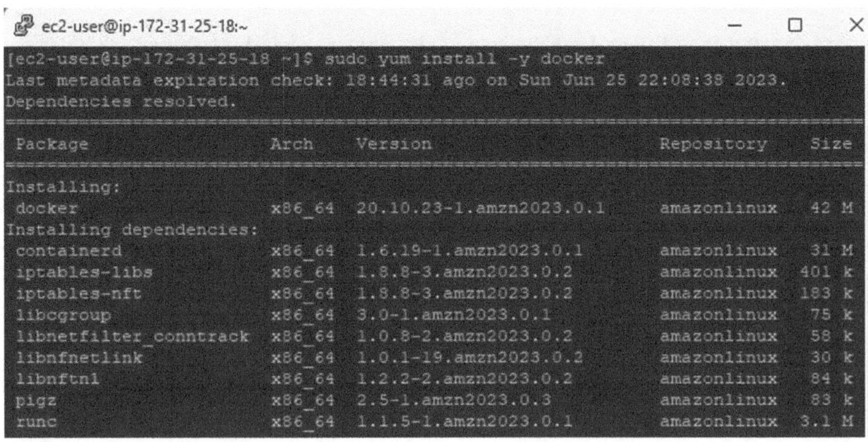

Figure 1.8 Installation of Docker.

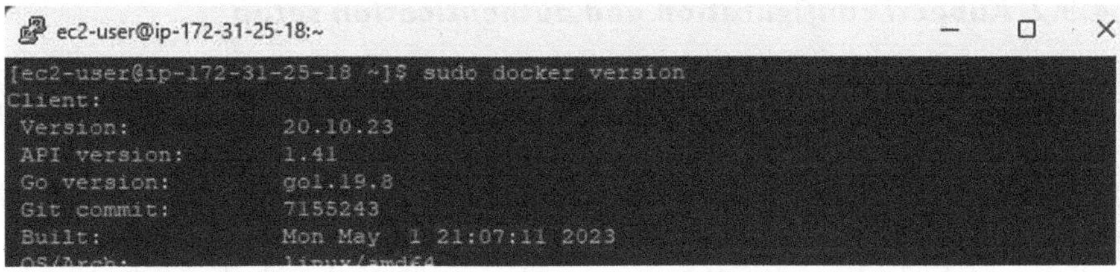

Figure 1.9 Docker version.

Docker version command, as shown in Figure 1.9. With Docker successfully installed, we were then ready to start creating and deploying containers on the EC2 instance using Docker commands.

1.4.3 Container live migration process

To facilitate container live migration, we opted to utilize Amazon EKS for container orchestration. We meticulously followed the AWS documentation to set up an EKS cluster and configure the requisite resources.

1.4.3.1 Establishing an Amazon EKS cluster

To create an Amazon EKS cluster, we first access the AWS Management Console and navigate to the Amazon EKS service. We then click on "Add cluster" followed by "Create." It's important to provide a meaningful cluster name and, if desired, enable tagging and logging options to aid in cluster management. For proper networking, we carefully select the VPC, subnets, and security groups that meet our requirements. Before proceeding, we review the configuration thoroughly to ensure accuracy. Once satisfied with the settings, we click "Create" to start the EKS cluster creation process. Figure 1.10 displays the status of our cluster after its creation.

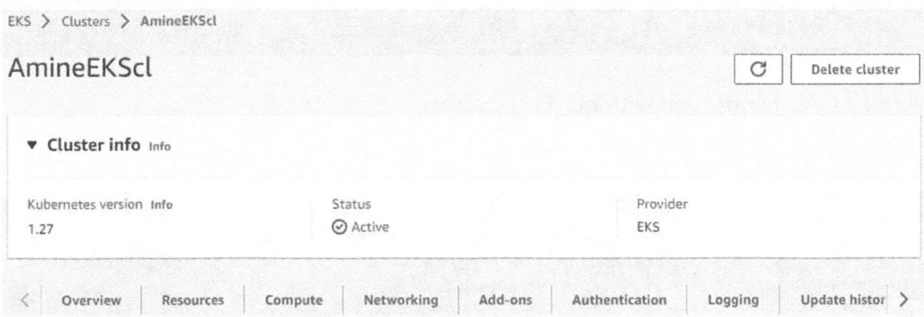

Figure 1.10 Our EKS cluster in active status.

1.4.3.2 Kubectl configuration and authentication setup

On our EC2 instance, we installed and configured the kubectl command-line tool to interact with Kubernetes clusters, as presented in Figure 1.11. First, we verified the installation by executing the command kubectl version –client to ensure that kubectl was installed correctly, as shown in Figure 1.12. Next, after confirming the installation, we configured kubectl to authenticate with our EKS cluster using the command aws eks update-kubeconfig –region us-east-1 –name AmineEKScl, as displayed in Figure 1.13. This command automatically updates the kubeconfig file on our EC2 instance with the necessary authentication details for accessing our EKS cluster. Finally, we verified the cluster connection by running kubectl cluster-info to ensure that kubectl could successfully connect to our EKS cluster, as illustrated in Figure 1.14.

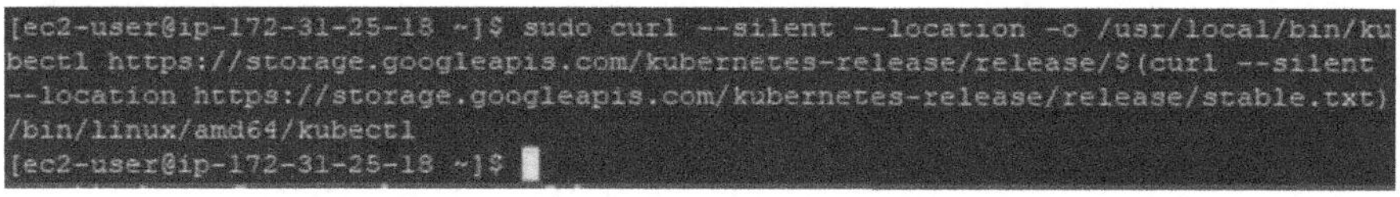

Figure 1.11 Installation of kubectl.

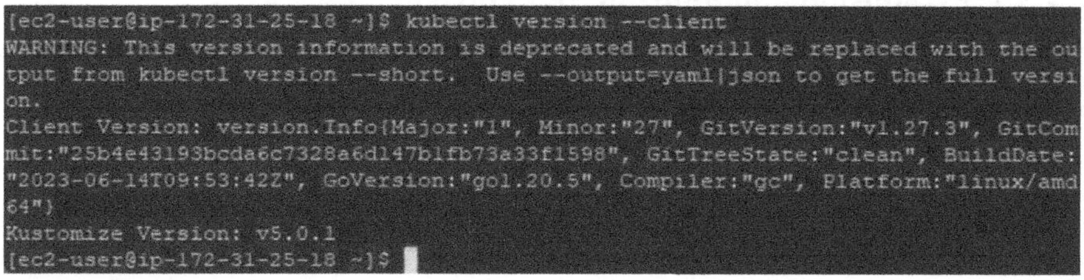

Figure 1.12 Kubectl version.

Figure 1.13 Authentication with our EKS cluster.

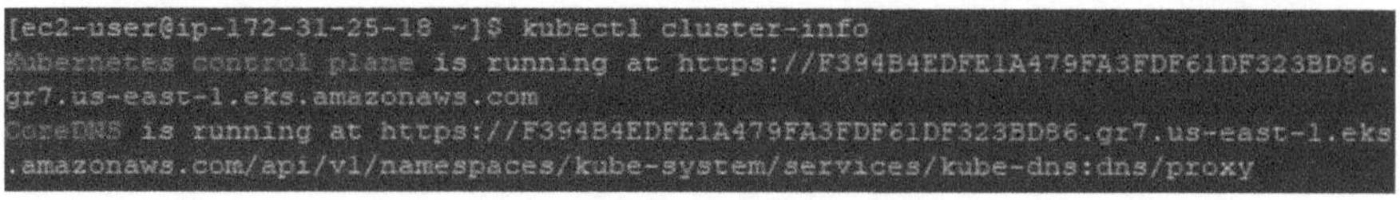

Figure 1.14 Successful kubectl connection to EKS cluster.

1.4.3.3 *Container deployment process*

Before initiating the migration, it is essential to deploy containers by constructing Kubernetes manifests (YAML files) that define the containers we intend to deploy and manage. These manifests outline the configuration of pods, replicas, container images, resource requirements, and any necessary service configurations for the application. For our specific use case, we chose to create a manifest file that specifies the deployment of the Nginx web server, as depicted in Figure 1.15.

Our YAML file defines a Kubernetes Deployment object, designed to manage the deployment and scaling of multiple instances of the Nginx web server. Named "nginx-deployment-amine," the deployment is set to run three replicas of the Nginx pod. It is configured with a selector that identifies pods labeled "app: nginx" as part of the deployment.

Within the template section, the configuration for the Nginx pod is specified. The pod, labeled "app: nginx," will run a single container named "nginx" using the official Nginx Docker image. The Nginx container will listen on port 80 and has resource requests defined, requiring a minimum of 0.5 GiB of memory and 500 mCPU units. By creating this deployment, Kubernetes ensures that three replicas of the Nginx web server are always

```
apiVersion: apps/v1
kind: Deployment
metadata:
  name: nginx-deployment-amine
spec:
  replicas: 3  # Number of replicas we want to run
  selector:
    matchLabels:
      app: nginx
  template:
    metadata:
      labels:
        app: nginx
    spec:
      containers:
        - name: nginx
          image: nginx
          ports:
            - containerPort: 80
          resources:
            requests:
              memory: "0.5Gi"
              cpu: "500m"
```

Figure 1.15 The Kubernetes manifest "Contsdeployment.yaml."

```
C:\Users\AMINE>kubectl apply -f D:\LATEX_PHD\comm2\Contsdeployment.yaml
deployment.apps/nginx-deployment-amine created

C:\Users\AMINE>kubectl get pods
NAME                                  READY   STATUS              RESTARTS   AGE
nginx-deployment-amine-657bc6789d-28hqx   0/1     ContainerCreating   0          15s
nginx-deployment-amine-657bc6789d-8gm42   0/1     ContainerCreating   0          15s
nginx-deployment-amine-657bc6789d-rct4k   0/1     ContainerCreating   0          15s

C:\Users\AMINE>kubectl get deployments
NAME                     READY   UP-TO-DATE   AVAILABLE   AGE
nginx-deployment-amine   0/3     3            0           2m21s
```

Figure 1.16 Deploying containers using "Contsdeployment.yaml" YAML file.

	Name	Namespace	Type	Age	Pod count	Status
○	coredns	kube-system	deployments	Created July 26, 2023, 16:50 (UTC+01:00)	0	0 Read
○	nginx-deployment-amine	default	deployments	Created 2 minutes ago	0	0 Read

Figure 1.17 Initial deployment overview in AWS Management Console.

running, managing scaling, and maintaining the desired state according to the defined configuration.

After defining our YAML file, we deployed the containers by applying the Kubernetes manifests using kubectl, as illustrated in Figure 1.16. The scheduler will then seek nodes with sufficient available resources to accommodate the new requests, migrating the pods accordingly, as shown in Figure 1.17.

1.4.3.4 Performing container live migration

To initiate container live migration in Kubernetes, including Amazon EKS, we don't trigger it directly with a manual command. Instead, we indirectly prompt migration by altering our deployment's resource requirements, which can lead the Kubernetes scheduler to reschedule the pods. The cluster's built-in scheduler typically manages container live migration, automatically handling the placement and movement of containers across nodes. For instance, updating resource requirements in our Kubernetes manifest can cause the scheduler to move the container to a different node.

For our deployment, we set up three nodes to serve as our computing units for managing containers, as displayed in Figure 1.18. These nodes

```
C:\Users\AMINE>kubectl get nodes
NAME                             STATUS   ROLES    AGE     VERSION
ip-172-31-21-109.ec2.internal    Ready    <none>   8m35s   v1.27.3-eks-a5565ad
ip-172-31-46-32.ec2.internal     Ready    <none>   8m27s   v1.27.3-eks-a5565ad
ip-172-31-82-243.ec2.internal    Ready    <none>   8m29s   v1.27.3-eks-a5565ad
```

Figure 1.18 The nodes of our EKS cluster.

are individual worker machines that form the infrastructure of our cluster, AmincEKScl. Each node is a VM responsible for running containers, as demonstrated in Figure 1.19, providing the necessary resources like CPU, memory, and storage to run applications. Managed by the control plane, these nodes communicate with it to receive instructions for running and managing pods, as evidenced in Figure 1.20.

Nodes (3) Info

Filter Nodes by property or value < 1 >

Node name	Instance type	Node group	Created	Status
ip-172-31-21-109.ec2.internal	t3.medium	AmineNP	Created 4 minutes ago	Ready
ip-172-31-46-32.ec2.internal	t3.medium	AmineNP	Created 4 minutes ago	Ready
ip-172-31-82-243.ec2.internal	t3.medium	AmineNP	Created 4 minutes ago	Ready

Figure 1.19 Nodes in the EKS cluster in AWS Management Console.

Figure 1.20 Detailed view of a node in AWS Management Console.

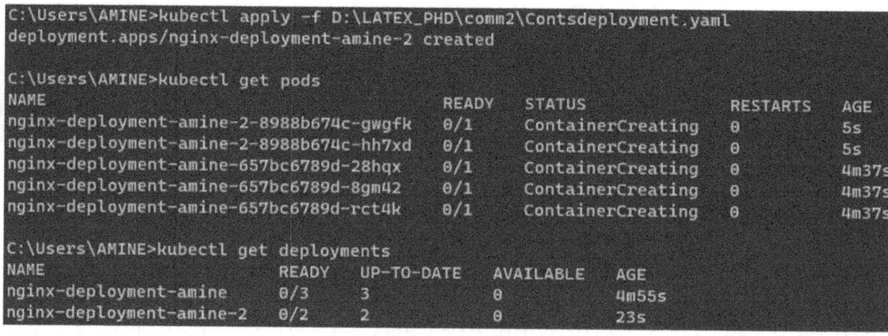

```
C:\Users\AMINE>kubectl apply -f D:\LATEX_PHD\comm2\Contsdeployment.yaml
deployment.apps/nginx-deployment-amine-2 created

C:\Users\AMINE>kubectl get pods
NAME                                        READY   STATUS              RESTARTS   AGE
nginx-deployment-amine-2-8988b674c-gwgfk    0/1     ContainerCreating   0          5s
nginx-deployment-amine-2-8988b674c-hh7xd    0/1     ContainerCreating   0          5s
nginx-deployment-amine-657bc6789d-28hqx     0/1     ContainerCreating   0          4m37s
nginx-deployment-amine-657bc6789d-8gm42     0/1     ContainerCreating   0          4m37s
nginx-deployment-amine-657bc6789d-rct4k     0/1     ContainerCreating   0          4m37s

C:\Users\AMINE>kubectl get deployments
NAME                       READY   UP-TO-DATE   AVAILABLE   AGE
nginx-deployment-amine     0/3     3            0           4m55s
nginx-deployment-amine-2   0/2     2            0           23s
```

Figure 1.21 New container deployment post "Contsdeployment.yaml" modification.

Pods, the smallest units in the Kubernetes model, are used to deploy and scale applications. They encapsulate application containers, storage resources, and unique network IP addresses, creating a cohesive unit for running and managing containerized applications. After our initial deployment, we decided to modify the YAML file, changing some parameters to create new deployments, as depicted in Figure 1.21. These changes are detected by the scheduler, which then automatically begins the migration by rescheduling containers to different nodes, as displayed in Figure 1.22.

In our approach, we created four different deployments by modifying our YAML file, "Contsdeployment.yaml," as represented in Table 1.1, allowing the scheduler to initiate the migration of containers between nodes.

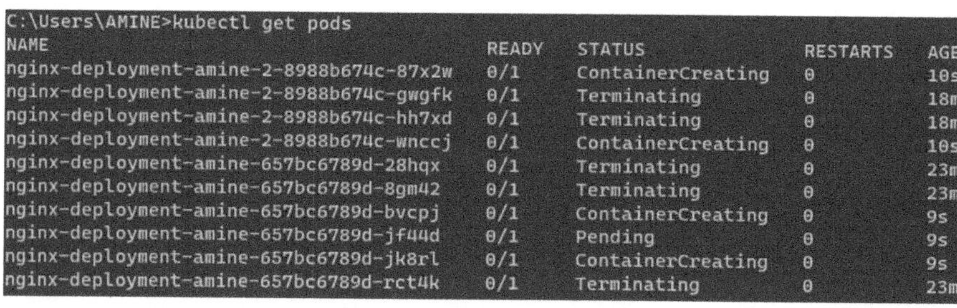

```
C:\Users\AMINE>kubectl get pods
NAME                                        READY   STATUS              RESTARTS   AGE
nginx-deployment-amine-2-8988b674c-87x2w    0/1     ContainerCreating   0          10s
nginx-deployment-amine-2-8988b674c-gwgfk    0/1     Terminating         0          18m
nginx-deployment-amine-2-8988b674c-hh7xd    0/1     Terminating         0          18m
nginx-deployment-amine-2-8988b674c-wnccj    0/1     ContainerCreating   0          10s
nginx-deployment-amine-657bc6789d-28hqx     0/1     Terminating         0          23m
nginx-deployment-amine-657bc6789d-8gm42     0/1     Terminating         0          23m
nginx-deployment-amine-657bc6789d-bvcpj     0/1     ContainerCreating   0          9s
nginx-deployment-amine-657bc6789d-jf44d     0/1     Pending             0          9s
nginx-deployment-amine-657bc6789d-jk8rl     0/1     ContainerCreating   0          9s
nginx-deployment-amine-657bc6789d-rct4k     0/1     Terminating         0          23m
```

Figure 1.22 The pods of our deployments.

Table 1.1 Deployment requirements

Deployment	*Replicas*	*Memory (GB)*	*mCPU (m)*
1	3	0.5	500
2	2	1	1000
3	3	1	700
4	2	0.5	700

1.4.4 Impact monitoring and measurement

To collect performance metrics from our EKS cluster, we utilized AWS CloudWatch. This tool provides valuable insights into CPU utilization, network traffic, and other metrics. By using CloudWatch, we can monitor the performance of our nodes and pods before, during, and after the live migration process, helping us analyze any potential impact on our application's performance. After enabling CloudWatch monitoring in the AWS Management Console, we can observe resource consumption, particularly CPU usage.

The CPU utilization data for node ip-172-31-21-109.ec2.internal indicates stable performance with occasional fluctuations. A significant spike in CPU utilization occurred at 2023-07-28 14:15, as indicated in Figure 1.23, reaching 2.126666667, corresponding to the migration event. Post-migration, CPU utilization stabilized, indicating successful migration and resource allocation.

For node ip-172-31-46-32.ec2.internal, CPU utilization showed consistent behavior with minor fluctuations. Notably, at 2023-07-28 14:20, there was a sharp increase, peaking at 2.166630615, coinciding with the migration. Afterward, CPU usage stabilized at a lower level, ensuring smooth cluster performance, as evidenced in Figure 1.24.

Node ip-172-31-82-243.ec2.internal exhibited steady performance with minor variations. During the migration event at 2023-07-28 14:20, CPU utilization peaked at 2.225003455. Following the migration, CPU utilization

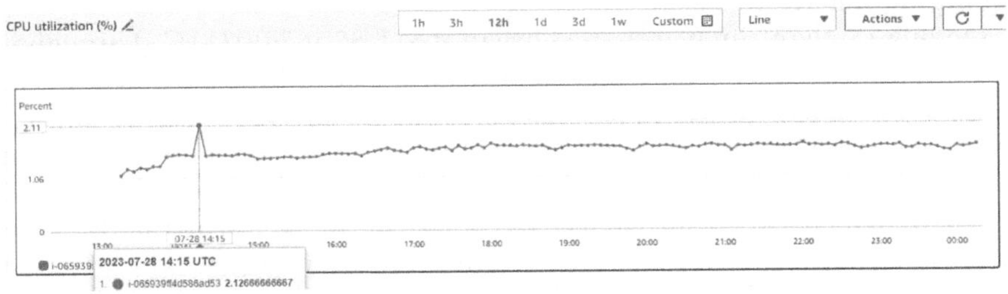

Figure 1.23 CPU utilization of the first node.

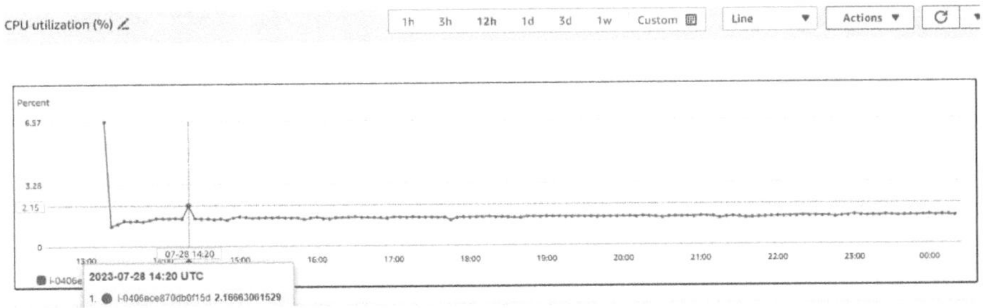

Figure 1.24 CPU utilization of the second node.

returned to a stable level, demonstrating successful migration and efficient resource management.

Overall, the CPU utilization analysis for all three nodes indicates that the migration process had a noticeable impact on each node's performance. Understanding these patterns is essential for monitoring and optimizing the cluster's overall performance during migration events. By correlating migration events with CPU utilization, administrators can make informed decisions regarding resource allocation, ensuring the stability and efficiency of the EKS cluster. Fine-tuning resource allocation and closely monitoring CPU utilization, post-migration can lead to a well-optimized and high-performing EKS cluster.

1.4.5 Cost evaluation and analysis

After evaluating the impact of migration on performance metrics, particularly CPU usage, we also decided to analyze the estimated costs to understand the financial implications of container migration in AWS.

By examining the total estimated charges in USD for every 6 hours after the migration on July 28, 2023, we gain valuable insights into the cost implications of container migration within the EKS cluster. Before the migration, from July 22 to July 27, the total estimated charges remained stable at around $0.48 per 6-hour interval, as shown in Figure 1.25. However, following the migration, there was a noticeable increase in estimated charges, indicating a significant impact on costs.

On July 28, the estimated charges began at $11.55 at 5:00 UTC. Throughout the day, these charges escalated, reaching $13.96 at 17:00 UTC and $15.38 at 23:00 UTC. This upward trend suggests that the migration event significantly impacted overall expenses, likely due to resource allocation, potential scaling, or temporary inefficiencies during the migration. The following day, July 29, also saw an increase in estimated charges, starting at $15.38 at 5:00 UTC and rising to $17.87 at 11:00 UTC. This continued rise in costs could be attributed to the ongoing stabilization and optimization of the cluster post-migration.

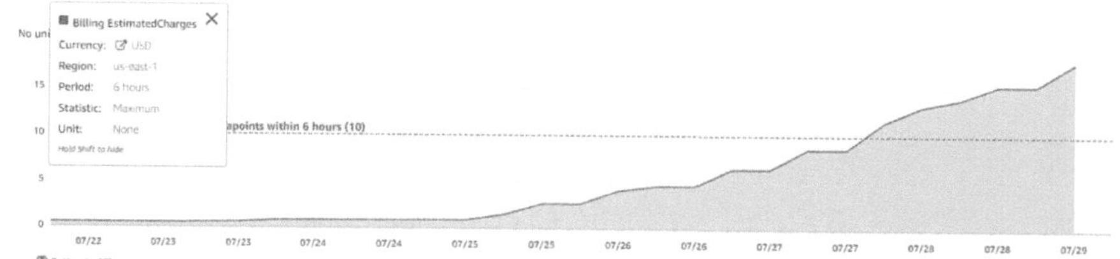

Figure 1.25 Total estimated charges.

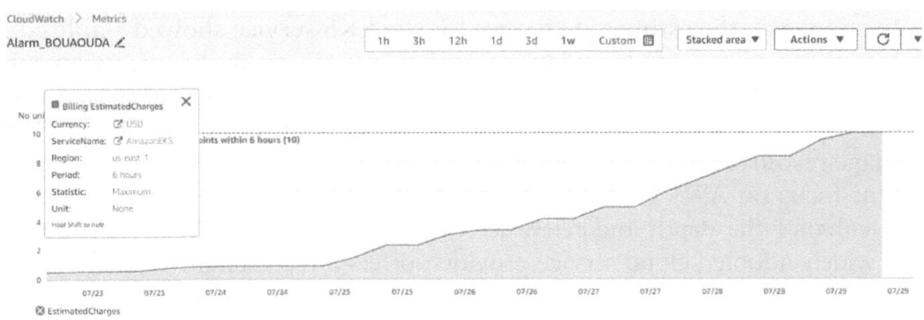

Figure 1.26 Estimated charges of the EKS service.

Overall, the cost analysis reveals that the container migration on July 28, 2023, had a noticeable effect on the estimated charges. The increased costs during and after the migration event highlight the need for careful resource management and monitoring to ensure optimal utilization and cost efficiency. Administrators should closely evaluate resource allocation and scale resources as needed to maintain a balance between performance and expenses in the EKS cluster. Ongoing optimization efforts and fine-tuning of the cluster configuration can help stabilize costs post-migration.

In addition to analyzing total estimated charges, we also examined the cost of using the EKS service. The estimated charges for the EKS service showed a clear pattern of increase over time, as demonstrated in Figure 1.26. From July 23 to July 27, costs remained relatively stable, ranging from \$0.48 to \$4.90 per 6-hour interval. During this period, the EKS service operated with consistent resource demands, resulting in steady and predictable charges.

On July 28, the estimated charges for the EKS service experienced a significant spike, starting at \$6.00 at 2:00 UTC and reaching \$8.40 at 20:00 UTC. This increase indicates higher resource consumption within the EKS cluster, likely triggered by the container migration process. The post-migration phase on July 29 continued to show higher estimated charges compared to the pre-migration period, ranging from \$8.40 to \$10.00 per 6-hour interval. This suggests that the EKS cluster was still undergoing adjustments after the migration, with fluctuations in costs attributed to the cluster's efforts to optimize performance and resource allocation.

The sudden increase in estimated charges on the migration day and the elevated costs in the subsequent days imply that the container migration introduced additional resource requirements. While the migration might have been necessary to enhance the cluster's capabilities or accommodate workload changes, it also resulted in higher short-term expenses. To optimize the EKS service's estimated charges post-migration, it is crucial to fine-tune resource allocation based on actual workload demands. Continuously monitoring cluster performance, identifying resource bottlenecks, and optimizing autoscaling policies can help balance cost efficiency and service quality.

In summary, the estimated charges for the EKS service showed stable costs before the migration, followed by a significant increase on the migration date, and continued higher charges post-migration. While the migration may have positively impacted cluster performance, administrators should carefully manage resources to ensure long-term cost efficiency.

Our focus on AWS in this study is based on several strategic considerations that enhance the depth and relevance of our investigation. AWS, as a leading and widely adopted cloud service provider, offers a representative ecosystem for exploring the intricacies of container live migration. Leveraging AWS technologies such as Amazon EC2, Amazon EKS, and AWS CloudWatch enables us to conduct a comprehensive analysis within a technologically diverse and established cloud infrastructure. Importantly, the methodologies and insights presented here are not exclusive to AWS; rather, they serve as a robust foundation for broader discussions applicable to other cloud platforms. The technologies employed, such as EC2, are ubiquitous across various cloud providers, ensuring the generalizability of our findings. While our study unfolds within the AWS context, the principles explored extend beyond, offering valuable insights applicable to the dynamic landscape of container live migration across diverse cloud platforms.

1.5 CONCLUSION AND FUTURE WORK

This study presents a methodology for evaluating the impact and cost of container live migration in cloud environments, utilizing AWS cloud services. By following a systematic approach that includes creating an Amazon EC2 instance, setting up Docker and containers, executing container live migration via orchestration tools, continuously monitoring performance metrics, and analyzing associated costs, we have gained valuable insights into the effectiveness and efficiency of container live migration. These findings provide cloud users with practical information and perspectives, aiding in decision-making processes and identifying optimization techniques for deploying containerized applications in the cloud. Future research can focus on further enhancing the impact, cost, and overall efficiency of container live migration in cloud environments.

ACKNOWLEDGMENT

We declare that this research project did not receive any funding. The work presented in this chapter was conducted without external financial support.

REFERENCES

1. Bernstein, D.: Containers and cloud: From LXC to Docker to Kubernetes. IEEE Cloud Computing **1**(3), 81–84 (2014). https://doi.org/10.1109/MCC.2014.51
2. Hardikar, S., Ahirwar, P., Rajan, S.: Containerization: Cloud computing based inspiration technology for adoption through Docker and Kubernetes. In: 2021 Second International Conference on Electronics and Sustainable Communication

Systems (ICESC), pp. 1996–2003 (2021). https://doi.org/10.1109/ICESC51422.2021.9532917
3. Liu, Y., Lan, D., Pang, Z., Karlsson, M., Gong, S.: Performance evaluation of containerization in edge-cloud computing stacks for industrial applications: A client perspective. IEEE Open Journal of the Industrial Electronics Society **2**, 153–168 (2021). https://doi.org/10.1109/OJIES.2021.3055901
4. Ma, L., Yi, S., Carter, N., Li, Q.: Efficient live migration of edge services leveraging container layered storage. IEEE Transactions on Mobile Computing **18**(9), 2020–2033 (2019). https://doi.org/10.1109/TMC.2018.2871842
5. Stoyanov, R., Kollingbaum, M.J.: Efficient live migration of Linux containers. In: Yokota, R., Weiland, M., Shalf, J., Alam, S. (eds.) High Performance Computing, pp. 184–193. Springer, Cham (2018)
6. Xu, B., Wu, S., Xiao, J., Jin, H., Zhang, Y., Shi, G., Lin, T., Rao, J., Yi, L., Jiang, J.: Sledge: Towards efficient live migration of Docker containers. In: 2020 IEEE 13th International Conference on Cloud Computing (CLOUD), pp. 321–328 (2020). https://doi.org/10.1109/CLOUD49709.2020.00052
7. Puliafito, C., Vallati, C., Mingozzi, E., Merlino, G., Longo, F., Puliafito, A.: Container migration in the fog: A performance evaluation. Sensors **19**(7) (2019). https://doi.org/10.3390/s19071488
8. Bouaouda, A., Afdel, K., Abounacer, R.: Forecasting the energy consumption of cloud data centers based on container placement with ant colony optimization and bin packing. In: 2022 5th Conference on Cloud and Internet of Things (CIoT), pp. 150–157 (2022). https://doi.org/10.1109/CIoT53061.2022.9766522
9. Bouaouda, A., Afdel, K., Abounacer, R.: Meta-heuristic and heuristic algorithms for forecasting workload placement and energy consumption in cloud data centers. Advances in Science, Technology and Engineering Systems Journal **8**(1), 1–11 (2023). https://doi.org/10.25046/aj080101
10. Torre, R., Urbano, E., Salah, H., Nguyen, G.T., Fitzek, F.H.P.: Towards a better understanding of live migration performance with Docker containers. In: European Wireless 2019; 25th European Wireless Conference, pp. 1–6 (2019)
11. He, T., N. Toosi, A., Buyya, R.: Performance evaluation of live virtual machine migration in SDN-enabled cloud data centers. Journal of Parallel and Distributed Computing **131**, 55–68 (2019). https://doi.org/10.1016/j.jpdc.2019.04.014
12. Felter, W., Ferreira, A., Rajamony, R., Rubio, J.: An updated performance comparison of virtual machines and Linux containers. In: 2015 IEEE International Symposium on Performance Analysis of Systems and Software (ISPASS), pp. 171–172 (2015). https://doi.org/10.1109/ISPASS.2015.7095802
13. Duggan, M., Shaw, R., Duggan, J., Howley, E., Barrett, E.: A multitime-steps-ahead prediction approach for scheduling live migration in cloud data centers. Software: Practice and Experience **49**(4), 617–639 (2019). https://doi.org/10.1002/spe.2635. https://onlinelibrary.wiley.com/doi/pdf/10.1002/spe.2635
14. Tyj, N., G, D.V.: Adaptive deduplication of virtual machine images using AKKA stream to accelerate live migration process in cloud environment. Journal of Cloud Computing **8** (2019). https://doi.org/10.1186/s13677-019-0125-z
15. Beloglazov, A., Buyya, R.: Energy efficient resource management in virtualized cloud data centers. In: 2010 10th IEEE/ACM International Conference on Cluster, Cloud and Grid Computing, pp. 826–831 (2010). https://doi.org/10.1109/CCGRID.2010.46
16. Govindaraj, K., Artemenko, A.: Container live migration for latency critical industrial applications on edge computing. In: 2018 IEEE 23rd International Conference on Emerging Technologies and Factory Automation (ETFA), vol. 1, pp. 83–90 (2018). https://doi.org/10.1109/ETFA.2018.8502659
17. Agarwal, A., Raina, S.: Live migration of virtual machines in cloud. International Journal of Scientific and Research Publications **2**(6), 1–5 (2012)

18. Noshy, M., Ibrahim, A., Ali, H.A.: Optimization of live virtual machine migration in cloud computing: A survey and future directions. Journal of Network and Computer Applications **110**, 1–10 (2018). https://doi.org/10.1016/j.jnca.2018.03.002
19. Benjaponpitak, T., Karakate, M., Sripanidkulchai, K.: Enabling live migration of containerized applications across clouds. In: IEEE INFOCOM 2020 - IEEE Conference on Computer Communications, pp. 2529–2538 (2020). https://doi.org/10.1109/INFOCOM41043.2020.9155403
20. Rybina, K., Patni, A., Schill, A.: Analysing the migration time of live migration of multiple virtual machines. CLOSER **14**, 590–597 (2014)
21. Huang, Q., Gao, F., Wang, R., Qi, Z.: Power consumption of virtual machine live migration in clouds. In: 2011 Third International Conference on Communications and Mobile Computing, pp. 122–125 (2011). https://doi.org/10.1109/CMC.2011.62
22. Voorsluys, W., Broberg, J., Venugopal, S., Buyya, R.: Cost of virtual machine live migration in clouds: A performance evaluation. In: Cloud Computing: First International Conference, CloudCom 2009, Beijing, China, December 1–4, 2009. Proceedings 1, pp. 254–265 (2009). Springer.
23. Shang, Y., Lv, D., Liu, J., Dong, H., Ren, T.: Container memory live migration in wide area network. In: Qiu, M. (ed.) Smart Computing and Communication, pp. 68–78. Springer, Cham (2021).
24. Zaher, C.: For CTO's: the no-nonsense way to accelerate your business with containers. Technical report, Canonical Limited 2017. Ubuntu, Kubuntu (February 2017)
25. Pahl, C., Brogi, A., Soldani, J., Jamshidi, P.: Cloud container technologies: A state-of-the-art review. IEEE Transactions on Cloud Computing **7**(3), 677–692 (2019). https://doi.org/10.1109/TCC.2017.2702586
26. Ruan, B., Huang, H., Wu, S., Jin, H.: A performance study of containers in cloud environment. In: Wang, G., Han, Y., Martínez Pérez, G. (eds.) Advances in Services Computing, pp. 343–356. Springer, Cham (2016)
27. Mathew, S., Varia, J.: Overview of Amazon web services. Amazon Whitepapers **105**, 1–22 (2014)
28. Gupta, B., Mittal, P., Mufti, T.: A review on Amazon Web Service (AWS), Microsoft Azure & Google Cloud Platform (GCP) services. EAI, (2021). https://doi.org/10.4108/eai.27-2-2020.2303255
29. Muhammed, A.S., Ucuz, D.: Comparison of the IoT platform vendors, Microsoft Azure, Amazon Web Services, and Google Cloud, from users' perspectives. In: 2020 8th international symposium on digital forensics and security (ISDFS), pp. 1–4 (2020). https://doi.org/10.1109/ISDFS49300.2020.9116254
30. Dutta, P., Dutta, P.: Comparative study of cloud services offered by Amazon, Microsoft & Google. International Journal of Trend in Scientific Research and Development **3**(3), 981–985 (2019)
31. Al-Sayyed, R.M., Hijawi, W., Bashiti, A.M., AlJarah, I., Obeid, N., Adwan, O.Y.: An investigation of Microsoft Azure and Amazon Web Services from users' perspectives. International Journal of Emerging Technologies in Learning **14**(10) (2019)
32. Stephen, A., Benedict, S., Kumar, R.A.: Monitoring IaaS using various cloud monitors. Cluster Computing **22**(Suppl 5), 12459–12471 (2019)
33. Ostermann, S., Iosup, A., Yigitbasi, N., Prodan, R., Fahringer, T., Epema, D.: A performance analysis of ec2 cloud computing services for scientific computing. In: Cloud Computing: First International Conference, CloudComp 2009 Munich, Germany, October 19–21, 2009 Revised Selected Papers 1, pp. 115–131 (2010). Springer

34. He, K., Fisher, A., Wang, L., Gember, A., Akella, A., Ristenpart, T.: Next stop, the cloud: Understanding modern web service deployment in EC2 and azure. In: Proceedings of the 2013 conference on internet measurement conference, pp. 177–190 (2013)
35. Pereira Ferreira, A., Sinnott, R.: A performance evaluation of containers running on managed Kubernetes services. In: 2019 IEEE International Conference on Cloud Computing Technology and Science (CloudCom), pp. 199–208 (2019). https://doi.org/10.1109/CloudCom.2019.00038
36. Ifrah, S., Ifrah, S.: Deploy a containerized application with Amazon EKS. Deploy Containers on AWS: With EC2, ECS, and EKS, 135–173 (2019)
37. Talha, M., Sohail, M., Hajji, H.: Analysis of research on Amazon AWS cloud computing seller data security. International Journal of Research in Engineering Innovation **4**(3), 131–136 (2020)

Chapter 2

A deep learning (DL) framework for gastrointestinal (GI) abnormality classification and localization within WCE images

Yassine Oukdach, Zakaria Kerkaou, Mohamed El Ansari, Lahcen Koutti, and Ahmed Fouad El Ouafdi

2.1 INTRODUCTION

Medical imaging analysis has garnered significant research attention recently, primarily due to numerous challenges associated with image analysis, interpretation, and disease identification. Computer-aided diagnostic-based artificial intelligence (AI) algorithms have been the most commonly used tools by researchers in addressing these challenges. These AI-driven algorithms have proven invaluable in enhancing the speed and accuracy of diagnoses, enabling early disease detection and personalized treatment strategies in different parts of the human body, including skin cancer recognition [1], brain tumor detection [2], prostate cancer [3], breast cancer [4], and gastrointestinal (GI) disease [5–11]. GI tract abnormalities, in particular, contribute significantly to cancer-related deaths worldwide [12]. In 2023, the United States recorded around 348,840 new cancer cases and 172,010 deaths due to cancer among both genders. Colon and rectal cancers were the most prevalent, with an estimated 153,020 new cases and 52,550 fatalities [13, 14]. Furthermore, colorectal cancer stands as the third leading cause of cancer-related mortality across all cancer types. The digestive system is among the most intricate aspects of the human body to examine, owing to its complex structure and sensitivity. In many cases, this examination necessitates surgical procedures to identify abnormalities and their source. However, numerous tools have been developed to tackle these challenges, including small bowel enteroscopy, colonoscopy, and duodenoscopy, to name a few. The latter two are effective for visualizing the surrounding bowel [15]. However, these methods have specific limitations when selecting a diagnostic approach for suspected small bowel diseases. Among these tools, only enteroscopy allows for the simultaneous assessment of the small bowel mucosa and the treatment of issues. The invasive nature of this procedure carries risks of adverse events, limits patient acceptance, is highly time-consuming, and is not widely accessible. Moreover, the intricate structure and mobility of the small bowel add complexity to the performance of enteroscopy. However, wireless capsule endoscopy (WCE) [16] technology has emerged as a new, non-invasive examination tool, addressing the limitations associated with the enteroscopy procedure.

DOI: 10.1201/9781032664293-3

WCE is a new imaging diagnostic procedure, introduced in 2000 by the Food and Drug Administration (FDA). It features a small, 11 × 11 mm capsule equipped with a camera. The process commences by having the patient swallow the capsule, which then navigates the entire digestive system. While traveling through the digestive tract, the miniature camera captures thousands of images, including various segments of the GI tract, such as the small bowel. In a single 8–10 hour examination, this capsule generates approximately 50,000–60,000 images. These images are subsequently transmitted to an external device for in-depth analysis by expert gastroenterologists, enabling the identification of any existing abnormalities [17]. Interpreting these images is a laborious and time-consuming task for doctors. Therefore, the implementation of an automated tool for the analysis of GI abnormalities not only reduces the time for healthcare professionals but also enhances accuracy and consistency in diagnosis, especially in cases located in deep regions or displaying small-sized anomalies. Computer vision tasks for image analysis, relying on machine learning (ML) and deep learning (DL) algorithms, have shown a significant impact in the last decade. They excel in detecting and localizing GI abnormalities based on images generated by both colonoscopy and WCE technology. ML algorithms, including Support Vector Machine (SVM) [18], K-Nearest Neighbors (KNN) [19], and Random Forest (RF) [20], have been employed to classify features extracted by classical algorithms such as local binary pattern (LBP) [21], Histogram of Oriented Gradients (HOG) [22], and Scale-Invariant Feature Transform (SEFT) [23], among others. The primary challenge with hand-crafted methods lies in their limited applicability to specific tasks, as they necessitate manual feature extraction design. Furthermore, WCE technology generates images with diverse features in terms of shape, texture, and color. This diversity makes it challenging for hand-crafted methods to adapt to the variations in features. However, DL-based convolutional neural networks (CNNs) have played a major role in detecting and localizing abnormalities in WCE images [24]. The CNN's ability to automatically extract and learn hierarchical and discriminative features from raw image data through convolution layers and pooling operations makes them well-suited for various tasks related to GI disease identification, including classification [25], detection [26], and segmentation [27]. Despite this, CNNs are effective, but they have several drawbacks, including their requirement for large training datasets and the limitation that convolution-based filters capture primarily local image information. These limitations pose significant challenges to CNN's performance. However, as the majority of datasets generated by WCE technology are small and imbalanced, transfer learning has emerged as an effective way to address the challenges posed by small datasets. The process involves exploiting pre-trained models, such as ResNet, MobileNet, and Xception, which were trained on large datasets like ImageNet, COCO, and VOC datasets, and making them suitable for a specific task. This approach is often highly effective in adapting well-established knowledge from one domain to another, significantly enhancing the performance of models when dealing with limited and imbalanced data. However, the field of detecting and localizing GI diseases remains a challenging area for both researchers and medical professionals, necessitating further opportunities for progress. Most datasets derived from WCE images are limited in size and suffer from significant imbalances. Additionally, WCE images are typically captured in conditions with poor lighting, leading to issues like reduced contrast, intricate backgrounds, and substantial variations in shape and texture. These factors add complexity to the analysis and interpretation of WCE images. Figure 2.1 displays some of the images generated by WCE images. This chapter introduces a comprehensive framework for both detecting and localizing abnormalities in WCE images. Our approach leverages two distinct methods

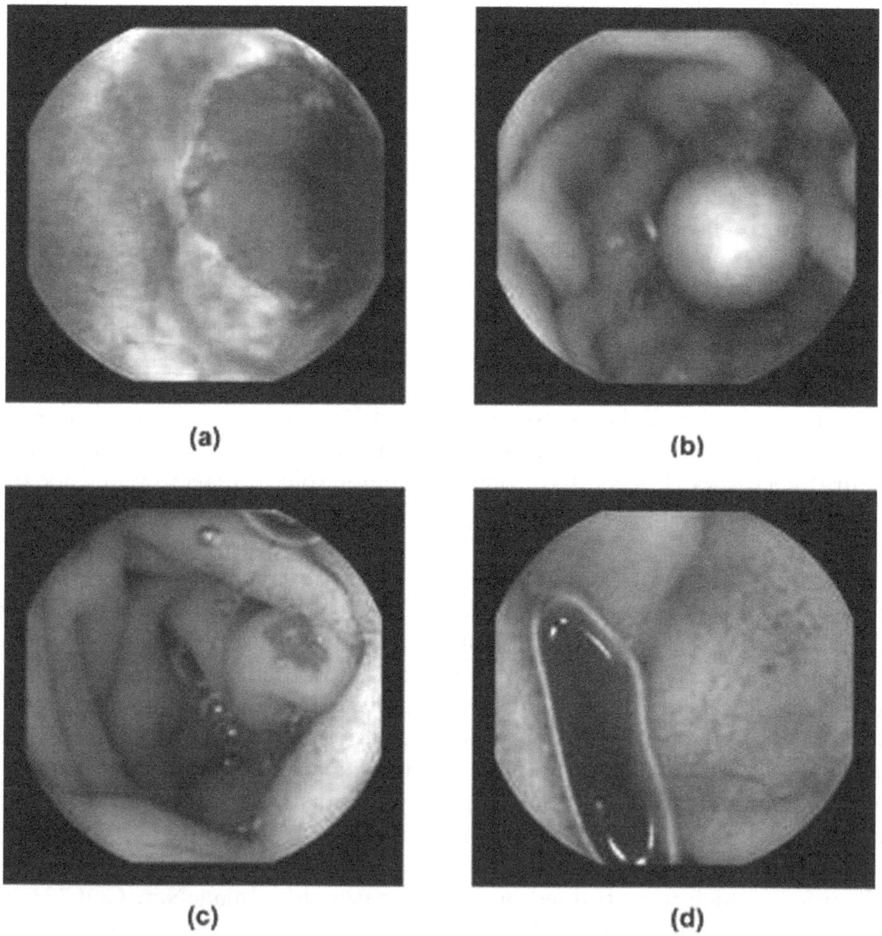

Figure 2.1 Examples of WCE images; (a) inflammatory; (b) polyp; (c) vascular; (d) healthy part.

for feature extraction. The first method employs a custom-designed, efficient CNN to classify cases into inflammatory, polyp, vascular, and normal categories. Meanwhile, the second method utilizes a modified DeepLabv3+ network to precisely localize abnormalities within WCE images.

The remainder of this chapter is structured as follows: Section 2.2 reviews state-of-the-art methods, and Section 2.3 details our proposed approach for efficient detection and localization of GI abnormalities. Section 2.5 discusses the dataset, experimental results, and analysis. Lastly, Section 2.6 presents the concluding remarks of this study.

2.2 RELATED WORKS

GI abnormalities, such as polyps, ulcers, and bleeding, have gained significant attention in the fields of computer vision and image processing. Automated detection and localization of abnormalities in the human digestive system pose significant challenges for researchers and healthcare professionals. Initially, feature detection and description in WCE images relied on handcrafted techniques, which involved manually defining features for extraction and utilizing ML algorithms to distinguish between normal and abnormal regions. Barbosa et al. [28] introduced statistical texture features based on color curvelet covariance for the automated detection of lesions in WCE images. Charfi et al. [29] proposed a new method for identifying colon abnormalities in WCE images. This approach utilizes LBPs, variance, and discrete wavelet transform features to compute texture-related information from the extracted features, which is subsequently employed in classification by an SVM. In their study, Hajabdollahi et al. [30] introduced an uncomplicated and efficient method for localizing bleeding regions within WCE images. They employed color transformations, such as hue, saturation, intensity, and LAB, for feature extraction and utilized a multi-perceptron learning approach for classification, leading to enhanced performance. Explore other studies that have utilized handcrafted methods for feature descriptions in WCE images [31–35]. However, handcrafted methods for feature extraction from WCE images have numerous inherent limitations. They involve manual feature design, making them subjective and potentially time-consuming. These methods are often task-specific, challenging to adapt, and may struggle with handling complex features present in WCE images, such as intricate textures and colors. Additionally, they lack the adaptability and automatic learning capabilities of DL, making them less suitable for large and diverse datasets. In clinical practice, these limitations may hinder their effectiveness in detecting abnormalities within WCE images, emphasizing the need for more advanced and flexible techniques. Deep CNN models, in contrast, have surpassed the limitations of handcrafted methods by automatically extracting a comprehensive set of features from WCE images [24]. Jain et al. [36, 37] introduced two separate studies focusing on both detecting and localizing abnormalities. In their initial work, they presented an efficient attention-based CNN designed to categorize images into four distinct groups: polyp, vascular, inflammatory, and normal. Subsequently, the authors applied a combination of Grad-CAM++ and a custom SegNet to accurately identify anomalous regions within abnormal images. Three parallel CNNs were used in their second study, contributing to enhanced classification performance through the statistical correlation of rich feature sets and the inclusion of the random forest algorithm as a classifier. Goel et al. [38] utilized a dual-branch CNN architecture in their work. One branch serves as the backbone network, while the other employs resolution-preserving dilated convolution operations. The backbone CNN is responsible for extracting diverse features at various scales, while the dilated convolution branch extends the receptive field and aids in capturing prominent features. Finally, the fusion of features from the backbone CNN and the dilated CNN combines to retrieve the overarching dominant features. In their study, Mohapatra et al. [39] introduced an approach to categorize diseases of the alimentary canal, encompassing conditions like Barrett's, Esophagitis, Hemorrhoids, Polyps, and Ulcerative colitis. Their method involves the integration of empirical wavelet transform (EWT) and CNN model. The study makes use of the publicly accessible HyperKvasir dataset for experimental purposes. The procedure initiates with several stages of image preprocessing, followed by the application of EWT. EWT is employed to dissect the images into distinct modes, simplifying the extraction of specific patterns within the images. These decomposed images are then fed into the deep CNN for the disease classification process. In [40], the authors employed RGB images, the Hessian, and the Laplacian of images as inputs for three parallel CNNs used in feature extraction. The obtained set of features is fused and then fed into a fully connected layer for classification. However, CNN models are effective in learning features from images due to their hierarchical structure. Nevertheless, they demand a substantial amount of training data, and regrettably, the majority of medical datasets, especially GI datasets, are private or limited in size. Transfer learning has emerged as an effective strategy for addressing the challenges presented by limited dataset sizes. The process involves transferring the knowledge acquired by pre-trained models from a specific task and fine-tuning them for particular tasks, such as classification, detection, and segmentation [32, 41, 42]. In our previous work [5], we fine-tuned the pre-trained model ResNet50 for feature extraction. We used the HSV color space instead of RGB images as inputs for the model. The best-selected features were fed into a multi-layer perceptron (MLP) module to categorize the WCE images into polyp, ulcer, and normal. In their work, Caroppo et al. [43] extracted features from the intermediate layers of ResNet, VGG, and Inception neural networks. These features were combined and enhanced using the maximum relevance-minimum redundancy (MRMR) technique, and ML algorithms were then utilized to distinguish between bleeding and non-bleeding regions within WCE images. In [44], authors utilized VGG and AlexNet models for feature extraction. The features extracted from these models were chosen through the use of the genetic algorithm (GA). They employed a serial approach to combine these features and fed them into ML algorithms for the classification of images into categories, including gastritis, ulcer, esophagitis, bleeding, and healthy in their private dataset. However, WCE technology generates images with diverse features, such as color, texture, and shape. These features can exhibit significant variations in different types of GI abnormalities. Furthermore, multiple diseases, such as polyps, ulcers, and inflammatory, can be small in size and may exist in deep regions of the GI tract, making their detection and localization challenging. Therefore, many efforts have been employed to detect and localize small anomalies in both colonoscopy and WCE images [9, 45, 46]. Souaidi et al. [26] introduced an approach named the multiscale pyramidal fusion single-shot multi-box detector network (MP-FSSD) to tackle the detection of small polyp areas within WCE images. Their method capitalizes on deep transfer learning to extract highly representative features and contextual information using the FSSD model. To accomplish this, they integrated an edge-pooling layer early in the network, conducted size transformations on feature maps from various layers and scales, and employed a concatenation module to merge these transformed feature maps. These modified feature maps were subsequently utilized in multi-box detectors to produce the ultimate detection results. The study in [47] introduces a DL framework called SR-AttNet for colorectal polyp segmentation. This framework employs an encoder-decoder architecture with both undilated and dilated filters to capture information from both nearby and distant regions and to comprehend image depth. To enhance the model, the authors implemented four-fold skip connections between each spatial encoder and decoder and used a Feature-to-Mask pipeline to manage both dilated and undilated features for the final prediction. Furthermore, their approach includes an attention mechanism named SR-Attention, inspired by the Stretch-Relax concept, which generates spatial features with high variance to create effective attention masks for feature selection.

2.3 METHODOLOGY

In this section, we will outline our proposed method for classifying and localizing GI diseases. The proposed system can be divided into three main parts. First, we have implemented several preprocessing steps, including data augmentation, resizing, and normalization. After data preparation, an efficient, simple, dilated CNN is employed for feature extraction, and an MLP is used to classify WCE images into four categories: Normal, inflammatory, polyp, and vascular. In the third phase of the proposed approach, a modified DeepLab version 3+ is fine-tuned through a transfer learning mechanism for GI disease localization. The details of each step are discussed in the following subsections, while Figure 2.2 presents the overall structure of the proposed approach.

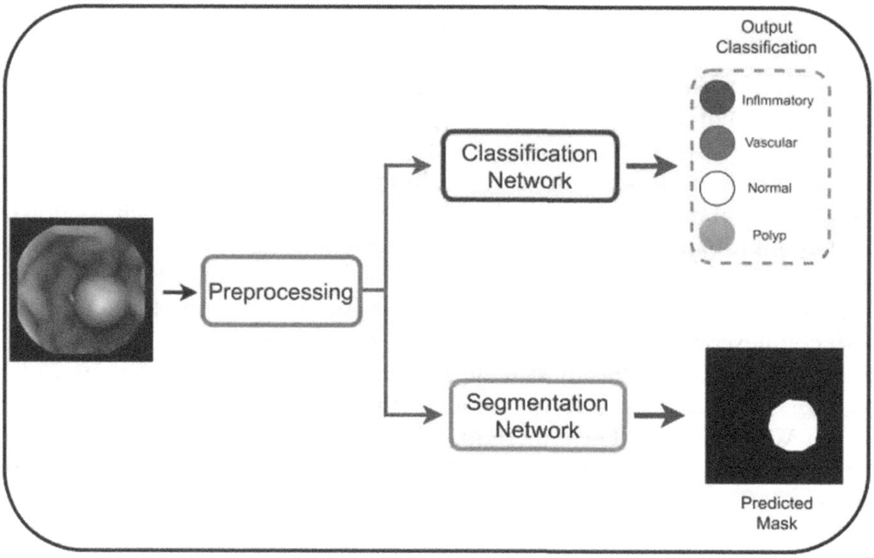

Figure 2.2 The structure of the proposed framework for GI abnormalities classification and localization in WCE images.

2.3.1 Data preprocessing

In this study, we utilized the KID dataset, characterized by class imbalance, which necessitated augmentation techniques to upsample the minority class. The augmentation methods employed encompassed flipping, cropping, zooming within a range of 0.2, rotations within the range of –20 to 20 degrees, and the addition of Gaussian noise with a standard deviation of 0.6. Following the data balancing process, all images were resized to 256 × 256 pixels and normalized, resulting in feature values within the [0–1] range. Figure 2.3 displays samples of the prepared data, while Table 2.1 presents the number of images for each class before and after the augmentation process.

2.3.2 The proposed dilated CNN for GI classification

The proposed CNN for GI disease classification involves multiple blocks of dilated convolutions, pooling, batch normalization, and ReLU activation layers. Dilated convolution, also known as atrous convolution, is a variant of the standard convolution operation. It differs from traditional convolution in that it introduces gaps or dilations between the values being convolved. This spacing between the values is controlled by a parameter called the dilation rate. Figure 2.4 illustrates the main difference between regular convolution and dilated convolution. In standard convolution, a filter slides over the input image, and at each position, it computes a weighted sum of the values within the filter's receptive field. The receptive field is a small region of the input

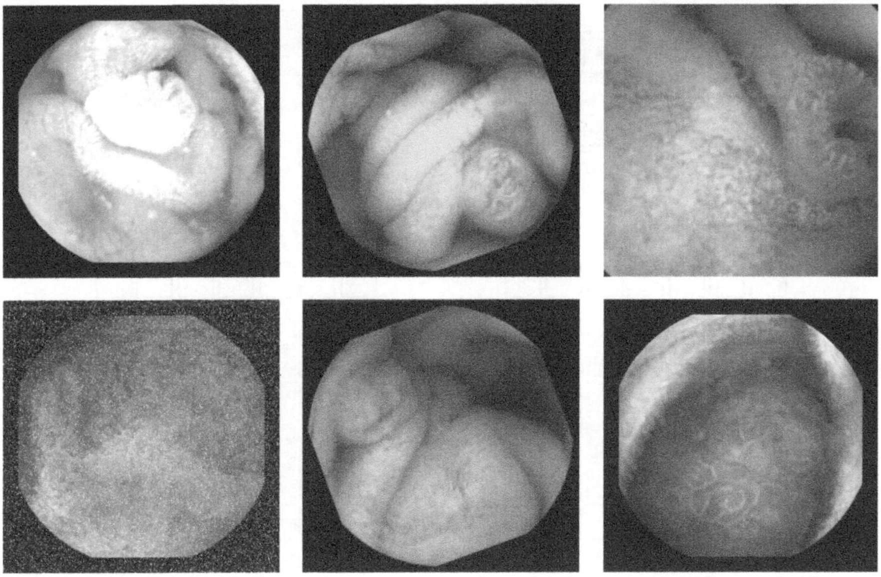

Figure 2.3 Examples of images in the dataset after applying various augmentation techniques, such as zooming, flipping, rotation, adding Gaussian noise, and more.

Table 2.1 Overview of images in the KID dataset before and after augmentation across various classes

Images in the dataset	*Before augmentation*	*After augmentation*
Inflammatory	227	1267
Vascular	303	1244
Polyp	44	1288
Normal	1327	1327

image defined by the filter's size. In dilated convolution, the dilation rate specifies the gap between the values in the receptive field, effectively increasing the receptive field without increasing the filter size. As depicted in Figure 2.5, the proposed dilated CNN for GI abnormalities consists of six sets. Each set includes a dilated convolution, batch normalization, and pooling layers. The number of filters starts at 16 and increases by 2 in each subsequent set. The dilation rate starts at 1 in the initial set and increments by 2 in each successive set. This strategy enables the proposed CNN to capture multiscale information, ranging from fine-grained details in the early set to broader context in the later set, enhancing its ability to understand both local and global features in the WCE images. Furthermore, it aids in mitigating over-fitting, enhancing the model's performance in GI abnormality detection.

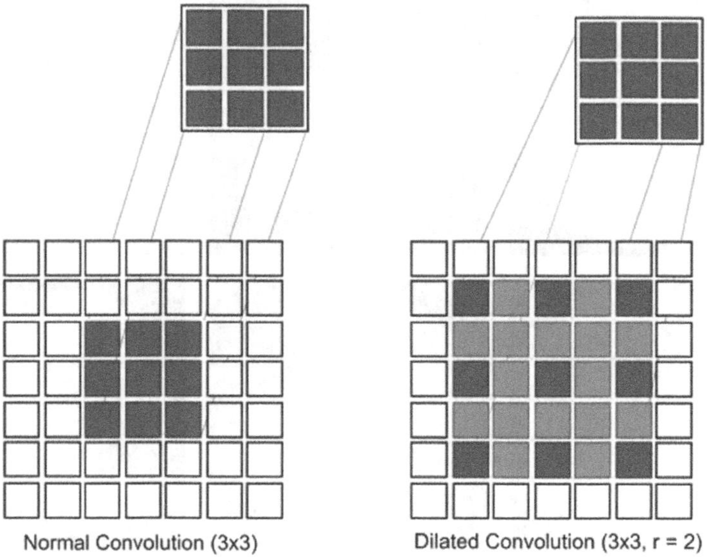

Figure 2.4 The main difference between the regular convolution and the dilated one.

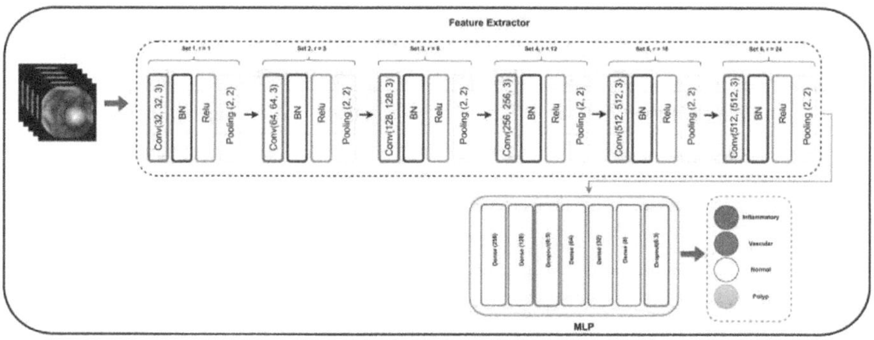

Figure 2.5 The proposed dilated network for GI abnormalities classification.

2.3.3 The proposed network for GI localization

In this section, we will elucidate the proposed encoder-decoder approach for localizing GI abnormalities. Our method leverages DeepLabv3+ [48], a widely employed framework in state-of-the-art image segmentation. The core elements of DeepLabv3+ encompass a deep, pre-trained Xception model as the backbone in the encoder, coupled with the atrous spatial pyramid pooling (ASPP). The decoder comprises a sequence of Conv2D and Upsampling layers. Figure 2.6 shows the overall diagram of the proposed framework for

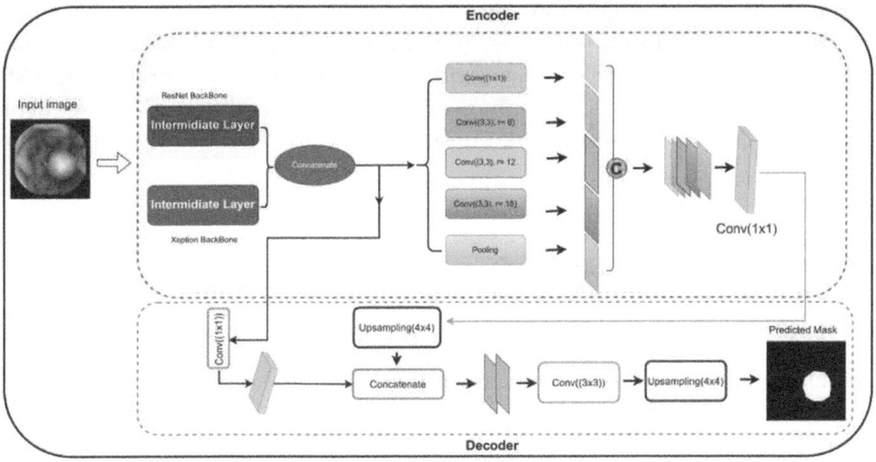

Figure 2.6 The pipeline of the proposed system for GI abnormality segmentation in WCE images.

GI localization, whereas the details of each component are discussed in the upcoming phases.

2.3.3.1 *Encoder architecture*

The encoder component in DeepLabv3+ initiates with a deep, pre-trained Xception model acting as the foundation for feature extraction, which is then followed by the atrous spatial pyramid pooling (ASSP) module. In the proposed approach, we have modified the original feature extraction backbone. We extract features from the intermediate layers of both ResNet50 and Xception. The features of ResNet50 and Xception are fused, and the feature set obtained is conveyed to the ASSP module for resampling the feature maps at varying atrous rates. The ASSP module consists of a max pooling layer and three Conv2D layers with a (1 × 1) kernel size, followed by an additional three Conv2D layers with (3 × 3) kernels, each employing different atrous rates—6, 12, and 18, respectively. By applying parallel convolutions with these different rates to the feature maps from the backbone, our network effectively captures multiscale information. This approach proves especially effective since WCE images commonly exhibit abnormalities of varying scales.

2.3.3.2 *Decoder architecture*

The decoder component generates segmentation maps at the original image resolution using low-scale feature maps from the encoder. As illustrated in Figure 2.3, the feature maps from the encoder are initially upscaled by a factor of 4 to produce the semantic segmentation map. These maps are then merged with intermediate low-level features from the ResNet and Xception backbones, maintaining the same spatial dimensions. By combining spatially rich low-level features from the encoder with high-level features from ASPP, segmentation performance is enhanced. A 3 × 3 convolution is subsequently applied, followed by another 4× upscaling to produce the final prediction mask.

2.4 EXPERIMENT RESULTS

2.4.1 Dataset

The experiments are conducted using the KID dataset, a multi-class benchmark that encompasses various categories representing both abnormal and normal segments of the GI tract. Within the KID dataset used in this work, there are three distinct abnormalities: Inflammatory (227 images), polyp (44 images), and vascular (303 images). The healthy part contains four categories: normal colon (169 images), normal esophagus (282 images), normal small bowel (728 images), and normal stomach (599). Only the normal small bowel is included in this study. The inflammatory diseases encompass a group of conditions marked by inflammation within the digestive system. One of the most common and widely recognized inflammatory GI diseases is inflammatory bowel disease (IBD), which includes Crohn's disease and ulcerative colitis. IBD can lead to chronic inflammation and damage in various parts of the GI tract, resulting in symptoms like abdominal pain, diarrhea, and rectal bleeding. GI vascular diseases involve disorders affecting the blood vessels within the digestive system. These conditions can lead to symptoms like bleeding, abdominal pain, and tissue damage, while polyps in the GI tract are growths that resemble groups of cells on the colon lining; resulting from uncontrolled cell proliferation, they often share a color similar to the surrounding intestinal walls. Their unique structure distinguishes them from healthy tissue. The proposed system is trained and validated using the KID dataset. To address the dataset's inherent imbalance, we apply various augmentation techniques, as outlined in the image processing step, to all classes, ensuring a balanced data distribution. Additionally, we resize all images to 254 × 254 pixels and partition them into three sets: 80% for training, 10% for validation, and 10% for testing.

2.5 PERFORMANCE OF THE PROPOSED CLASSIFICATION MODEL

The classification model uses a dilated CNN at different atrous rates for classifying GI abnormalities into four categories: Inflammatory, vascular, polyp, and normal. The proposed system is built with Keras and TensorFlow as the backend, running on a machine equipped with an Intel i7 processor, 64 GB

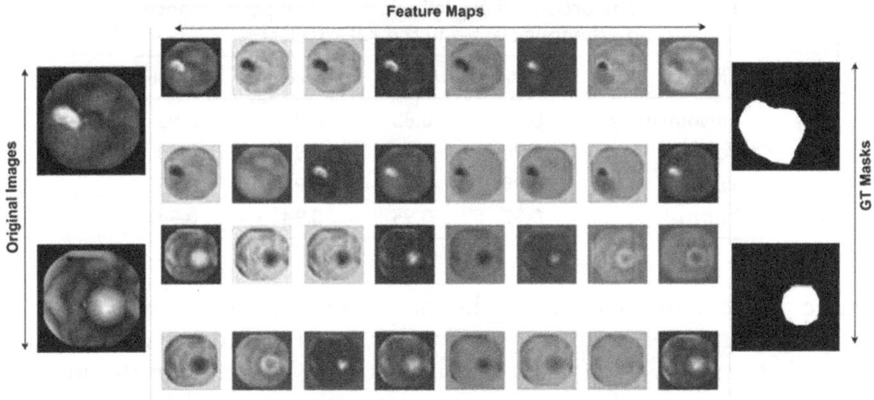

Figure 2.7 Visualization of the predicted features of the proposed CNN classifier on the test set.

of RAM, and a 24 GB NVIDIA GeForce RTX 3090 GPU. It trains on RGB images for 100 epochs using the Adam optimizer with a learning rate of 0.001 and employs categorical cross-entropy as the loss function. The model's effectiveness in classification tasks is evaluated through the assessment of various metrics, such as precision (Prec), recall (Rec), accuracy (Acc), F1-score (F-1), and area under the curve (AUC). In the proposed analysis, as shown in Figure 2.7, we utilized feature maps from the last dilated convolutional layer of the proposed method to visualize the model's attention in the test set. Comparing these feature maps with the ground truth labels and predicted features, the visualization clearly illustrates that the model has the ability to identify abnormal regions, showcasing its ability to detect areas of interest within the images. Table 2.2 presents that the proposed model achieves its highest performance with the Adam optimizer and a batch size of 32. Adam's superiority can be attributed to its adaptive learning rates, incorporation of momentum, bias correction, ability to handle sparse gradients, and regularization. These factors collectively enable faster convergence and make Adam a versatile and user-friendly choice for training the proposed method, even though the relative performance of optimizers can vary depending on the

Table 2.2 Comparative study between the use of Adam and SGD algorithms for training

Adam	*SGD*	*#Batch size = 32*	*#Batch size = 16*	*Precision*	*Recall*	*F-1 score*	*Accuracy*
✓	×	✓	×	94.35	93.78	93.45	93.39
✓	×	×	✓	91.33	90.12	90.65	91.56
×	✓	✓	×	85.33	84.78	84.65	83.85
×	✓	×	✓	86.22	85.01	84.11	84.77

Table 2.3 The proposed model's classification performance on the four classes in the KID dataset

Classes	*Precision*	*Recall*	*F1-score*	*Accuracy*
Inflammatory	0.98	0.85	0.91	0.93
Polyp	0.86	0.96	0.90	0.93
Vascular	0.99	0.99	0.99	0.95
Normal	0.92	0.95	0.94	0.94

specific problem and dataset. Furthermore, to comprehensively evaluate the model's performance, we present in Table 2.3 and Figure 2.8 a performance report that assesses its effectiveness in distinguishing among the four distinct classes. The classification report indicates that the model exhibits strong performance, with high precision in categories like 'inflammatory' and 'vascular,' ensuring accurate positive predictions. The model also demonstrates impressive recall in classes like polyp and 'normal,' meaning it effectively identifies a high proportion of relevant instances. The balanced F1-scores

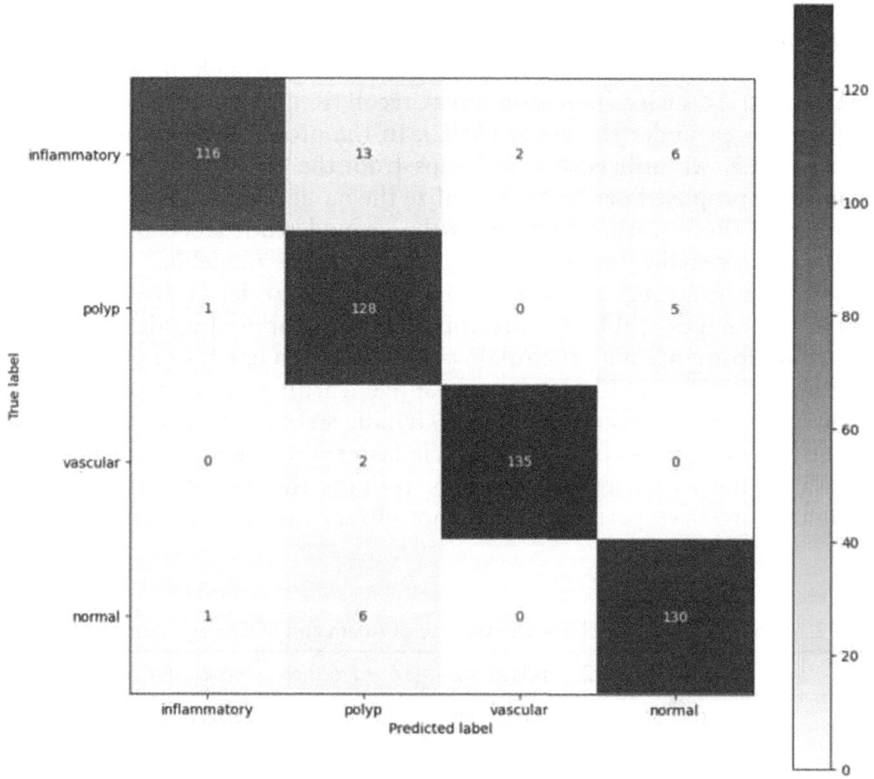

Figure 2.8 Visualization of the confusion matrix for the proposed system's classifier.

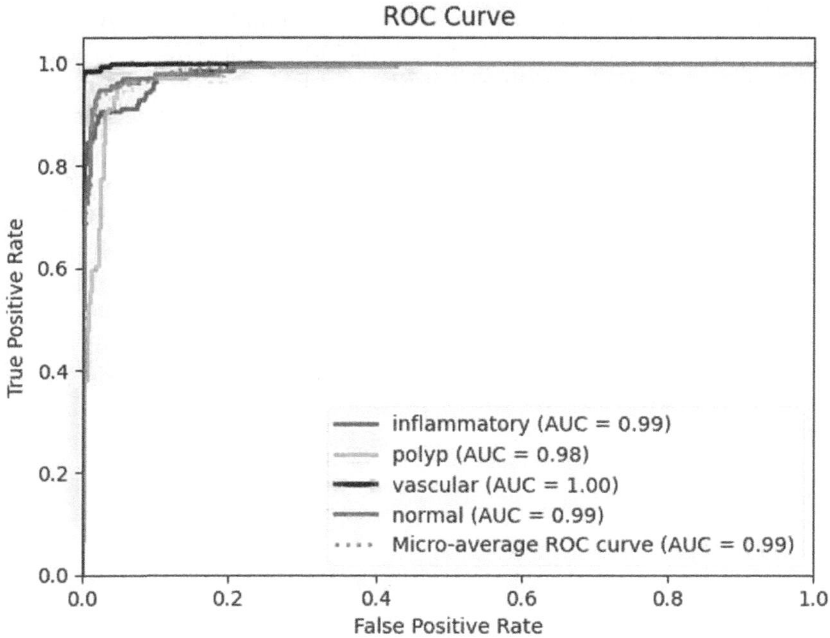

Figure 2.9 ROC curves of the proposed model.

across all classes underscore the model's overall effectiveness. The confusion matrix reveals the proposed model's proficiency in distinguishing between the vascular, normal, and polyp classes. However, it faces a minor challenge when differentiating the inflammatory class from the polyp and healthy segments, possibly due to similarities in the abnormal regions of these categories. This similarity in abnormal regions may contribute to the model's occasional difficulty in accurate classification. Additionally, we present a Receiver Operating Characteristic (ROC) curve in Figure 2.9, providing further insights into the model's performance and its ability to discriminate between these classes.

2.5.1 The proposed network for GI localization

The proposed model has been trained on a binary KID dataset for localizing GI abnormalities. We combined all abnormal images with their corresponding masks. This approach presents a considerable challenge for the model in accurately identifying disease regions and predicting the true mask, given the significant diversity among abnormal categories in their features. Furthermore, the limited dataset size and its imbalanced nature can influence the model's performance.

However, we addressed these challenges by training the proposed model with data augmentation strategies, which were applied during the classification phase. To assess the model's performance, we employed standard metrics commonly used in segmentation tasks, including the Dice coefficient (D), Jaccard index (J), precision (P), recall (R), and F1-score (F). Additionally, we utilized a combination of loss functions, including the Dice coefficient and binary cross-entropy (BCE), to guide the training process. The metrics used and the loss function are presented in the following equations, where TP, FP, FN, y, and $\hat{y}$ represent the true positive, false positive, false negative, ground truth, and predicted values, respectively:

$$J = \frac{TP}{TP + FP + FN} \tag{2.1}$$

$$D = \frac{2 \cdot TP}{2 \cdot TP + FP + FN} \tag{2.2}$$

$$\mathrm{P} = \frac{TP}{TP + FP} \tag{2.3}$$

$$\mathrm{R} = \frac{TP}{TP + FN} \tag{2.4}$$

$$\mathrm{BCE} = -(y \log(\hat{y}) + (1 - y)\log(1 - \hat{y})) \tag{2.5}$$

$$\text{Combined Loss} = \lambda \times \mathrm{BCE} - (1 - \lambda) \times \mathrm{D} \tag{2.6}$$

A comparative study was conducted with various backbones in the proposed segmentation model. The primary objective was to assess the most effective backbone for feature extraction before feeding them into the ASSP module and the decoder section. The study aims to assess the model's performance through a comparison of three configurations: using ResNet as the backbone, using Exception as the backbone, and employing their feature concatenation from the intermediate layer within the backbone. The objective is to determine the most effective configuration among these options. As shown in Table 2.4, the proposed feature extraction process outperforms ResNet and Xception in all metrics except the Dice coefficient, where the ResNet displays a dice of 70.94%. These improvements may be due to the

Table 2.4 Results of the comparative study

Backbones	*P*	*R*	*F-1*	*J*	*D*
ResNet	0.7094	0.7409	0.5702	0.5516	**0.7094**
Xception	0.5389	0.7387	0.4991	0.4806	0.5389
The proposed approach	**0.9498**	**0.8938**	**0.8700**	**0.8443**	0.4403

Table 2.5 Results of the comparative study

Thresholds	*P*	*R*	*F-1*	*J*	*D*
0.1	0.9499	0.8933	0.8698	0.8439	0.4452
0.3	0.9497	0.8942	0.8702	0.8446	0.4370
0.5	0.9496	0.8947	0.8704	0.8450	0.4319
0.7	0.9495	0.8951	0.8452	0.8452	0.4268
0.9	0.9493	0.8958	0.8708	0.8457	0.4193

successful concatenation of features from the intermediate layers of ResNet and Xception. The poorest results were obtained with the use of the Xception model in the backbone; this may be attributed to its limited feature extraction capabilities, and it might not fully meet the unique requirement of image segmentation. Nonetheless, the study affirms that the fusion of features from the intermediate layers of these models played a pivotal role in enhancing the model's performance. The model's performance at different thresholds is displayed in Table 2.5, while Figure 2.10 displays the prediction results of the GI abnormalities localization; the results demonstrate the ability of the proposed approach to locate various diseases in the GI tract. However, while the proposed model exhibits promising results in both the classification and segmentation of GI abnormalities, it still encounters several challenges, particularly in classification performance due to difficulties in distinguishing between classes. The imbalanced dataset and the considerable variation in image features contribute significantly to inaccurate mask predictions in localization performance. Therefore, it is imperative to explore alternative augmentation techniques or data generation methods to expand our training dataset. Furthermore, the separation and precise localization of diseases could offer valuable insights for future segmentation work.

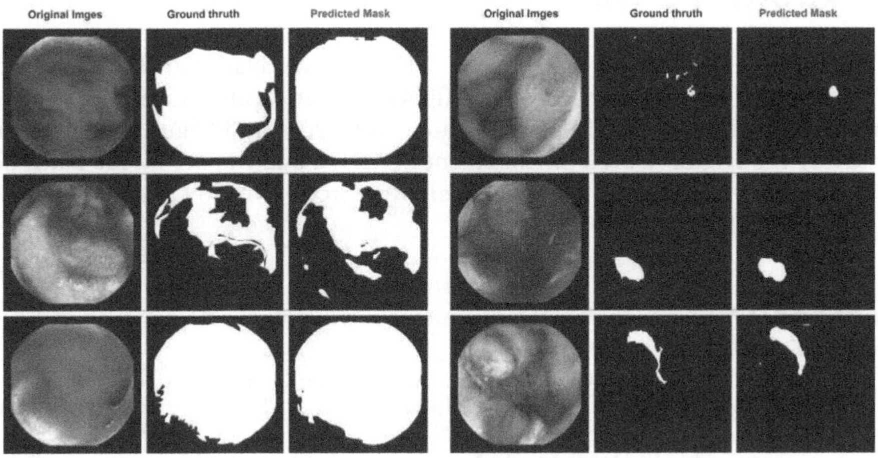

Figure 2.10 The model's performance for abnormalities localization in WCE images.

2.6 CONCLUSION

In this chapter, we introduce a hybrid approach for both GI classification and the localization of abnormalities. The classification phase employs an efficient and straightforward dilated CNN for feature extraction. The proposed CNN extracts features from WCE images at various rates, capturing fine-grained and high-level information crucial for accurate classification. The set features obtained are fed to an MLP module to categorize WCE images into four categories that are inflammatory, vascular, polyp, and normal. Meanwhile, the segmentation branch demonstrates the simplicity and efficiency of the DeepLabv3+ architecture. The original architecture was evaluated with various backbones, and the model's performance confirms the significance of the proposed feature extraction approach. This approach combines features from the intermediate layers of ResNet and Exception networks, resulting in a notable enhancement of GI intestinal localization.

The experiments were performed on the KID dataset, which contains several abnormalities and normal classes. The obtained results validate the hybrid approach's significant capability in classifying and locating diseases within WCE images. However, the proposed approach still faces several challenges, including confusion in classifying certain cases and difficulties in adapting to the significant variation in features within images. Therefore, for future work, it is essential to address these challenges. Possible avenues include improving the classification model's ability to handle complex cases and enhancing its robustness to feature variations within images. Furthermore, for the segmentation aspect, research could focus on refining the localization of abnormalities within the GI tract. Exploring advanced techniques for fine-tuning the segmentation model and adapting it to various feature variations would be crucial.

ACKNOWLEDGMENT

This work was supported by the Ministry of National Education by Vocational Training; in part by the Higher Education and Scientific Research through the Ministry of Industry, Trade, and Green and Digital Economy; in part by the Digital Development Agency (ADD); and in part by the National Center for Scientific and Technical Research (CNRST) under Project ALKHAWARIZMI/2020/20.

REFERENCES

1. Tschandl, P., Rinner, C., Apalla, Z., Argenziano, G., Codella, N., Halpern, A., Janda, M., Lallas, A., Longo, C., Malvehy, J., *et al.*: Human–computer collaboration for skin cancer recognition. Nature Medicine 26(8), 1229–1234 (2020).
2. Mohsen, H., El-Dahshan, E.-S.A., El-Horbaty, E.-S.M., Salem, A.-B.M.: Classification using deep learning neural networks for brain tumors. Future Computing and Informatics Journal 3(1), 68–71 (2018)

3. Goldenberg, S.L., Nir, G., Salcudean, S.E.: A new era: Artificial intelligence and machine learning in prostate cancer. Nature Reviews Urology 16(7), 391–403 (2019).
4. Bentaher, N., Kabbadj, Y., Salah, M.B.: Enhancing breast masses detection and segmentation: A novel u-net-based approach. In: 2023 10th International Conference on Wireless Networks and Mobile Communications (WINCOM), pp. 1–6 (2023). https://doi.org/10.1109/WINCOM59760.2023.10322998
5. Oukdach, Y., Kerkaou, Z., El Ansari, M., Koutti, L., El Ouafdi, A.F.: Gastrointestinal diseases classification based on deep learning and transfer learning mechanism. In: 2022 9th International Conference on Wireless Networks and Mobile Communications (WINCOM), pp. 1–6 (2022). IEEE.
6. Lafraxo, S., El Ansari, M., Koutti, L.: Computer-aided system for bleeding detection in WCE images based on CNN-GRU network. Multimedia Tools and Applications, 1–26 (2023).
7. Garbaz, A., Lafraxo, S., Charfi, S., El Ansari, M., Koutti, L.: Bleeding classification in wireless capsule endoscopy images based on Inception-ResNet-V2 and CNNS. In: 2022 IEEE Conference on Computational Intelligence in Bioinformatics and Computational Biology (CIBCB), pp. 1–6 (2022). IEEE.
8. Souaidi, M., Lafraxo, S., Kerkaou, Z., El Ansari, M., Koutti, L.: A multiscale polyp detection approach for GI tract images based on improved DenseNet and single-shot multibox detector. Diagnostics 13(4), 733 (2023).
9. Lafraxo, S., Souaidi, M., El Ansari, M., Koutti, L.: Semantic segmentation of digestive abnormalities from WCE images by using AttResU-Net architecture. Life 13(3), 719 (2023).
10. Belabbes, M.A., Koutti, L., Charfi, S.: Computer-aided diagnosis of polyps, ulcer and bleeding on wireless capsule endoscopy images using deep learning: A systematic review. 2022 IEEE Information Technologies & Smart Industrial Systems (ITSIS), pp. 1–5 (2022).
11. Oukdach, Y., Kerkaou, Z., Ansari, M.E., Koutti, L., Ouafdi, A.F.E.: Conv-vit: Feature fusion-based detection of gastrointestinal abnormalities using CNN and ViT in WCE images. In: 2023 10th International Conference on Wireless Networks and Mobile Communications (WINCOM), pp. 1–6 (2023). https://doi.org/10.1109/WINCOM59760.2023.10322944
12. Bray, F., Ferlay, J., Soerjomataram, I., Siegel, R.L., Torre, L.A., Jemal, A.: Global cancer statistics 2018: GLOBOCAN estimates of incidence and mortality worldwide for 36 cancers in 185 countries. CA: A Cancer Journal for Clinicians 68(6), 394–424 (2018).
13. Siegel, R.L., Miller, K.D., Wagle, N.S., Jemal, A.: Cancer statistics, 2023. CA: A Cancer Journal for Clinicians 73(1), 17–48 (2023).
14. Siegel, R.L., Wagle, N.S., Cercek, A., Smith, R.A., Jemal, A.: Colorectal cancer statistics, 2023. CA: A Cancer Journal for Clinicians 73(3), 233–254 (2023).
15. Deepak, P., Axelrad, J.E., Ananthakrishnan, A.N.: The role of the radiologist in determining disease severity in inflammatory bowel diseases. Gastrointestinal Endoscopy Clinics 29(3), 447–470 (2019).
16. Wang, A., Banerjee, S., Barth, B.A., Bhat, Y.M., Chauhan, S., Gottlieb, K.T., Konda, V., Maple, J.T., Murad, F., Pfau, P.R., *et al.*: Wireless capsule endoscopy. Gastrointestinal Endoscopy 78(6), 805–815 (2013).
17. Costamagna, G., Shah, S.K., Riccioni, M.E., Foschia, F., Mutignani, M., Perri, V., Vecchioli, A., Brizi, M.G., Picciocchi, A., Marano, P.: A prospective trial comparing small bowel radiographs and video capsule endoscopy for suspected small bowel disease. Gastroenterology 123(4), 999–1005 (2002).
18. Cortes, C., Vapnik, V.: Support-vector networks. Machine Learning 20, 273–297 (1995).
19. Cover, T., Hart, P.: Nearest neighbor pattern classification. IEEE Transactions on Information Theory 13(1), 21–27 (1967).

20. Breiman, L.: Random forests. Machine Learning 45, 5–32 (2001).
21. Constantinescu, A.F., Ionescu, M., Rogoveanu, I., Ciurea, M.E., Streba, C.T., Iovanescu, V.F., Artene, S.A., Vere, C.C.: Analysis of wireless capsule endoscopy images using local binary patterns. Applied Medical Informatics 36(2), 31–42 (2015).
22. Ellahyani, A., Jaafari, I.E., Charfi, S., Ansari, M.E.: Detection of abnormalities in wireless capsule endoscopy based on extreme learning machine. Signal, Image and Video Processing 15, 877–884 (2021).
23. Souaidi, M., Ansari, M.E.: Multi-scale analysis of ulcer disease detection from WCE images. IET Image Processing 13(12), 2233–2244 (2019).
24. Jia, X., Xing, X., Yuan, Y., Xing, L., Meng, M.Q.-H.: Wireless capsule endoscopy: A new tool for cancer screening in the colon with deep-learning-based polyp recognition. Proceedings of the IEEE 108(1), 178–197 (2019).
25. Dheir, I.M., Abu-Naser, S.S.: Classification of anomalies in gastrointestinal tract using deep learning (2022).
26. Souaidi, M., El Ansari, M.: A new automated polyp detection network MP-FSSD in WCE and colonoscopy images based fusion single shot multibox detector and transfer learning. IEEE Access 10, 47124–47140 (2022).
27. Jha, D., Ali, S., Tomar, N.K., Johansen, H.D., Johansen, D., Rittscher, J., Riegler, M.A., Halvorsen, P.: Real-time polyp detection, localization and segmentation in colonoscopy using deep learning. Ieee Access 9, 40496–40510 (2021).
28. Barbosa, D.J., Ramos, J., Correia, J.H., Lima, C.S.: Automatic detection of small bowel tumors in capsule endoscopy based on color curvelet covariance statistical texture descriptors. In: 2009 Annual International Conference of the IEEE Engineering in Medicine and Biology Society, pp. 6683–6686 (2009). IEEE.
29. Charfi, S., Ansari, M.E.: Computer-aided diagnosis system for colon abnormalities detection in wireless capsule endoscopy images. Multimedia Tools and Applications 77, 4047–4064 (2018).
30. Hajabdollahi, M., Esfandiarpoor, R., Soroushmehr, S., Karimi, N., Samavi, S., Najarian, K.: Segmentation of bleeding regions in wireless capsule endoscopy images an approach for inside capsule video summarization. arXiv preprint arXiv:1802.07788 (2018).
31. Yuan, Y., Li, B., Meng, M.Q.-H.: WCE abnormality detection based on saliency and adaptive locality-constrained linear coding. IEEE Transactions on Automation Science and Engineering 14(1), 149–159 (2016).
32. Sharif, M., Attique Khan, M., Rashid, M., Yasmin, M., Afza, F., Tanik, U.J.: Deep CNN and geometric features-based gastrointestinal tract diseases detection and classification from wireless capsule endoscopy images. Journal of Experimental & Theoretical Artificial Intelligence 33(4), 577–599 (2021).
33. Ghosh, T., Fattah, S.A., Wahid, K.A., Zhu, W.-P., Ahmad, M.O.: Cluster based statistical feature extraction method for automatic bleeding detection in wireless capsule endoscopy video. Computers in Biology and Medicine 94, 41–54 (2018).
34. Romain, O., Histace, A., Silva, J., Ayoub, J., Granado, B., Pinna, A., Dray, X., Marteau, P.: Towards a multimodal wireless video capsule for detection of colonic polyps as prevention of colorectal cancer. In: 13th IEEE International Conference on Bioinformatics and Bioengineering, pp. 1–6 (2013). IEEE.
35. Li, B., Meng, M.Q.-H.: Automatic polyp detection for wireless capsule endoscopy images. Expert Systems with Applications 39(12), 10952–10958 (2012).
36. Jain, S., Seal, A., Ojha, A., Yazidi, A., Bures, J., Tacheci, I., Krejcar, O.: A deep CNN model for anomaly detection and localization in wireless capsule endoscopy images. Computers in Biology and Medicine 137, 104789 (2021).

37. Jain, S., Seal, A., & Ojha, A.: A convolutional neural network with meta-feature learning for wireless capsule endoscopy image classification. *Journal of Medical and Biological Engineering*, *43*(4), 475–494 (2023).
38. Goel, N., Kaur, S., Gunjan, D., Mahapatra, *S. J.*: Dilated CNN for abnormality detection in wireless capsule endoscopy images. *Soft Comput* 26, 1231–1247 (2022).
39. Mohapatra, S., Pati, G.K., Mishra, M., Swarnkar, T.: Gastrointestinal abnormality detection and classification using empirical wavelet transform and deep convolutional neural network from endoscopic images. Ain Shams Engineering Journal 14(4), 101942 (2023).
40. Segúı, S., Drozdzal, M., Pascual, G., Radeva, P., Malagelada, C., Azpiroz, F., Vitria`, J.: Generic feature learning for wireless capsule endoscopy analysis. Computers in Biology and Medicine 79, 163–172 (2016).
41. Rustam, F., Siddique, M.A., Siddiqui, H.U.R., Ullah, S., Mehmood, A., Ashraf, I., Choi, G.S.: Wireless capsule endoscopy bleeding images classification using CNN based model. IEEE Access 9, 33675–33688 (2021).
42. Lan, L., Ye, C., Wang, C., Zhou, S.: Deep convolutional neural networks for WCE abnormality detection: CNN architecture, region proposal and transfer learning. IEEE Access 7, 30017–30032 (2019).
43. Caroppo, A., Leone, A., Siciliano, P.: Deep transfer learning approaches for bleeding detection in endoscopy images. Computerized Medical Imaging and Graphics 88, 101852 (2021).
44. Ayyaz, M.S., Lali, M.I.U., Hussain, M., Rauf, H.T., Alouffi, B., Alyami, H., Wasti, S.: Hybrid deep learning model for endoscopic lesion detection and classification using endoscopy videos. Diagnostics 12(1), 43 (2021).
45. Souaidi, M., El Ansari, M.: Multi-scale hybrid network for polyp detection in wireless capsule endoscopy and colonoscopy images. Diagnostics 12(8), 2030 (2022).
46. Khan, M.A., Khan, M.A., Ahmed, F., Mittal, M., Goyal, L.M., Hemanth, D.J., Satapathy, S.C.: Gastrointestinal diseases segmentation and classification based on duo-deep architectures. Pattern Recognition Letters 131, 193–204 (2020).
47. Alam, M.J., Fattah, S.A.: SR-AttNet: An interpretable stretch–relax attention based deep neural network for polyp segmentation in colonoscopy images. Computers in Biology and Medicine 160, 106945 (2023).
48. Chen, L.-C., Zhu, Y., Papandreou, G., Schroff, F., Adam, H.: Encoder-decoder with atrous separable convolution for semantic image segmentation. In: Proceedings of the European Conference on Computer Vision (ECCV), pp. 801–818 (2018).

Chapter 3

Breast cancer segmentation using U-Net and transfer learning approaches

Noura Bentaher, Younes Kabbadj, and Mohamed Ben Salah

3.1 INTRODUCTION

One of the most prevalent cancers affecting women globally is breast cancer [1]. Breast cancer cells have the potential to travel to lymph nodes and possibly harm other bodily organs, including the lungs. Invasive ductal carcinoma, or dysfunctional milk-producing ducts, is more frequently the initial cause of breast cancer. But it can also start in other breast tissues or cells, such as the glandular structures known as lobules [2]. Additionally, changes in environment, lifestyle, and hormones have been proven by researchers to increase the risk of breast cancer [2].

The size of the tumor has been long considered a crucial prognostic indicator. A precise preoperative assessment of the size of breast cancer is necessary for both surgical resection and the creation of a chemotherapy treatment plan [1]. Additionally, tracking the tumor's volume change during the course of treatment is a crucial source of information for response evaluation standards in solid tumors. For this reason, during the clinical course, precise measurements of size and volume are essential [1]. Medical imaging can be used to reliably and non-invasively collect anatomic information since it is superior at assessing the size and volume of tumors [1]. A low-dose x-ray of the breasts is taken to see the internal breast structures; this process is referred to in medicine as mammography. It is among the best methods of medical imaging for identifying breast cancer [2]. Comparing modern mammography equipment to older models, the radiations that are received by the breast are significantly lower. It has been shown to be one of the most trustworthy screening instruments and a crucial technique for the early diagnosis of breast cancer in recent years [2]. For every breast, two separate views are obtained for the mammograms: the mediolateral oblique (MLO) view and the craniocaudal (CC) view (Figure 3.1).

Experts in radiology examine mammograms daily to look for unusual lesions and identify any worrying areas in the breast, including their location, shape, and type. Despite being considered essential and requiring more precision and accuracy, this method is still costly and vulnerable to mistakes because of the rising number of screening mammography performed every day. Doctors can obtain comprehensive information about suspected tumor locations for additional diagnosis and pathology results with the aid of medical

DOI: 10.1201/9781032664293-4

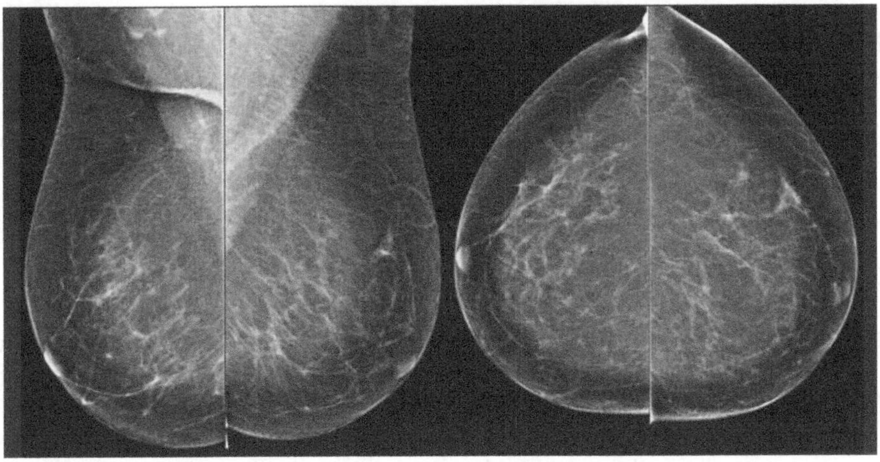

Figure 3.1 MLO view (medial-lateral oblique view) and CC view (cranial-caudal view) [3].

image segmentation tasks. As a result, an automated system can make use of the large volume of mammograms and do this task automatically [4].

Recently, there has been an increased focus on computer-aided diagnosis (CAD) [5–8] for breast cancer imaging due to the rapid advancements in deep learning algorithms and artificial intelligence. This renewal has a focus on cutting-edge systems that could greatly improve clinical care [4].

The encoder-decoder architecture is used most frequently in models created for the segmentation of medical and biological image data [9]. In this type of architecture, the encoder is used to downsample the input image to a lower-dimensional representation and capture the context by extracting features. However, the decoder is used to upsample this representation, returning to the original size of the image, then generating the segmentation map based on information of localization and the fine-grained details of the structures of interest. The popular models for medical and biomedical image segmentation that use the encoder-decoder architecture include SegNet, U-Net, and variational autoencoders (VAEs).

In this work, we will use U-Net architecture that was presented by Ronneberger et al. [10]. This architecture consists of two primary parts: a symmetric expanding path targeted at accurate localization and a contracting path intended to capture context. Furthermore, skip links are integrated within the architecture to improve the relationship between precise localization and context extraction by connecting these two channels. Using data augmentation allows the U-Net training technique to learn effectively from a small number of annotated photos. Many U-Net-inspired techniques have been created for various applications [9].

Our aim is to explore the depths of medical imaging with the goal of transforming the precise accuracy of breast mass diagnosis. We carefully examine the effectiveness of transfer learning and training from scratch approaches for breast mass segmentation in mammography images. Our goal with this extensive study is to not only compare but also shed light on the way to improve efficiency and accuracy. Our research bridges the gap between theory and application, offering revolutionary insights into early and accurate breast cancer diagnosis, due to the potent tools of U-Net and pre-trained models.

3.2 RELATED WORKS

Breast mass detection and segmentation methods that are automated have already been demonstrated. Ojala et al. [11] provided a method to separate the breast area from digital mammograms; nevertheless, this segmentation may contain inaccuracies due to bright objects outside of the breast area. The entire breast, pectoral muscles, and nipple extraction typically make up a segmented breast. Sameti et al. [12] separated a mammography into various mass candidate regions and then computed the discrete texture features for each mass candidate area. Gray-level co-occurrence matrices (GLCM), which necessitate large computing loads, were used to construct the features. The usefulness of the textural information that mass areas contained in contrast to the mass margins was assessed. Bellotti et al. [13] defined areas of interest (ROIs) using textural characteristics that were derived from the GLCM. Texture cues could be used to separate lesions with masses from normal regions. Khuzi et al. [14] also used GLCM, which was built in four orientations for every ROI, to extract the textural characteristics. Oktay et al. [15] suggested attention gates (AGs), and instead of using a single U-Net for segmenting the pancreas, the network topology linked with AGs produced better segmentation results. Dhungel et al. [16] suggested a series of deep learning techniques that include a convolutional neural network (CNN) to retain the right candidates, a random forest (RF) to minimize false positives, and a deep belief network to identify suspect regions. Li et al. [17] were able to enhance the outcomes for the benign and malignant categorization of mammograms. The authors employed a CNN that was modified from the DenseNet model for the classification. In addition to applying rotations of 90°, 180°, and 270°, vertical and horizontal mirroring, and an 80% scalar reduction, the authors also used data augmentation techniques. Ultimately, the method's accuracy was 94.55%. Al-Masni et al. [18] created a CAD system that combines simultaneous detection and classification, utilizing the You Only Look Once (YOLO) architecture. The YOLO architecture is exclusively utilized in the identification of masses in the study of Al-Antari et al. [19], processing the ROI acquired in the preceding stage in a full-resolution convolutional network (FrCN). Wang et al. [20] suggested an encoder-decoder architecture to segment masses according to regions of interest. The decoder module uses multi-scale feature fusion to recover the segmentation mask, while the encoder uses a nested atrous spatial pyramid pooling module to extract the image's multi-scale features.

3.3 BACKGROUND

3.3.1 U-Net

The U-Net approach, which was first presented by Ronneberger, Fischer, and Brox [10], has proven to be incredibly effective at biomedical image segmentation. Its name comes from the way it is structured symmetrically, with both an expanding and a contracting path. Numerous sectors have benefited greatly from this innovative design, including brain lesion segmentation, cell segmentation, and the segmentation of medical imaging like CT or MRI scans. Figure 3.2 illustrates the U-Net architecture [10].

3.3.2 ResNet

"Residual blocks" are the building blocks of residual networks (ResNets), and each one has a shortcut link running parallel to the main branch. In order to prevent learning a residual function with reference to the input, the final shortcut and main branch are introduced. The terms "convolutional block" and "identity block" refer to the two categories of residual blocks as depicted in Figure 3.3. To minimize the spatial size of the feature maps, the "convolutional block" consists of a convolutional layer with a stride of 2, a batch normalization layer on its shortcut path, and a stride of 2 for the first convolutional layer on its main path. Conversely, "identity blocks" include identity

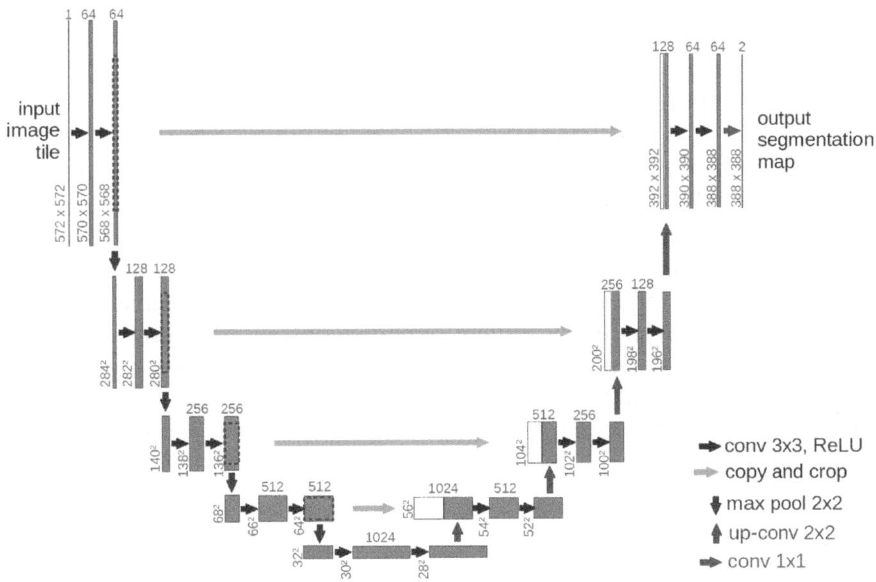

Figure 3.2 U-Net architecture [10].

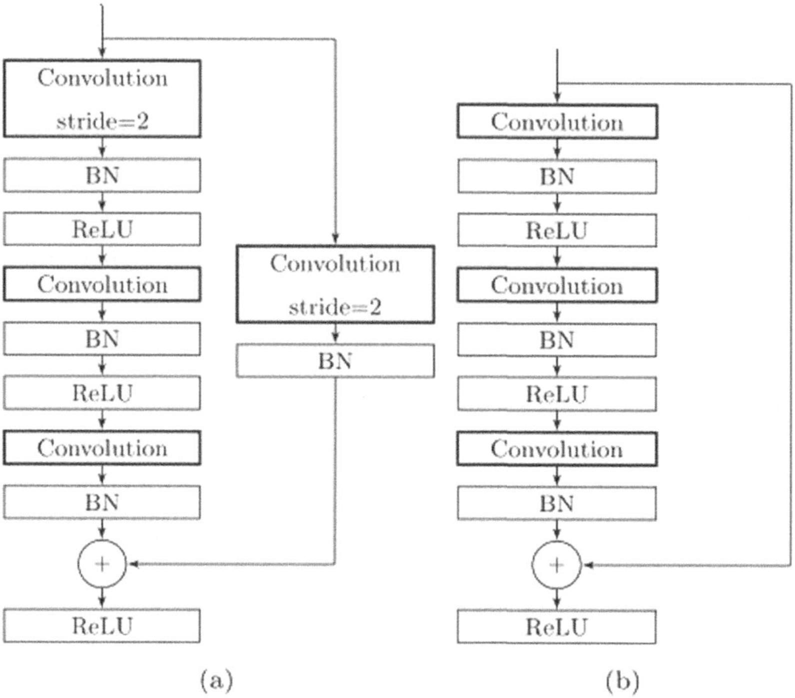

Figure 3.3 The ResNet-50 (a) diagram of the convolutional block and (b) diagram of the identity block [21].

shortcuts that add the stage's input to the output without altering the feature maps' size (stride 1) [21].

The architecture of the ResNet starts with a convolutional layer (stride of 2 and kernel size of 7 × 7), a batch normalization layer, a ReLU, and a max-pooling layer. The remaining part of the network is composed of four stages, each of which consists of a single convolutional block followed by multiple identity blocks; the number of identity blocks varies among the four stages and is arranged as follows: 2, 3, 5, and 2, respectively. The number of kernels of convolutional layers in each stage remains constant. To convert the two-dimensional feature maps into a one-dimensional feature vector, a global average pooling layer is added. This layer converts the two-dimensional feature maps into a one-dimensional feature vector, which is then adjusted by a fully connected (FC) layer comprising neurons based on the number of classes [21].

3.3.3 Transfer learning

Transfer learning is a machine learning strategy that transfers information extracted from relevant data given by a CNN to solve different problems.

Transfer learning is based on the principle of developing learning by using knowledge from related tasks learned in new tasks through transfer. Pre-trained deep learning networks are already well trained on other datasets and can be improved to achieve high accuracy on much smaller datasets. Because the learning process in transfer learning is based on patterns discovered through the solution of another problem, it avoids the need for time-consuming computations. Moreover, it facilitates the development of an accurate model. Transfer learning is not a model or approach for machine learning; rather, it is seen as a design methodology. Usually, pre-trained models can be used using this design technique. On deep convolutional neural networks, these pre-trained models are built. This technique for deep learning involves first training the CNN with a sizable training dataset for classification issues. A key component of a successful training process is having data available for initial training, as the CNN model is capable of learning to extract significant characteristics from images. This model's suitability for transfer learning will be assessed based on CNN's capacity to identify and select the best display aspects [22].

3.3.4 Loss function

3.3.4.1 *Jaccard distance and Dice loss*

Enhancing the alignment between the predicted segmentation and the ground truth's foreground is the objective of the two losses: Dice loss and the Jaccard distance loss. Through an emphasis on optimizing the overlap among these regions, these functions considerably improve medical image segmentation performance [23]. The Dice loss function and the Jaccard distance loss function have the following specific expressions:

$$L_{dice} = 1 - \frac{2\sum_{i=1}^{k} p_i g_i}{\sum_{i=1}^{K} p_i^2 + \sum_{i=1}^{k} g_i^2} \quad and$$

$$L_{jaccard} = 1 - \frac{\sum_{i=1}^{k} p_i g_i}{\sum_{i=1}^{k} p_i^2 + \sum_{i=1}^{k} g_i^2 - \sum_{i=1}^{k} p_i g_i} \tag{3.1}$$

where p_i denotes the pixel i of the predicted binary segmentation map and g_i denotes the pixel i of the ground truth binary segmentation map. K is the number of pixels in the image.

3.3.5 Optimizers

Optimizers refer to algorithms or techniques employed to minimize an error function (also known as a loss function) or enhance production efficiency. These mathematical functions are contingent upon the model's learnable parameters, namely weights and biases. Optimizers play a vital role in determining how the neural network's weights and learning rate should be adjusted to minimize the loss. The choice of these algorithms significantly impacts the accuracy of the deep learning model, as well as its training speed [24].

3.3.6 Callbacks

Training a deep learning model is a very challenging process because it can take many hours, and sometimes we need to change some parameters at later stages while training which is impossible, and also it is impossible to predict the good parameters for the model. Those parameters can include the number of epochs and learning rate.

A callback in Keras aids us in an accurate training of the model. It is an object that we can pass to the model while using the fit method and can call it during different points of the training. There are different kinds of callbacks that can be used during the model training. The early stopping is used to interrupt the training process when the validation loss is no longer improving. The Model Checkpointing helps in saving the current weight of the model at different points during the training. The Reduce On Plateau can be used to reduce the learning rate when the validation loss has stopped improving [25].

3.4 METHODOLOGY

3.4.1 Training from scratch approach

In our initial approach, we implemented the original U-Net architecture as proposed by Ronneberger et al. [10]. The input layer was configured with a shape of $224 \times 224 \times 3$, representing the mammography image and its corresponding mask label. The output of this architecture comprised an image with pixel values of 0 or 1, achieved through the sigmoid activation function in the final convolution layer of the decoder. The depth of the resulting image was 1, indicating the binary segmentation. During the training phase, the Jaccard distance and Dice loss functions were employed, tailored specifically for segmentation tasks. These functions consider the spatial relationships between pixels, aiming to maximize the overlap between the predicted mask and the ground truth. The Adam optimizer was utilized with an initial learning rate set to 0.0001. To optimize the training process, a ModelCheckpoint Callback was defined to save the best model state.

The model underwent rigorous training over 300 epochs, resulting in a total of 23,784,241 trainable parameters.

3.4.2 Transfer learning approach

In our second approach, we leveraged the foundational U-Net architecture with various pre-trained models serving as encoders. These models included ResNet, DenseNet, MobileNet, Inception, and EfficientNet, all sourced from the Keras library. The segmentation models were defined using the segmentation models module [26], integrating the specified backbones to augment U-Net. The input size for these models was standardized to 224 × 224 × 3, while the output consisted of a mask with dimensions 224 × 224 × 1. Hyperparameters such as the Adam optimizer, a learning rate of 0.0001, and Jaccard distance and Dice loss functions remained consistent with the initial U-Net model. Training these models spanned 150 epochs, with the ModelCheckpoint Callback ensuring the preservation of the best-performing versions.

3.5 PERFORMANCE ANALYSIS AND DISCUSSIONS

3.5.1 Dataset preparation

The INbreast database contains 107 cases with mass (abnormal cases) and 303 normal cases; in order to have a balanced dataset, we used only 107 cases from the 303 cases (Figure 3.4). For every image, there is a mask. For the normal cases, we create masks that contain pixels of value zero. All images are resized to 224 × 224 and divided into three sets: train, validation, and test. The Keras ImageDataGenerator is also used to augment the size of our data by applying a rotation, a shift in width and height, a zoom, and a horizontal and vertical flip; the fill mode is set to "nearest" in this case.

3.5.2 Evaluation metrics

3.5.2.1 Precision and recall

Precision and recall are two common measures that help us understand the types of errors we make. Precision, or a positive predictive value, gives us a measure of how confident we can be in a positive prediction from our model. Recall, sensitivity, or true positive rate (TPR) gives us a measure of how many values are actually "true" that we have detected [27]. Precision and recall are defined as follows:

$$\text{Precision} = \frac{TP}{TP+FP} \tag{3.2}$$

$$\text{Recall} = \frac{TP}{TP+FN} \tag{3.3}$$

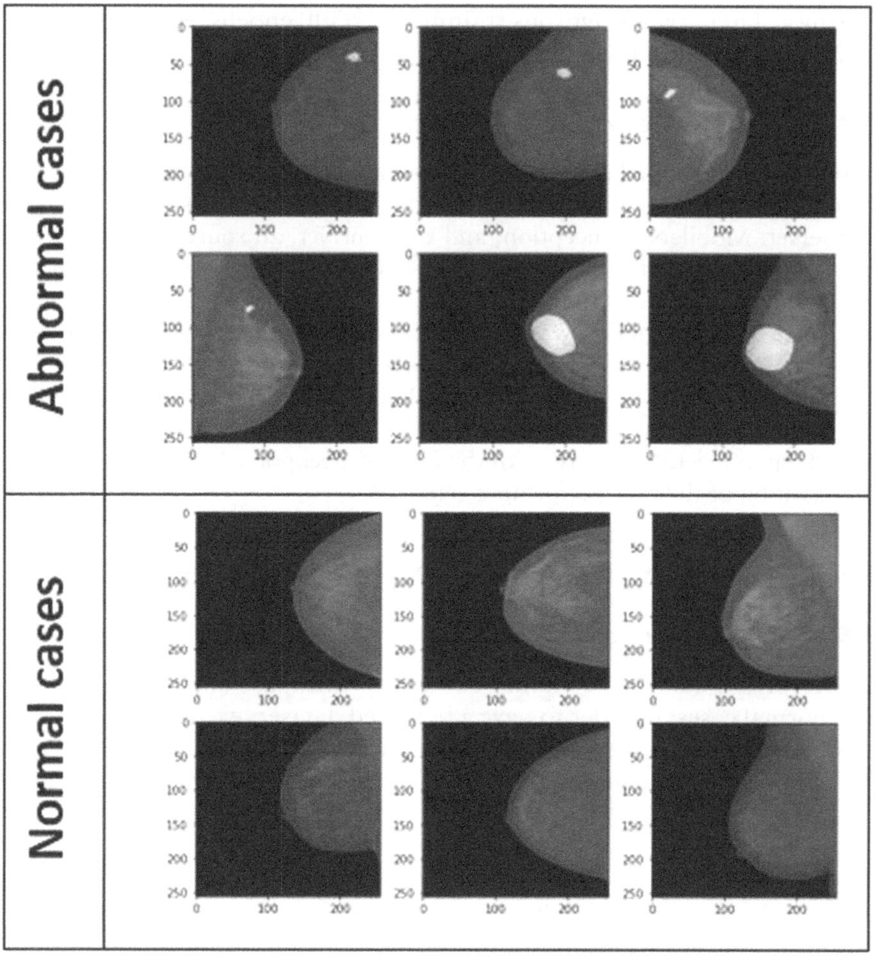

Figure 3.4 INbreast database.

3.5.2.2 Intersection over Union

The Intersection over Union (IoU) metric, also known as the Jaccard Index, is basically a method of quantifying the percentage of overlap between the target mask and the predicted output. This metric is closely related to the Dice coefficient, which is often used as a loss function during training. The IoU metric measures the number of pixels common to both the target and predictive masks divided by the total number of pixels common to both masks [28].

$$IoU = \frac{\text{target} \cap \text{prediction}}{\text{target} \cup \text{prediction}} \tag{3.4}$$

3.5.2.3 *Dice similarity coefficient*

Dice similarity coefficient also called the overlap index is a validation metric for the segmentation of white matter lesions in medical images. The value of a Dice similarity coefficient (DSC) ranges from 0, indicating no spatial overlap between two masks of binary segmentation results, to 1, indicating complete overlap [29].

$$DSC = \frac{2 * (\text{target} \cap \text{prediction})}{\text{target} + \text{prediction}} \tag{3.5}$$

3.5.3 Experimental results

The first approach (U-Net trained from scratch) achieved a precision of 97%, a recall of 76%, an f1-score of 85%, and an IoU-score of 74%. As a result, two abnormal cases with a small mass region are missed and predicted as false negatives. A total of 12 normal cases are correctly predicted with a black mask (true negative). The rest of the cases are abnormal and correctly predicted. The corresponding segmentations are very close to the ground truth. ResNet-UNet trained from scratch also achieved good results compared with the other models (DenseNet, MobileNet, Inception, and EfficientNet), with 91% as precision, 66% as recall, 76% as f1-score, and IoU-score of 62%. In the prediction, three abnormal cases with a small mass region are missed and predicted as false negatives. For other cases the rate of FP and the rate of FN are very small.

After the comparison between these two approaches, we can conclude that the use of transfer learning in the segmentation with U-Net doesn't show good results depending on the high rate of FP and FN that appear in the predicted mask for the two approaches that are trained with ImageNet weights, unlike training from scratch that predicted just two incorrect masks over 22 masks.

3.5.4 Comparative analysis

In this experiment, we tried to compare the original U-Net segmentation results with the combination of U-Net with standard models (ResNet, DenseNet, MobileNet, Inception, and EfficientNet), including the use of transfer learning in order to show the impact of ImageNet weights in medical imaging segmentation, especially breast mass segmentation.

The results showed that ResNet-UNet and U-Net have performed well in terms of precision, recall, the Dice coefficient, and IoU. Based on the scores of recall and f1-score, we can assume that those models were able to correctly identify the pixels in the region of interest. Also, we can say that the models have learned useful representations of the image and were able to generate an accurate mask as presented in Figure 3.6 and Figure 3.8. ResNet architecture is known for the high quality of features extracted due to the skip connection and residual blocks. That is why it avoids any loss in information and

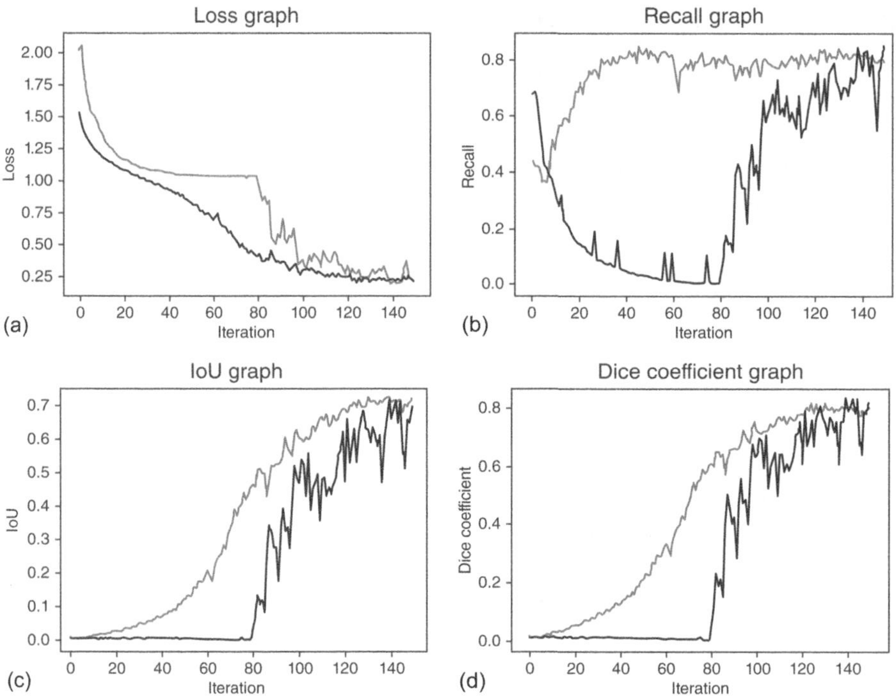

Figure 3.5 ResNet-UNet learning curves when trained from scratch: (a) loss curve on 150 epochs, (b) recall curve on 150 epochs, (c) IoU curve on 150 epochs, and (d) Dice coefficient curve on 150 epochs.

also avoids such problems as gradient vanishing. The union between U-Net and ResNet has a big potential to predict a mask with a high overlap with the ground truth. From Table 3.1, we can see that training from scratch has achieved good results in all previous implementations, compared with transfer learning Table 3.2. The utilization of ImageNet weights has not yielded favorable results in terms of the Dice coefficient and IoU. This can be interpreted

Table 3.1 Results of training models from scratch

Models	*Precision (%)*	*Recall (%)*	*Dice coef. (%)*	*IoU (%)*	*Epochs*
Inception-UNet	78	54	58	44	150
EfficientNet-UNet	75	54	59	47	150
MobileNet-UNet	65	71	62	49	150
DenseNet-UNet	66	62	63	49	150
ResNet-UNet	**91**	**66**	**76**	**62**	**150**
U-Net	**97**	**76**	**85**	**74**	**300**

Table 3.2 Results of training models using transfer learning

Models	*Precision (%)*	*Recall (%)*	*Dice coef. (%)*	*IoU (%)*	*Epochs*
Inception-UNet	76	69	70	57	150
EfficientNet-UNet	57	57	57	40	150
MobileNet-UNet	69	57	62	48	150
DenseNet-UNet	57	67	61	47	150
ResNet-UNet	74	63	68	51	150
U-Net	88	46	61	44	300

as a significant difference between the predicted mask and the ground truth as illustrated in Figure 3.7 and Figure 3.9. Therefore, it can be concluded that using weights from a domain unrelated to the medical field is not advantageous in segmentation tasks.

Additionally, more reliable and accurate models have been made possible by the use of the Jaccard index and Dice loss as loss functions. During training, the model can be trained to generate segmentations that are closer to the original mask by using these loss functions to determine how close the predicted segmentation is to the ground truth. Through the reduction of these loss functions, the model may function better and yield more precise outcomes.

A thorough understanding of the development and performance of the ResNet-UNet model may be gained from the learning curves (Figure 3.5), which were created from scratch across 150 epochs of training. Initially, the loss curve's declining trend shows that the model is becoming better at producing segmentation masks that are more accurate, demonstrating that it can closely match predictions to the real ground truth. Simultaneously, the increasing recall curve emphasizes how sensitive the model is at finding breast masses, which lowers the possibility of false negative results. In medical imaging, this sensitivity is essential since it makes sure the model detects all possible cases of breast masses, reducing the possibility that important diagnostic data might be missed. The model's increasing ability to correctly identify the borders of segmented sections is demonstrated by the rising IoU curve, which also highlights the model's capacity to exactly represent the complex forms and contours of breast masses in mammography images. At the same time, the model's improved segmentation accuracy is highlighted by the rising Dice coefficient curve, which shows that it can reliably produce predictions that nearly match the actual data. Collectively, these encouraging curves indicate the ResNet-UNet model's impressive advancements in segmentation quality overall, sensitivity, precision, and spatial accuracy. These developments are highly promising for the detection of breast cancer, as precise and accurate segmentation is essential for prompt and effective medical therapies. This highlights the potential benefit of this model for enhancing patient outcomes and clinical decision-making.

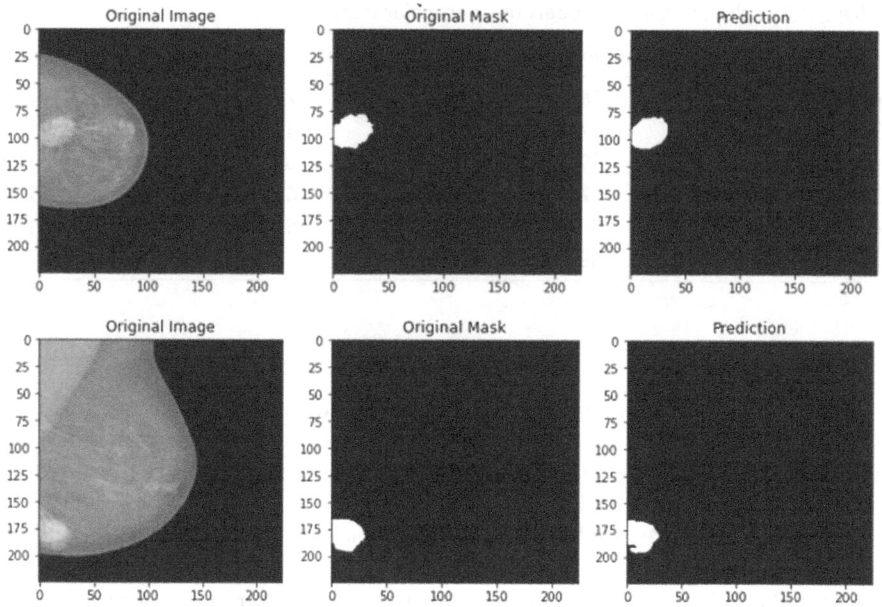

Figure 3.6 Predictions of abnormal masks with a small rate of false negative using U-Net trained from scratch.

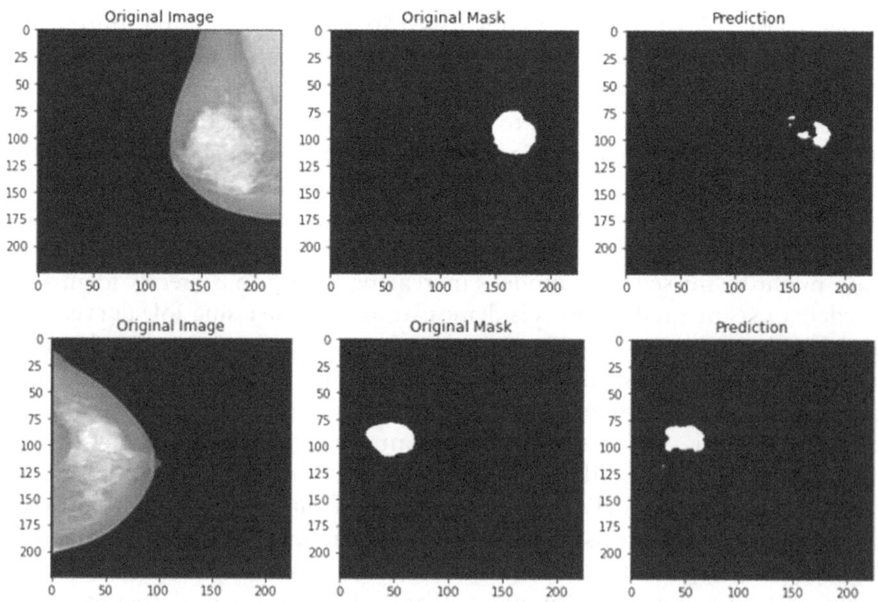

Figure 3.7 Predictions of abnormal masks with an important rate of false negative using U-Net trained using ImageNet weights.

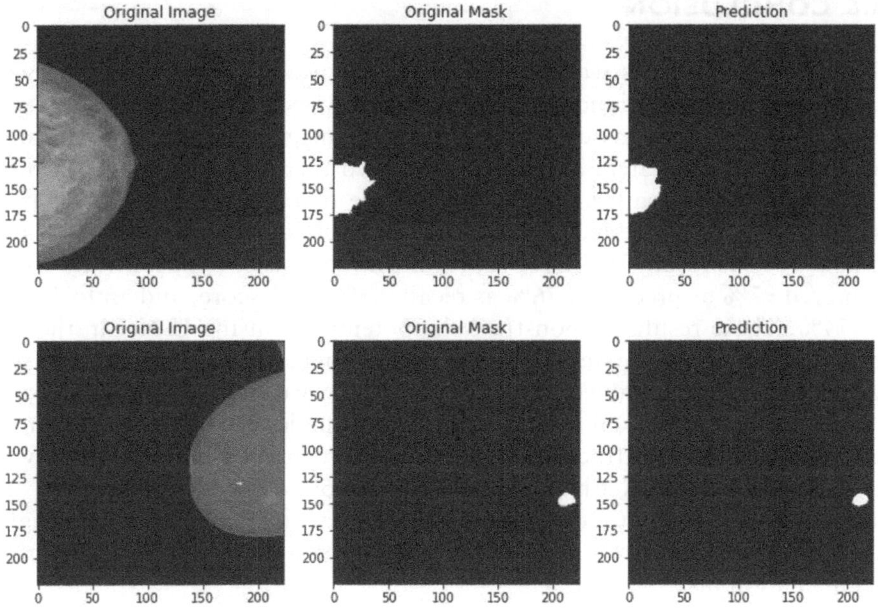

Figure 3.8 Predictions of abnormal masks using Res-UNet trained from scratch with a small rate of false positive and false negative.

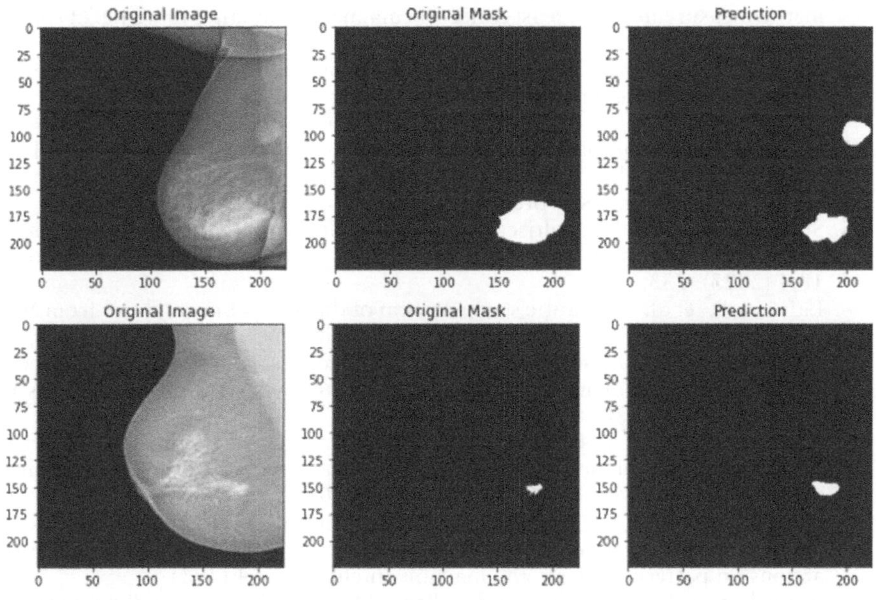

Figure 3.9 Prediction of abnormal cases with a serious number of false negative and false positive pixels using Res-UNet trained with ImageNet weights.

3.6 CONCLUSION

The main goal of this work was to study the use of transfer learning with U-Net and pre-trained models combined with U-Net architecture in the segmentation of breast masses in mammography images. We started by training all models from scratch, and then we trained them using transfer learning. In terms of performance, the results indicate that training the models from scratch was the most effective approach, ResNet-UNet achieved 91% as precision, 66% as recall, 76% as f1-score, and an IoU-score of 62%, while U-Net achieved 97% as precision, 76% as recall, 85% as f1-score, and an IoU-score of 74%. These results demonstrate the potential of using U-Net in the segmentation of breast masses in mammography images, and its capacity to provide accurate and reliable masks. ResNet-UNet was a good approach, based on the power of residual blocks that prevent the loss in spatial information and ensure a good extraction of features and context. In the training with transfer learning, U-Net had achieved a recall of 46% and IoU of 44%; subsequently, we can say that the use of transfer learning and ImageNet weights doesn't display good results and it is not a good approach, especially in the segmentation task.

REFERENCES

1. Yue, W., et al. "Deep learning-based automatic segmentation for size and volumetric measurement of breast cancer on magnetic resonance imaging." Frontiers in oncology 12 (2022): 984626.
2. Gardezi, S.J.S., et al. "Breast cancer detection and diagnosis using mammographic data: Systematic review." Journal of medical internet research 21.7 (2019): e14464.
3. Teoh, M. S. "Mammography," http://penangbreastcare.com/mammo.html
4. Baccouche, A., et al. "Connected-UNets: A deep learning architecture for breast mass segmentation." NPJ breast cancer 7.1 (2021): 151.
5. Souaidi, M., et al. "A multiscale polyp detection approach for GI tract images based on improved DenseNet and single-shot multibox detector." Diagnostics 13.4 (2023): 733.
6. Lafraxo, S., et al. "Semantic segmentation of digestive abnormalities from WCE images by using AttResU-Net architecture." Life 13.3 (2023): 719.
7. Lafraxo, S., M. El Ansari, and L. Koutti. "Computer-aided system for bleeding detection in WCE images based on CNN-GRU network." Multimedia tools and applications 83 7 (2024): 21081–21106.
8. Lafraxo, S., M. El Ansari, and S. Charfi "MelaNet: An effective deep learning framework for melanoma detection using dermoscopic images." Multimedia tools and applications 81.11 (2022): 16021–16045.
9. Minaee, S., Y.Y. Boykov, F. Porikli, A.J. Plaza, N. Kehtarnavaz, and D. Terzopoulos. "Image segmentation using deep learning: A survey." IEEE transactions on pattern analysis and machine intelligence 44 (2021).
10. Ronneberger, O., P. Fischer, and T. Brox. "U-Net: Convolutional networks for biomedical image segmentation". In: International conference on medical image computing and computer-assisted intervention. Springer, 2015, pp. 234–241.

11. Ojala, T., J. Näppi, and O. Nevalainen. "Accurate segmentation of the breast region from digitized mammograms." Computerized medical imaging and graphics 25.1 (2001): 47–59.
12. Sameti, M., et al. "Texture feature extraction for tumor detection in mammographic images." In: 1997 IEEE Pacific Rim conference on communications, computers and signal processing, PACRIM. 10 years networking the Pacific Rim, 1987–1997. Vol. 2, IEEE, 1997.
13. Bellotti, R., et al. "A completely automated CAD system for mass detection in a large mammographic database." Medical physics 33.8 (2006): 3066–3075.
14. Khuzi, A.M., et al. "Identification of masses in digital mammogram using gray level co-occurrence matrices." Biomedical imaging and intervention journal 5.3 (2009).
15. Oktay, O., et al. "Attention U-Net: Learning where to look for the pancreas." arXiv preprint arXiv:1804.03999 (2018).
16. Dhungel, N., G. Carneiro, and A.P. Bradley. "Deep learning and structured prediction for the segmentation of mass in mammograms." In: International conference on medical image computing and computer-assisted intervention. Cham: Springer International Publishing, 2015.
17. Li, H., et al. "Benign and malignant classification of mammogram images based on deep learning." Biomedical signal processing and control 51 (2019): 347–354.
18. Al-Masni, M.A., et al. "Simultaneous detection and classification of breast masses in digital mammograms via a deep learning YOLO-based CAD system." Computer methods and programs in biomedicine 157 (2018): 85–94.
19. Al-Antari, M.A., et al. "A fully integrated computer-aided diagnosis system for digital X-ray mammograms via deep learning detection, segmentation, and classification." International journal of medical informatics 117 (2018): 44–54.
20. Wang, R., et al. "Multi-level nested pyramid network for mass segmentation in mammograms." Neurocomputing 363 (2019): 313–320.
21. Tsochatzidis, L., P. Koutla, L. Costaridou, and I. Pratikakis. "Integrating segmentation information into CNN for breast cancer diagnosis of mammographic masses." Computer methods and programs in biomedicine 200 (2021).
22. Sahinbas, K., and F.O. Catak. "Transfer learning-based convolutional neural network for COVID-19 detection with X-ray images." Data science for COVID-19. Elsevier. 2021, pp. 451–466.
23. Wei, Z., H. Song, L. Chen, Q. Li, and G. Han. "Attention- based DenseUnet network with adversarial training for skin lesion segmentation", 2019, 10.1109/ACCESS.2019.2940794
24. Musstafa, M. "Optimizers in deep learning", 2021. https://medium.com/mlearning-ai/optimizers-in-deep-learning-7bf81fed78a0.
25. Mohanty, A. Understanding Callbacks in Keras. Analytics Vidhya. 2020. https://medium.com/analytics-vidhya/understanding-callbacks-in-keras-98c935095219.
26. Iakubovskii, P. Segmentation Models [Computer software]. GitHub. 2019. https://github.com/qubvel/segmentation_models.
27. Bressler, N., and S. Tannor. A guide to evaluation metrics for classification models. 2021. https://deepchecks.com/a-guide-to-evaluation-metrics-for-classification-models/
28. Jordan, J. "Evaluating image segmentation models," 2018. https://www.jeremyjordan.me/evaluating-image-segmentation-models/
29. Zou, K.H., et al. "Statistical validation of image segmentation quality based on a spatial overlap index1: Scientific reports." Academic radiology 11.2 (2004): 178–189.

Chapter 4

Transforming graphics processing unit-accelerated machine learning (ML) environments with Docker Cloud containers

Amine Bouaouda, Karim Afdel, and Rachida Abounacer

4.1 INTRODUCTION

Machine learning and deep learning have spearheaded transformative changes across numerous industries, spanning healthcare, finance, autonomous vehicles, and natural language processing [1–4]. Fueled by the analysis of extensive datasets, these algorithms demand substantial computational power for both training and inference. Graphics processing units (GPUs) have emerged as a pivotal technology, facilitating the acceleration of these computations and enabling the training of intricate models within reasonable timeframes [5, 6]. However, realizing the full potential of GPUs necessitates a meticulously configured and optimized environment, a task that can be formidable [6].

This chapter delves into the utilization of Docker containers as a solution to the challenges associated with creating and managing machine learning environments with GPU support. Docker containers offer a lightweight and portable approach to package applications alongside their dependencies, libraries, and system configurations [7–9]. By encapsulating the entire environment within a container, researchers and practitioners can ensure consistent execution across diverse systems, eliminating the compatibility hurdles that often arise when deploying machine learning models on various hardware setups [10]. The objective of this chapter is to showcase the advantages and effectiveness of Docker containers in addressing the complexities of GPU-enabled machine learning environments.

4.2 BACKGROUND

The landscape of creating GPU-supported environments for machine learning has evolved to meet the growing demands of the field. Researchers have embraced various techniques to establish such environments, each with its merits and drawbacks. We provide an overview of four prominent techniques: Bare-metal installation, Anaconda, virtual machines (VMs), and Docker containers.

The conventional approach of bare-metal installation involves directly installing the necessary software components and libraries on the host

 DOI: 10.1201/9781032664293-5

machine or server [11–13]. While this technique offers fine-grained control over the environment, facilitating precise configuration, it also introduces challenges such as software conflicts, version compatibility issues, and complexities in replicating the environment across multiple systems [11].

Anaconda, a widely adopted package manager and environment manager for Python, provides a streamlined method for managing machine learning dependencies. It enables the creation of isolated virtual environments, each with specific library versions, thereby preventing conflicts. Despite easing the burden of dependency management, Anaconda may encounter challenges related to version compatibility and package availability [14].

VMs offer an alternative by establishing isolated environments on a host system. VMs provide hardware-level isolation, enabling multiple environments to coexist without interference [9]. While effective for managing complex software stacks and offering a degree of separation from the host system, VMs introduce additional resource overhead and may suffer from performance bottlenecks [15, 16].

Docker containers, the focal point of this chapter, present a modern solution to these challenges. Docker enables the creation of images that package an application, its dependencies, and environment variables into a single unit [17]. These containers can be consistently deployed across various environments, ensuring reproducibility and minimizing compatibility concerns [7, 18]. By abstracting away the underlying system specifics, Docker containers offer a lightweight and efficient approach to GPU-enhanced machine learning environments.

In the subsequent sections, we will delve into a comparative analysis of these four techniques—bare-metal installation, Anaconda, VMs, and Docker containers—to illustrate the advantages and drawbacks of each in the context of GPU-accelerated machine learning environments.

The remaining sections of the chapter are structured as follows: We define the problem in Section 4.3. In Section 4.4, we present the application of our proposed solution and the results. Finally, the chapter concludes with the conclusion and future work in Section 4.5.

4.3 PROBLEM DEFINITION AND SOLUTION

Machine learning and deep learning have emerged as powerful tools for extracting insights and patterns from vast datasets. However, the computational demands of these algorithms have driven the need for high-performance computing environments, particularly those equipped with GPUs for efficient training and inference. Creating and managing such environments is not without challenges, as each approach carries its own set of advantages and disadvantages.

GPU-accelerated environments play a pivotal role in expediting the training of complex machine learning and deep learning models. Several techniques have been employed to create such environments, each with its trade-offs. In this section, we explore three common techniques—bare-metal installations, Anaconda, and VMs—highlighting their drawbacks and limitations, and present our proposed solution using Docker containers.

4.3.1 Bare-metal technique

The bare-metal technique involves installing all necessary software components and libraries directly on the host machine. While this approach offers fine-grained control and customization, it often results in time-consuming setup processes and intricate compatibility issues [11–13]. The provisioning of the required dependencies, configurations, and libraries can be labor-intensive and error-prone. Additionally, bare-metal setups can quickly become cluttered and challenging to manage, lacking the encapsulation and isolation offered by modern containerization solutions [11, 13].

4.3.2 Anaconda Technique

Anaconda provides a package management system that simplifies the creation of isolated Python environments [14]. While Anaconda alleviates some compatibility issues by enabling the management of different library versions, it can lead to increased resource consumption due to the duplication of libraries across environments [14]. Moreover, the installation and configuration of specialized GPU drivers within Anaconda environments can be convoluted, potentially hindering efficient GPU utilization. As a result, while Anaconda offers dependency management, it may incur performance overhead and complexity.

4.3.3 Virtual machine technique

VMs create isolated environments within a host system, offering hardware-level isolation [9, 15]. However, this isolation comes at the cost of resource overhead, as each VM requires its operating system instance [9]. While VMs provide robust separation and the ability to manage complex software stacks, they can experience performance bottlenecks due to the overhead of running multiple operating systems simultaneously [15, 16, 19]. Additionally, the process of setting up VMs can be time-consuming, and the management of multiple VMs can become unwieldy.

4.3.4 Solution proposed

To overcome the challenges posed by the aforementioned techniques, we propose the utilization of Docker containers as seen in Figure 4.1. Containers provide a lightweight and efficient means of packaging applications along with their dependencies and configurations. In the following subsections, we

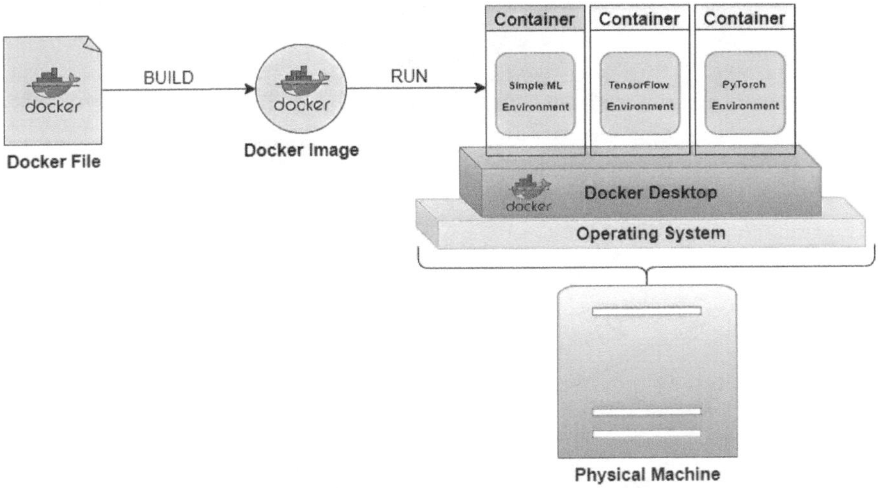

Figure 4.1 Overview of our proposed solution with Docker containers.

explore the concepts of containers, Docker, and Dockerfiles, highlighting how they address the limitations of previous techniques and provide an efficient solution for machine learning environments utilizing GPU technology.

4.3.4.1 Cloud containers

Containers are lightweight, standalone executable units that encapsulate an application, along with its runtime environment and all necessary dependencies [7, 9]. These isolated environments ensure consistent behavior across different computing environments, as shown in Figure 4.2, making it possible to run an application seamlessly regardless of the underlying system [9, 20]. Containers achieve this by leveraging operating system-level virtualization, where each container shares the host OS kernel while maintaining its own isolated file system, processes, and network space [10, 19, 21].

4.3.4.2 Docker

Docker is one of the most popular containerization platforms that simplifies the creation, deployment, and management of containers [7, 18]. It provides a unified platform to develop, ship, and run applications, ensuring that they run consistently across various environments [22]. Docker employs a client-server architecture, where the Docker client communicates with the Docker daemon running on the host system. Users interact with Docker through the command line interface (CLI) or graphical user interfaces to manage containers and images [23].

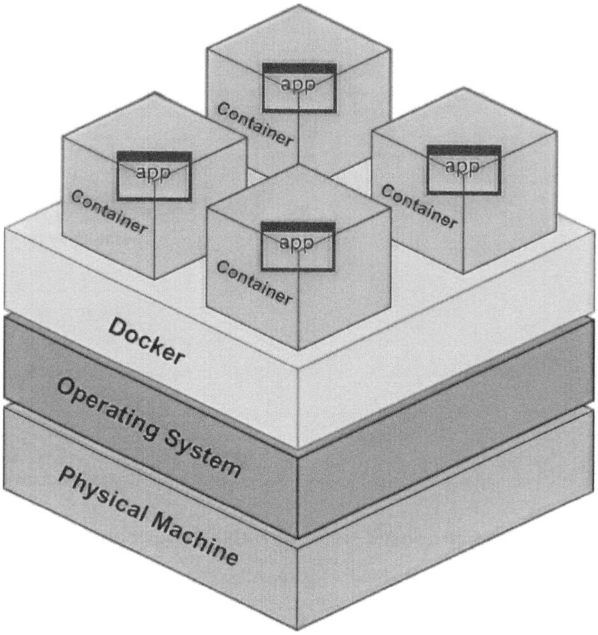

Figure 4.2 Docker container operation.

Docker images serve as the blueprints for containers [16, 22, 23]. These images are composed of a base file system, application code, runtime, libraries, and dependencies. Docker images are built using Dockerfiles, which contain a series of instructions that define the environment and configuration of the image. Once an image is built, it can be shared and deployed across different environments, ensuring that the application behaves consistently [18, 22].

4.3.4.3 Dockerfiles

Dockerfiles are plain-text configuration files that define the process of creating a Docker image. They contain a sequence of commands that instruct Docker how to assemble the image step by step [24]. Dockerfiles enable the automation of the image creation process and ensure that the entire configuration is version-controlled, making it easy to reproduce the same environment consistently [25]. A Dockerfile typically starts with a base image that serves as the foundation for the container. Subsequent instructions can include copying files into the image, setting environment variables, installing packages, and configuring the runtime environment. These instructions can be highly customized to suit the specific needs of the application [24, 25].

In summary, containers, Docker, and Dockerfiles collectively offer a powerful solution for creating and managing GPU-accelerated machine learning environments. Containers provide isolation, consistency, and portability, while Docker simplifies the containerization process. Dockerfiles further enhance the workflow by automating image creation and allowing for version-controlled environment definitions. The next section will delve into a comparative analysis of these techniques, highlighting their respective strengths and limitations.

Here is an overview of the structure of a Dockerfile:

Structure of a Dockerfile

```
# Use a base image suitable for the needs, such as Python
FROM base image
# Set up environment variables ENV VAR NAME=value
# Create a new directory and set it as the working
directory WORKDIR/app
# Copy files from the host to the container's filesystem
COPY source destination
# Run a command within the container RUN command
# Expose a port for communication EXPOSE port number
# Define the command to run when the container starts CMD
["command"]
```

- FROM base image: This instruction specifies the base image for your Docker image, containing the underlying operating system and pre-defined software. Choose a base image that suits your needs, such as one with Python, CUDA, or other dependencies.
- ENV VAR NAME=value: This instruction sets environment variables within the container, configuring aspects of the container's runtime behavior or application.
- WORKDIR/app: This instruction creates a new directory named "app" within the container, setting it as the working directory for subsequent commands. It's good practice for organizing files and executing commands.
- COPY source destination: This instruction copies files or directories from the host machine (where the Dockerfile is located) to the container's filesystem. Specify the source path on the host and the destination path within the container.
- RUN command: This instruction runs a command within the container during the image build process. Use it to install software, set up configurations, or perform tasks needed to set up the environment.
- EXPOSE port number: This instruction informs Docker that the container will listen on a specific network port at runtime. Although it doesn't publish the port to the host, it provides documentation for users of the image.
- CMD ["command"]: This instruction specifies the default command to run when the container starts. It's the primary command defining the container's behavior. If the user provides their command when running the container, it will override this default.

4.4 RESULTS AND DISCUSSION

In this section, we will demonstrate the process of creating Docker images for three different scenarios using provided Dockerfiles based on Docker documentation. Each Dockerfile corresponds to a specific machine learning environment.

1. Creating a GPU-supported environment with Miniconda3, featuring Jupyter Notebook.
2. Constructing a TensorFlow GPU environment with Jupyter Notebook.
3. Building a PyTorch environment with GPU support and JupyterLab.

4.4.1 Experience 1: Building a simple machine learning environment

In Experience 1, we created a GPU-supported environment using Miniconda3 as the base image. This environment includes Jupyter Notebook for interactive development.

Dockerfile 1

```
# Use the official Miniconda3 base image with GPU support
FROM continuumio/miniconda3
# Set up environment variables ENV CONDA ENV NAME=myenv
# Create a new conda environment and activate it
RUN conda create -y -n $CONDA ENV NAME python=3.10
# Set the working directory WORKDIR/app
# Activate the new environment and install necessary
packages
RUN echo" conda activate $CONDA ENV NAME" »~/.bashrc ENV
PATH/opt/conda/envs/$CONDA ENV NAME/bin:$PATH
# Install necessary packages
RUN conda install -y notebook matplotlib numpy seaborn
pandas scikit-learn
# Expose Jupyter Notebook port EXPOSE 8888
# Start Jupyter Notebook CMD ["jupyter"," notebook","
-ip=0.0.0.0"," -port=8888"," - allow-root"]
```

Dockerfile 1 is designed to create a self-contained environment that can be used for GPU-accelerated machine learning tasks, particularly using Jupyter Notebook for interactive development and analysis.

- Base image selection: The Docker image starts from the official Miniconda3 base image that supports GPU usage.
- Environment variable: An environment variable named CONDA ENV NAME is defined to hold the name of the conda environment to be created.

- Conda environment creation: A new conda environment is created with the specified name (myenv) and Python version (3.10). This environment will serve as an isolated workspace for the machine learning tasks.
- Working directory: The working directory inside the container is set to/ app, which is where the subsequent commands will be executed.
- Environment activation: The newly created conda environment is activated using the conda activate command. Additionally, the PATH is updated to include the environment's bin directory so that its executables are accessible.
- Package installation: Required machine learning packages are installed within the conda environment. The specified packages include Jupyter Notebook, Matplotlib, NumPy, Seaborn, Pandas, and Scikit-learn.
- Port exposing: Port 8888, which is the default port for Jupyter Notebook, is exposed to allow communication between the host system and the container.
- Container start command: The command to start Jupyter Notebook within the activated conda environment is specified using the CMD directive. The notebook is configured to listen on all available IPs (–ip=0.0.0.0) and use port 8888 (–port=8888). The –allow-root flag allows Jupyter to run as the root user.

With the Dockerfile ready, it's time to build the Docker image as shown in Figure 4.3. We opened a terminal and navigated to the directory containing the Dockerfile. We used the following command to build the image:

```
docker build -t my_ml_env.
```

The "-t" flag assigns a name and optional tag to the image, allowing us to easily reference it later. The "." at the end indicates that the build context is the current directory.

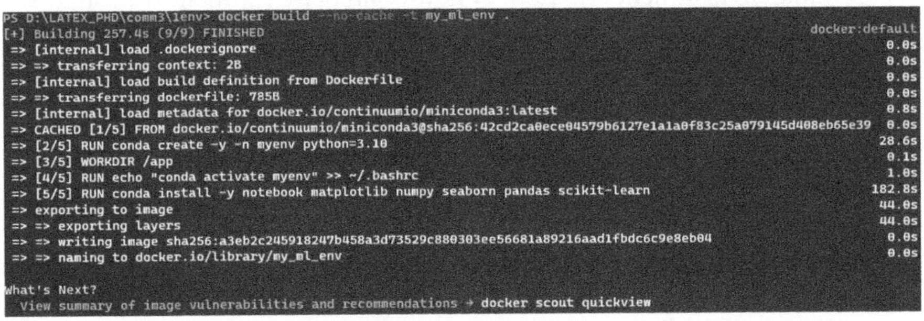

Figure 4.3 Building the first Docker image for a simple machine learning environment.

4.4.2 Experience 2: Building a TensorFlow environment

In Experiment 2, we decided to create a Docker image for a TensorFlow environment with GPU support. This environment features Jupyter Notebook for interactive development. We represent below the Dockerfile 2 to build the TensorFlow image.

Dockerfile 2

```
# Use the official Miniconda3 base image with GPU support
FROM continuumio/miniconda3
# Set up environment variables ENV CONDA ENV NAME=myenv
# Create a new conda environment and activate it
RUN conda create -y -n $CONDA ENV NAME python=3.10
# Set the working directory WORKDIR/app
# Activate the new environment and install necessary
packages
RUN echo" conda activate $CONDA ENV NAME" » ~/.bashrc ENV
PATH/opt/conda/envs/$CONDA ENV NAME/bin:$PATH
# Install necessary packages RUN conda install -y tensorflow
RUN conda install -y notebook matplotlib numpy seaborn
pandas scikit-learn
# Expose Jupyter Notebook port EXPOSE 8888
# Start Jupyter Notebook CMD ["jupyter"," notebook","
-ip=0.0.0.0"," -port=8888"," - allow-root"]
```

Dockerfile 2 is tailored to create an environment optimized for GPU-accelerated machine learning tasks with TensorFlow. It incorporates Jupyter Notebook for interactive model development and includes necessary libraries for data processing, visualization, and machine learning.

- Base image selection: The Docker image is based on the official Miniconda3 base image, which is equipped with GPU support.
- Environment variable: An environment variable named CONDA ENV NAME is defined to specify the name of the conda environment to be created.
- Conda environment creation: A new conda environment is created with the specified name (myenv) and Python version (3.10). This environment will be isolated and self-contained for machine learning tasks.
- Working directory: The working directory inside the container is set to/ app, where the following instructions will be executed.
- Environment activation: The newly created conda environment is activated using the conda activate command. The PATH is updated to include the environment's bin directory, ensuring access to its executables.
- Package installation: Essential machine learning packages are installed within the conda environment. Notably, TensorFlow is installed using the conda install command. Additionally, other necessary libraries such as Jupyter Notebook, Matplotlib, NumPy, Seaborn, Pandas, and Scikit-learn are also installed.

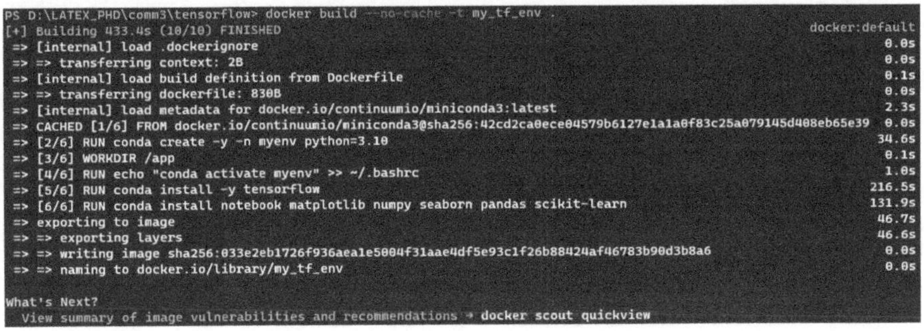

Figure 4.4 Building the second Docker image for a TensorFlow environment.

- Port exposing: Port 8888, the default port for Jupyter Notebook, is exposed to enable communication between the host system and the container.
- Container start command: The command to initiate Jupyter Notebook within the activated conda environment is provided using the CMD directive. The notebook is configured to listen on all available IPs (–ip=0.0.0.0) and utilize port 8888 (–port=8888). The –allow-root flag allows Jupyter Notebook to run as the root user.

With the Dockerfile prepared, we are now ready to build the Docker image. We opened our terminal and navigated to the directory containing the Dockerfile. In this location, we used the following command to initiate the image-building process as shown in Figure 4.4:

```
docker build -t my_tf_env.
```

4.4.3 Experience 3: Building a PyTorch environment

In this last experience, we created a Docker image for a PyTorch environment with GPU support. This environment includes JupyterLab for interactive development. Dockerfile 3 creates an environment optimized for GPU-accelerated deep learning tasks using PyTorch. It includes JupyterLab for interactive development and analysis of machine learning models, along with essential libraries for data manipulation and visualization.

Dockerfile 3

```
# Use the official PyTorch GPU base image FROM pytorch/
pytorch:latest # Set up environment variables
ENV CONDA ENV NAME=myenv
# Create a new conda environment and activate it
RUN conda create -y -n $CONDA ENV NAME python=3.8
# Set the working directory WORKDIR/app
# Activate the new environment and install necessary
packages RUN echo" conda activate $CONDA ENV NAME" »
~/.bashrc ENV PATH/opt/conda/envs/$CONDA ENV NAME/bin:$PATH
```

```
# Install necessary packages
RUN conda install -y jupyterlab matplotlib numpy seaborn
pandas scikit-learn
# Expose JupyterLab port EXPOSE 8888
# Start JupyterLab
CMD ["jupyter"," lab"," -ip=0.0.0.0"," -port=8888","
-allow-root"]
```

- Base image selection: The Docker image is based on the official PyTorch GPU base image, ensuring GPU support for deep learning tasks.
- Environment variable: An environment variable named CONDA ENV NAME is defined to specify the name of the conda environment to be created.
- Conda environment creation: A new conda environment is created with the specified name (myenv) and Python version (3.8). This environment will serve as an isolated workspace for machine learning tasks.
- Working directory: The working directory inside the container is set to/ app, which is where the subsequent commands will be executed.
- Environment activation: The newly created conda environment is activated using the conda activate command. Additionally, the PATH is updated to include the environment's bin directory so that its executables are accessible.
- Package installation: Essential machine learning packages are installed within the conda environment. The specified packages include JupyterLab, Matplotlib, NumPy, Seaborn, Pandas, and Scikit-learn.
- Port exposing: Port 8888, the default port for JupyterLab, is exposed to facilitate communication between the host system and the container.
- Container start command: The command to start JupyterLab within the activated conda environment is specified using the CMD directive. The lab is configured to listen on all available IPs (–ip=0.0.0.0) and use port 8888 (–port=8888). The –allow-root flag allows JupyterLab to run as the root user.

With the Dockerfile ready, it's time to build the Docker image. In the local machine, as seen in Figure 4.5, we opened a terminal and navigated to the

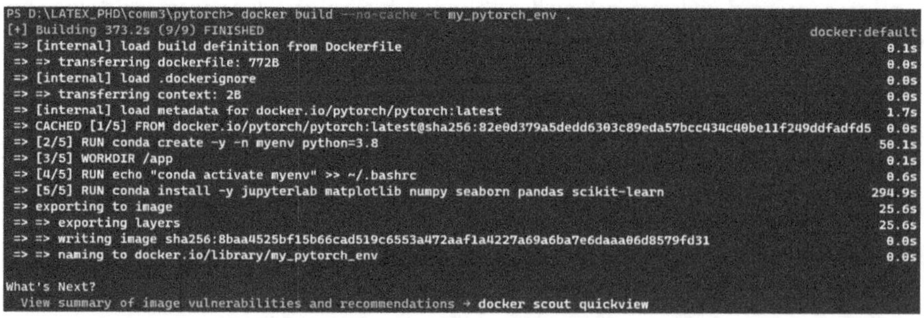

Figure 4.5 Building the third Docker image for a PyTorch environment.

directory containing the Dockerfile. Similar to the first Dockerfiles, we used the same command to build the PyTorch image:

```
docker build -t my_pytorch_env.
```

4.4.4 Running Docker containers

Now that we have created Docker images for different machine learning environments, it's time to run containers based on these images. To run a container based on the Miniconda3 image that we have built-in Experience 1, as shown in Figure 4.6, we used the following command:

```
docker run -gpus all -p 8888:8888 my_ml_env
```

- –gpus all ensures that the container can access the GPU.
- -p 8888:8888 maps port 8888 inside the container to port 8888 on our machine, allowing access to the Jupyter Notebook.
- my_ ml_ env specifies the name of the Docker image.

4.4.5 Accessing Jupyter Notebook or JupyterLab

With the container up and running, we can access Jupyter Notebook or JupyterLab in the web browser as shown in Figure 4.7. In the browser, we navigated to http://localhost:8888.

We can apply the same command to access the environments of the other Docker images (TensorFlow and PyTorch) by changing only the image name.

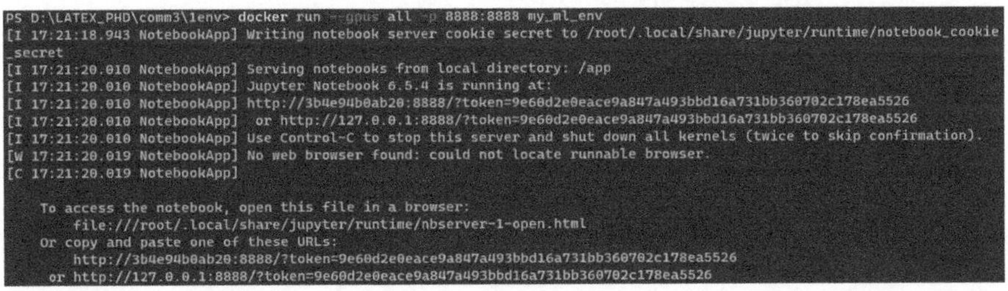

Figure 4.6 Running a Docker container to access the environment created in the first experience.

Figure 4.7 Jupyter Notebook after running the first Docker image on a container.

4.4.5.1 Testing Jupyter Notebook of the first Docker image

To assess the functionality of the Jupyter Notebook within the container, we opted to develop a fundamental Python script, highlighting the utilization of widely used data science libraries installed via Dockerfile. Noteworthy libraries encompassed in this script are Matplotlib, NumPy, Pandas, and Seaborn. The script generates synthetic data, structures it into a Pandas DataFrame, and produces a scatter plot using Seaborn. This demonstration vividly showcases the seamless visualization of data within the Docker environment. The outcome of applying this example is presented in Figure 4.8.

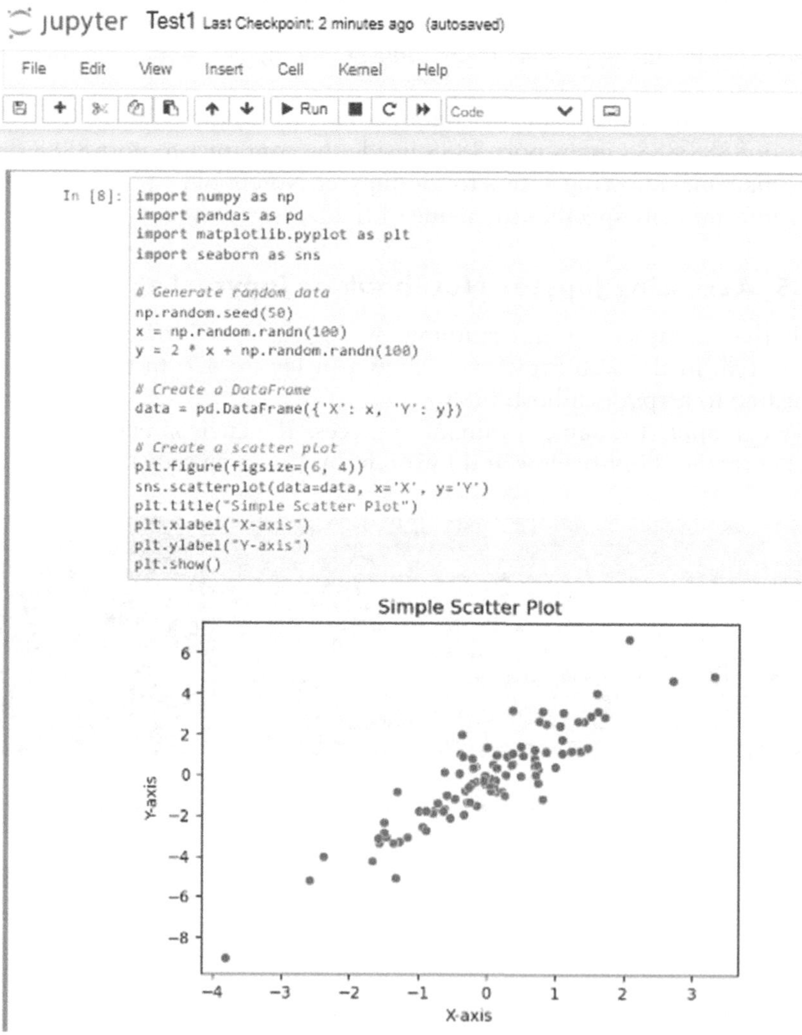

Figure 4.8 Example of testing Jupyter Notebook and the modules installed of the first Docker image.

4.4.5.2 *Testing Jupyter Notebook of the second Docker image*

Upon deploying the second Docker image onto a container, we proceeded to access the Jupyter Notebook to evaluate the environment. Subsequently, we conducted tests by installing modules and implementing a simple machine learning model. In Figure 4.9, a Python script illustrates the use of TensorFlow for a basic linear regression process. This script generates random data points, creates a linear regression model, and trains the model to predict the relationship between input (X) and output (Y). The visual representation depicts the model's predictions juxtaposed with the original data points for clarity.

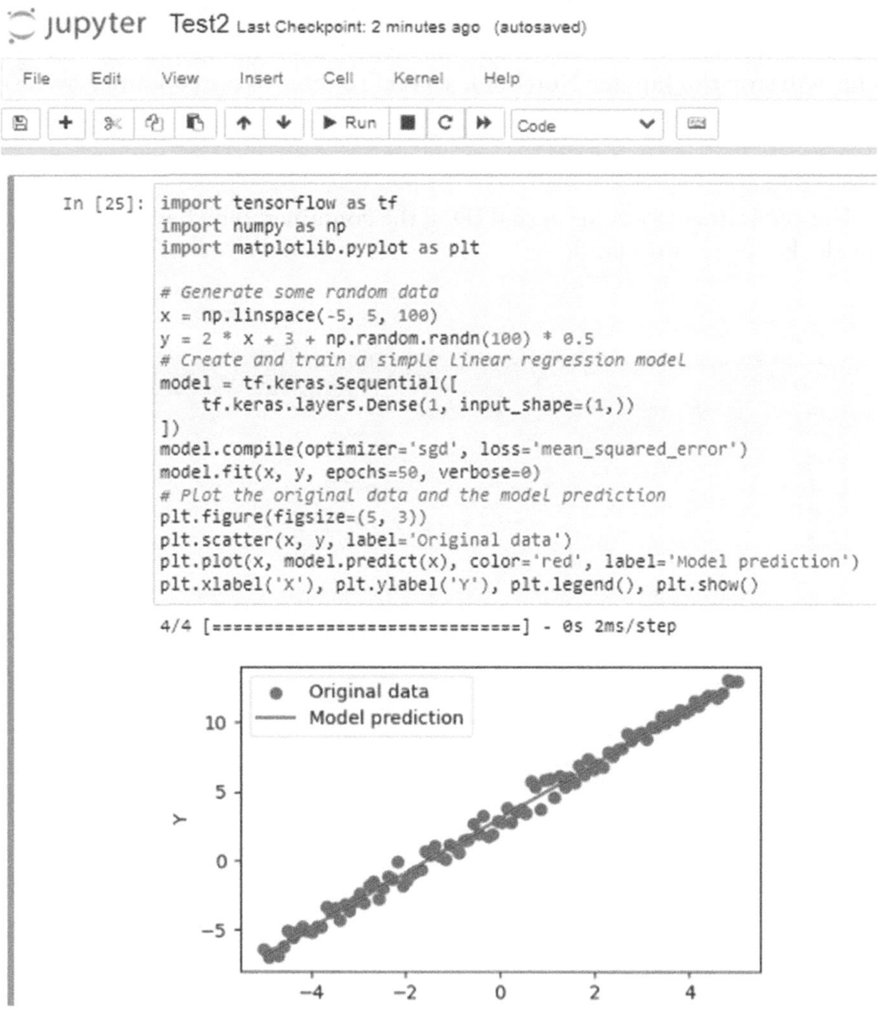

Figure 4.9 Example of testing Jupyter Notebook and TensorFlow of the second Docker image.

4.4.5.3 Testing JupyterLab of the third Docker image

To use the third Docker image, we launched a container to access JupyterLab and test the PyTorch installation.

In Figure 4.10, a Python script is showcased, designed to verify the presence of a GPU and print its name if available; otherwise, it gracefully resorts to utilizing the CPU. Subsequently, the script creates a PyTorch tensor, conducts a fundamental operation (multiplication), and prints the resultant output. This code exemplifies the seamless adaptability of PyTorch to leverage available hardware resources for efficient tensor operations.

4.4.6 Stopping and removing containers

To stop a container, we go to the terminal where it's running and press Ctrl+C. This will stop the Jupyter Notebook server. To remove the container, we used the following command:

```
docker rm <container id>.
```

The <container id> is the actual ID of the container, which we can get using the docker ps -a command.

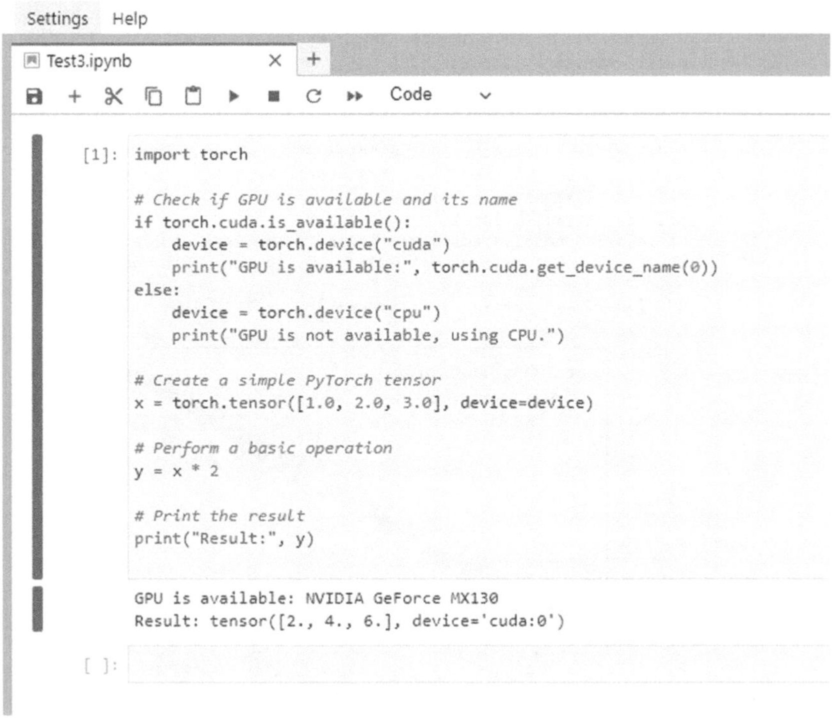

Figure 4.10 Example of testing JupyterLab and PyTorch of the third Docker image.

The adoption of containers, such as Docker, brings forth a plethora of advantages that streamline the workflow and overcome the limitations posed by traditional methods. According to our application, we deduced several advantages of building a GPU-supported environment for machine learning and deep learning on containers instead of using other techniques that we have mentioned before. These advantages include the following:

- Effortless installation: Simplifying the installation process, containers encapsulate the entire environment, along with its dependencies and configuration settings. This drastically reduces the time and complexity required for environment setup across diverse machines.
- Resource consumption: Containers ensure optimal resource utilization by sharing the host system's kernel, resulting in minimal overhead. Their lightweight nature diminishes resource requirements compared to VMs, enhancing overall efficiency.
- Isolation and reproducibility: Containers provide isolated runtime environments, ensuring consistent behavior across different systems. This isolation eliminates conflicts between libraries and packages, guaranteeing the reproducibility of experiments and results.
- Portability: Inherently portable, containers facilitate seamless deployment across diverse platforms and environments. This proves advantageous during transitions between development, testing, and production stages.
- Quick startup and execution: Containerized GPU environments exhibit rapid startup times, fostering swift iteration and experimentation. The encapsulated structure minimizes the overhead associated with starting and stopping environments, enabling faster iterations during development.

In the assessment of containerization against alternative methodologies, numerous distinct advantages come to the forefront. Contrasted with the conventional bare-metal technique, containerization adeptly abstracts intricate hardware and software intricacies, thereby eradicating the often formidable complexities associated with bare-metal installations. Furthermore, containerized environments elevate convenience levels, facilitating effortless sharing and deployment. Collaborative endeavors benefit significantly as these environments can be promptly shared and deployed, offering a marked departure from the time-consuming process of constructing bare-metal environments from the ground up.

Moreover, when juxtaposed against the Anaconda Technique, containers showcase superior capabilities concerning isolation and portability. By averting conflicts among dependencies and libraries, containers assure environment stability, thereby facilitating smoother workflow operations. The lightweight nature of containers results in efficient utilization of system resources, optimizing GPU performance and enhancing overall computational efficiency.

Last, compared to the VM technique, containers exhibit rapid initiation and economical resource consumption. Their prowess in dynamic scaling and efficient resource allocation surpasses the VM paradigm, rendering containerization an attractive choice for resource-conscious environments.

4.5 CONCLUSION AND FUTURE WORK

In conclusion, Docker containers have emerged as a pivotal asset in constructing GPU-supported machine learning and deep learning environments. Through the creation of various setups, including a simple machine learning environment, TensorFlow environment, and PyTorch environment, we have demonstrated the expeditious installation process, optimal resource consumption, and efficient execution that containers offer. By surpassing traditional methods such as bare-metal setups, Anaconda environments, and VMs, Docker containers establish themselves as the preferred solution for researchers and practitioners seeking swift, collaborative, and reproducible environments for their GPU-accelerated machine learning and deep learning projects. Looking ahead, future work could delve deeper into optimizing containerized GPU environments for specific machine learning frameworks. Further research might focus on enhancing resource allocation strategies to maximize GPU utilization and exploring advanced container orchestration techniques for deploying distributed machine learning workflows.

ACKNOWLEDGMENT

We declare that there is no funding to declare for this research project. The work presented in this chapter was carried out without any external financial support or funding.

REFERENCES

1. Sharma, N., Sharma, R., Jindal, N.: Machine learning and deep learning applications-a vision. Global Transitions Proceedings **2**(1), 24–28 (2021). https://doi.org/10.1016/j.gltp.2021.01.004
2. Shinde, P.P., Shah, S.: A review of machine learning and deep learning applications. In: 2018 Fourth International Conference on Computing Communication Control and Automation (ICCUBEA), pp. 1–6 (2018). https://doi.org/10.1109/ICCUBEA.2018.8697857
3. Dargan, S., Kumar, M., Ayyagari, M.R., Kumar, G.: A survey of deep learning and its applications: A new paradigm to machine learning. Archives of Computational Methods in Engineering 27, pp. 1–22 (2020).
4. Mosavi, A., Ardabili, S., Várkonyi-Kóczy, A.R.: List of deep learning models. In: Várkonyi-Kóczy, A.R. (ed.) Engineering for Sustainable Future, pp. 202–214. Springer, Cham (2020).

5. Wang, Z., Liu, K., Li, J., Zhu, Y., Zhang, Y.: Various frameworks and libraries of machine learning and deep learning: A survey. Archives of Computational Methods in Engineering 31, pp. 1–24 (2019).
6. Raschka, S., Patterson, J., Nolet, C.: Machine learning in Python: Main developments and technology trends in data science, machine learning, and artificial intelligence. Information **11**(4) (2020). https://doi.org/10.3390/info11040193
7. Hardikar, S., Ahirwar, P., Rajan, S.: Containerization: Cloud computing based inspiration technology for adoption through Docker and Kubernetes. In: 2021 Second International Conference on Electronics and Sustainable Communication Systems (ICESC), pp. 1996–2003 (2021). https://doi.org/10.1109/ICESC51422.2021.9532917
8. Bouaouda, A., Afdel, K., Abounacer, R.: Forecasting the energy consumption of cloud data centers based on container placement with ant colony optimization and bin packing. In: 2022 5th Conference on Cloud and Internet of Things (CIoT), pp. 150–157 (2022). https://doi.org/10.1109/CIoT53061.2022.9766522
9. Zaher, C.: For CTO's: The no-nonsense way to accelerate your business with containers. Technical report, Canonical Limited 2017. Ubuntu, Kubuntu (February 2017).
10. Pahl, C., Brogi, A., Soldani, J., Jamshidi, P.: Cloud container technologies: A state-of-the-art review. IEEE Transactions on Cloud Computing 7(3), 677–692 (2019). https://doi.org/10.1109/TCC.2017.2702586
11. Kominos, C.G., Seyvet, N., Vandikas, K.: Bare-metal, virtual machines and containers in OpenStack. In: 2017 20th Conference on Innovations in Clouds, Internet and Networks (ICIN), pp. 36–43 (2017). https://doi.org/10.1109/ICIN.2017.7899247
12. Zhang, X., Zheng, X., Wang, Z., Yang, H., Shen, Y., Long, X.: High-density multi-tenant bare-metal cloud. In: Proceedings of the Twenty-Fifth International Conference on Architectural Support for Programming Languages and Operating Systems. ASPLOS '20, pp. 483–495. Association for Computing Machinery, New York, NY, USA (2020). https://doi.org/10.1145/3373376.3378507. https://doi.org/10.1145/3373376.3378507
13. Cheng, K., Doddamani, S., Chiueh, T.-C., Li, Y., Gopalan, K.: Directvisor: Virtualization for bare-metal cloud. In: Proceedings of the 16th ACM SIGPLAN/SIGOPS International Conference on Virtual Execution Environments, pp. 45–58 (2020).
14. Sykora, P., Sinko, M., Vrskova, R., Kamencay, P., Hudec, R.: Comparison of convolutional neural network in python environment on CPU and GPU. In: 2020 ELEKTRO, pp. 1–4 (2020). https://doi.org/10.1109/ELEKTRO49696.2020.9130321
15. Helali, L., Omri, M.N.: A survey of data center consolidation in cloud computing systems. Computer Science Review **39**, 100366 (2021).
16. Potdar, A.M., Narayan, D., Kengond, S., Mulla, M.M.: Performance evaluation of Docker container and virtual machine. Procedia Computer Science **171**, 1419–1428 (2020).
17. Xu, B., Wu, S., Xiao, J., Jin, H., Zhang, Y., Shi, G., Lin, T., Rao, J., Yi, L., Jiang, J.: Sledge: Towards efficient live migration of Docker containers. In: 2020 IEEE 13th International Conference on Cloud Computing (CLOUD), pp. 321–328 (2020). https://doi.org/10.1109/CLOUD49709.2020.00052
18. Rad, B.B., Bhatti, H.J., Ahmadi, M.: An introduction to Docker and analysis of its performance. International Journal of Computer Science and Network Security (IJCSNS) **17**(3), 228 (2017).
19. Randal, A.: The ideal versus the real: Revisiting the history of virtual machines and containers. ACM Computing Surveys **53**(1) (2020). https://doi.org/10.1145/3365199

20. Bouaouda, A., Afdel, K., Abounacer, R.: Meta-heuristic and heuristic algorithms for forecasting workload placement and energy consumption in cloud data centers. Advances in Science, Technology and Engineering Systems Journal **8**(1), 1–11 (2023). https://doi.org/10.25046/aj080101
21. Bhardwaj, A., Krishna, C.R.: Virtualization in cloud computing: Moving from hypervisor to containerization—A survey. Arabian Journal for Science and Engineering **46**(9), 8585–8601 (2021).
22. Yadav, A.K., Garg, M.L., Ritika, M.: Docker containers versus virtual machine-based virtualization. In: Emerging Technologies in Data Mining and Information Security: Proceedings of IEMIS 2018, Volume 3, pp. 141–150 (2019). Springer.
23. Marathe, N., Gandhi, A., Shah, J.M.: Docker swarm and Kubernetes in cloud computing environment. In: 2019 3rd International Conference on Trends in Electronics and Informatics (ICOEI), pp. 179–184 (2019). IEEE.
24. Santoro, C., Messina, F., D'Urso, F., Santoro, F.F.: Wale: A Dockerfile-based approach to deduplicate shared libraries in Docker containers. In: 2018 IEEE 16th Intl Conf on Dependable, Autonomic and Secure Computing, 16th Intl Conf on Pervasive Intelligence and Computing, 4th Intl Conf on Big Data Intelligence and Computing and Cyber Science and Technology Congress (DASC/PiCom/DataCom/CyberSciTech), pp. 785–791 (2018). https://doi.org/10.1109/DASC/PiCom/DataCom/CyberSciTec.2018.00135
25. Reis, D., Piedade, B., Correia, F.F., Dias, J.P., Aguiar, A.: Developing Docker and Docker-compose specifications: A developers' survey. IEEE Access **10**, 2318–2329 (2022). https://doi.org/10.1109/ACCESS.2021.3137671

Chapter 5

A review of image-based deep learning approaches for atmospheric visibility estimation

Kabira Ait Ouadil, Soufiane Idbraim, Taha Bouhsine, Nidhal Carla Bouaynaya, Husam Alfergani, and Charles Cliff Johnson

5.1 INTRODUCTION

Meteorological visibility plays a vital role in road safety, air operations, and maritime activities. Good visibility is essential for drivers to accurately assess their surroundings and respond swiftly to road conditions. Drivers with good visibility can spot obstacles on the road, such as other vehicles, pedestrians, road signs, and traffic lights. Reduced visibility due to adverse weather conditions, such as fog, heavy rain, snow, or dust, makes early detection of these impediments more complex, increasing the risk of accidents and leading to flight delays or cancellations. Thus, estimating atmospheric visibility accurately is a challenging topic in meteorology, particularly in transport safety.

Previously, human eye observation was frequently employed to forecast visibility, but it is affected by a variety of subjective factors and can result in significant uncertainties. To accurately gauge visibility, it was replaced by specialized optical weather instruments, such as visibility meters and transmissometers, but this equipment is quite costly, and it can only be installed at a few weather stations to measure the visibility of specific scenes.

For ubiquitous and efficient visibility monitoring, the researchers used images captured by different cameras to estimate real-time visibility in any given scenario, analyzing the characteristics of the image (contrast, brightness, transmittance, etc.) using preprocessing algorithms.

Image-based visibility estimation approaches can be divided into two categories: physical model-based methods and deep learning (DL)-based methods. The first estimates visibility by defining the relationship between visibility and other data based on fundamental equations and laws, such as the Duntley, Lambert-Beer, and Koschmieder laws. For instance, Pomerleau et al. [1] tracked lane lines in the image and then estimated visibility by measuring the attenuation of contrast among consistent road characteristics. Babari et al. [2] suggested a model-driven approach to estimating visibility from ordinary outdoor cameras based on the mapping function between the contrast and visibility distance. Although this method is often interpretable, images acquired by cameras are affected by different factors, including climatic conditions, lighting conditions, and specific camera settings, making

DOI: 10.1201/9781032664293-6

the use of physical models for feature extraction more complex. Recently, DL-based methods have demonstrated outstanding success in a variety of domains, including computer vision, natural language processing (NLP), and reinforcement learning.

They have achieved advanced outcomes in a variety of tasks, outperforming traditional methods. To improve the feature extraction from weather images, DL techniques were introduced to estimate atmospheric visibility, leveraging the tremendous power of convolutional neural network (CNN) architectures to identify the complex relationships between extracted data and visibility labels. In addition to CNNs, transformers have recently been used.

Based on our previous work [3], this chapter focuses only on image-based DL approaches for estimating meteorological visibility, as presented in Figure 5.1. The objective of this survey is to summarize and classify datasets and DL architectures used in prior relevant publications. Further, the outcomes and limitations of the reviewed studies are discussed. The research gaps are also analyzed to assist future researchers in exploring new trends in this area.

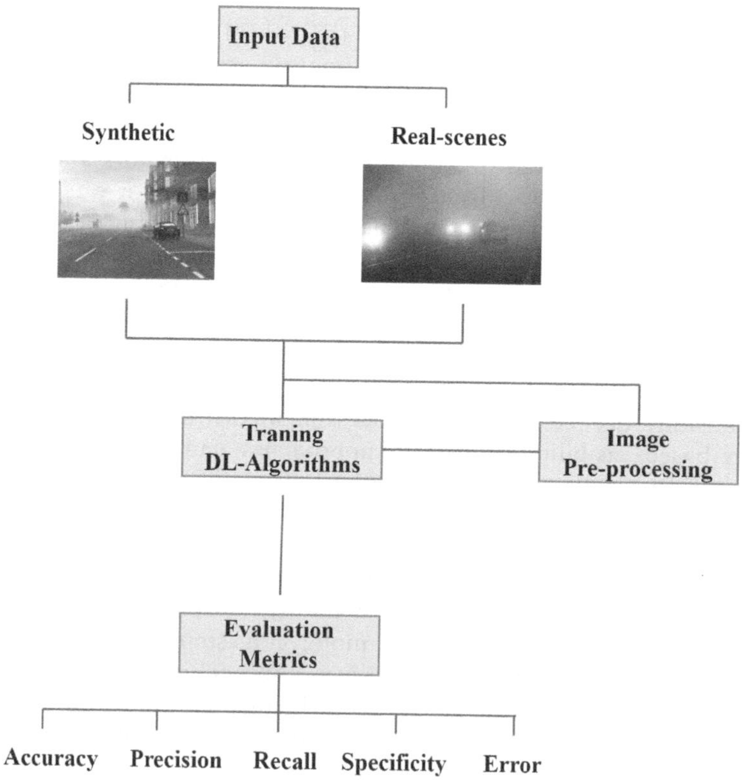

Figure 5.1 Image-based DL process for visibility estimation.

The content of this chapter was divided into five sections. The first section introduces the general context and background of visibility estimation. The second section includes the image preprocessing algorithms. In the third section, DL architectures are described. Furthermore, the available datasets are listed, and the results of the reviewed studies are discussed. Finally, conclusions and promising future research directions are presented.

5.2 IMAGE PREPROCESSING

Image preprocessing is an important step in image analysis and computer vision jobs. It entails the use of multiple techniques to improve image quality and prepare them for analysis and further processing. Figure 5.2 presents the three main image preprocessing techniques used to prepare weather image datasets.

5.2.1 Image enhancement and dehazing

Weather images captured by cameras are influenced by several factors, including camera quality, camera settings, and weather conditions, leading to noisy and low-quality photos.

Image enhancement is required to improve image quality and remove noise while retaining important image characteristics.

For example, Chaabani et al. [4] applied the Fourier transform to the Foggy ROad Sign Images (FROSI) [5] dataset to take the power spectrum of the

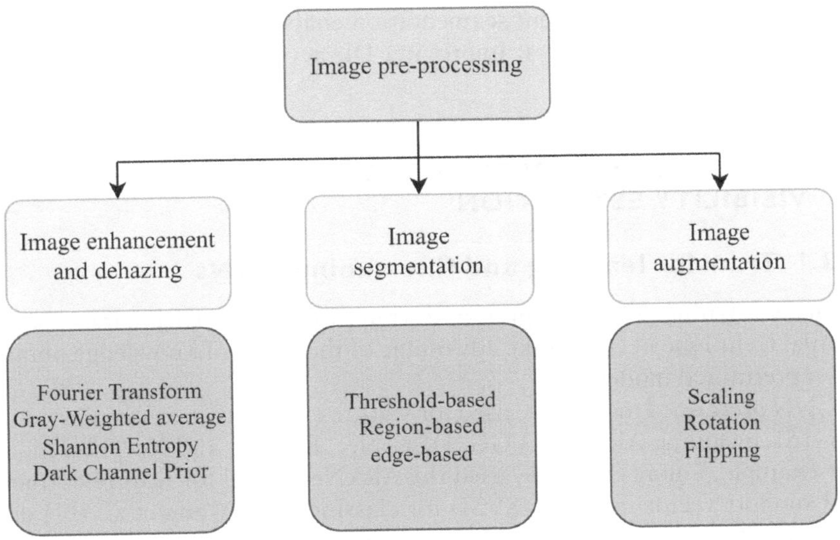

Figure 5.2 Image preprocessing techniques.

image and Shannon entropy to get the image texture. Palvanov et al. [6] introduced a multiple-stream model based on pre-processed images. They applied a fast Fourier transform to eliminate low frequencies (such as fog, clouds, and sky) and a spectral filter to obtain fog and low-contrast areas. Li et al. [7] worked with the Hong Kong Observatory (HKO) dataset gathered from the Central Pier Automatic Weather Station. The gray-weighted average is applied to the image database to remove interference areas such as sky and clouds, to facilitate the location of the most effective image subregions (containing landmarks).

Image dehazing methods were also utilized to assist DL models in obtaining more valuable fog information. The dark channel prior method is the most often employed [8, 9].

5.2.2 Image segmentation

Image segmentation separates objects or regions of interest (fog regions) from the background to extract meaningful information (landmarks) from an image.

For this purpose, various segmentation algorithms are applied, including threshold-, region-, and edge-based techniques. Palvanov et al. [10] used a Laplacian of Gaussian filter for edge detection and then cropped the region of interest to estimate visibility. Liu et al. [11] developed an Auto Seed Region Segmentation (ASRS) to separate the fog area in the input image in order to simplify the feature extraction phase. The threshold-based method [7, 12] was also used to segment the input image into subregions containing static features such as buildings.

Image segmentation is a decisive step since it has a significant impact on visibility estimation. Efficient segmentation enables accurate detection of fog zones containing landmarks, improving DL accuracy and decreasing inference time.

5.3 DEEP LEARNING METHODS FOR VISIBILITY ESTIMATION

5.3.1 Transfer learning and fine-tuning CNNs

Transfer learning (TL) is the preferred strategy for most articles; it has become a usual technique in DL to take advantage of the depth of knowledge obtained from pretrained models.

CNN pretrained models are used for feature extraction tasks in most studies [13–15], including AlexNet, VGG, DenseNet, ResNet, and Xception models. For example, Outay et al. [16] used the AlexNet model for feature extraction and support vector machine (SVM) for classification. Wang et al. [17] developed ResNet streams to extract features from multimodal datasets (visible and infrared images). TL is also performed by Lo et al. [7, 12, 18], where regions of interest (subregions) are provided as input. These subregions are fed into pretrained networks (VGG-16, VGG-19, DenseNet, and ResNet_50) for feature extraction, followed by the multi-support vector regression (SVR) model for visibility prediction. Liu et al. [19] introduced a deep integrated model by combining VGG16 and Xception. The ResNet pretrained model is the most often used for estimating visibility since it looks for promising outcomes.

5.3.2 Customized CNN architectures

VisNet was developed by Palvanov et al. [6], and it is regarded as a universal model for estimating daytime visibility. VisNet is a new, deeply integrated architecture that implements three CNN streams: the first takes the Fast Fourier Transform FFT-filtered image, the second receives the spectral-filtered image, and the third takes the original image. All streams are trained in parallel to take advantage of previously processed images in order to extract as much meaningful information as possible.

As seen in Figure 5.3, many studies have developed their own architectures. Qin et al. [20] proposed a novel CNN system that used multiscale mapping to identify features at various scales and an activation function called Modified_sigmoid for visibility estimation.

Xun et al. [21] established an end-to-end system called VISOR-Net that used the ordinal information and relative relation of images for visibility estimation. The VISOR-Net consists of two components: the Feature Extraction Regression Module (FERM) for predicting image visibility level, and the

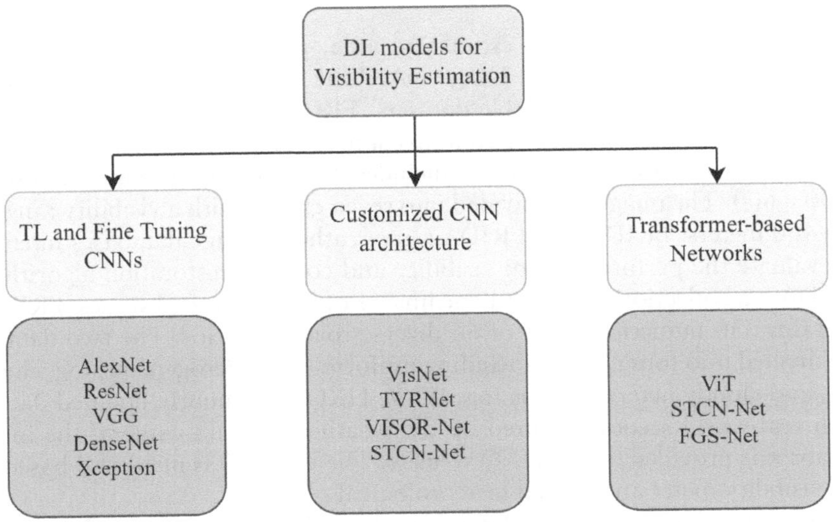

Figure 5.3 Classification of DL algorithms for visibility estimation.

second component is a pairs comparison with relative learning, where the ordinal information in each image batch is encoded as a collection of paired images in relative order. VISOR-Net is evaluated as both a classification and regression task on different datasets.

5.3.3 Transformer-based networks

In 2017, transformer architecture was introduced, and it has been extensively used in various NLP tasks, but it has also been expanded to other artificial intelligence (AI) domains, including computer vision and audio processing.

In the visibility estimation area, Bouhsine et al. [14] tested a Vision Transformer (ViT) model for ground-level visibility estimation on the Federal Aviation Administration (FAA) dataset. Liu et al. [9] proposed a novel hybrid DL system named STCN-Net that combines engineered and learned features using both CNN and Transformer. The DDT matrix is fed into the ResNet-18, while the original picture is fed into the Swin-Transformer (Swin-T). The coordinate attention is embedded in STCN-Net to incorporate different feature information from CNN and Swin-T. Recently, they developed FGS-Net [11], which has a similar architecture to STCN-Net, but they applied an ASRS segmentation technique for the accurate segmentation of fog regions before training the model.

Transformers have recently been utilized to estimate visibility and have shown good performance, particularly when combined with CNN architectures.

5.4 RESULTS

5.4.1 Datasets

In the atmospheric visibility estimation area, public datasets are extremely limited. FROSI, Foggy Road Image DAtabase (FRIDA) [22], FRIDA2 [23], and HKO are the only available datasets. FROSI and FRIDA are both synthetic datasets. FROSI dataset contains simple road scenes that are used for visibility classification problems. It includes 3528 photos with a resolution of 1400 × 600. The images are divided into seven classes with a visibility range of 50–400 meters. FRIDA and FRIDA2 are synthetic image databases intended to evaluate the performance of visibility and contrast restoration algorithms. FRIDA is a collection of 90 synthetic images of 18 urban road scenes. FRIDA2 contains 330 numeric images of 66 diverse road scenarios. The two datasets are divided into four classes, including uniform fog, heterogeneous fog, cloudy fog, and cloudy heterogeneous fog. While HKO is an hourly updated dataset with real-world scenes captured by the weather station's camera, the image database is provided by the HKO website. The visibility is measured based on the visibility meter and skilled meteorological observers.

FROSI is the most popular dataset used as a benchmark for evaluating the performance of DL models and algorithms. Most researchers compare their results with existing state-of-the-art methods using this dataset, but it is so small and has a relatively narrow visibility range (a few meters) and cannot be used for estimating real scenarios.

The key constraint of prior studies is the lack of databases with a high number of real images describing all visibility situations with a long visibility interval, hence the need to create a database that meets all these needs.

Most studies have developed their own dataset since the deficiency of public datasets, including a large number of real images describing all visibility situations with a long visibility interval.

5.4.2 Evaluation

Most papers considered visibility estimation as a classification problem. However, others estimate the continuous value of visibility instead of predicting visibility level. After the feature extraction step, various classifiers are used to guide the visibility, including SVM, SVR, artificial neural network (ANN), CNN, and others.

To evaluate the performance of the reviewed studies, different metrics were used: accuracy, precision, recall, specificity, and error, but accuracy is the most commonly used metric in all studies. Table 5.1 compares the more relevant DL techniques used in the prior studies for estimating visibility.

Table 5.1 Comparison of various techniques used for visibility estimation

Reference	*Technique*	*Dataset*	*Metric*
Outay et al. [16]	TL using AlexNet and SVM for range classification	FROSI	Acc = 99.02%
Palvanov et al. [6]	Customized DCNN (VisNet)	FOVI	Acc = 90.4%
You et al. [8]	Hybrid CNN-RNN model	Outdoors images	Acc = 82.2%
Wang et al. [17]	Multimodal DCNN	Multimodal dataset	Acc = 98.3%
Lo et al. [18]	TL using VGG16 and multi-SVR for visibility estimation	HKO	Acc = 85%
Li et al. [7]	TL based on feature fusion	HKO	Acc = 91.2%
Qin et al. [20]	Personalized DCNN (TVRNet) model	ARD dataset	Error = 0.0016
Bouhsine et al. [14]	TL for visibility range classification	FAA dataset	Acc = 98.47%
Liu et al. [9]	Combination of CNN and Swin-Transformer (STCN-Net)	Private	Acc = 99.4%
Xun et al. [21]	Novel architecture VISOR-Net	FHVI	Acc = 87.83%
Li et al. [24]	TL for visibility detection	Private	Acc = 93.86%

In this chapter, various DL-based techniques used for visibility estimation have been studied. These techniques are classified into three main categories, including TL and fine-tuning models, novel CNN architectures, and transformers. The first category is the most employed since it is easy to use, especially with the smallest datasets. New architectures achieved good results, but they are costly and take a long training time.

Lately, researchers have tended to work with transformer-based networks. Transformers can be combined with CNNs and have achieved outstanding results; however, these architectures are very complex and time-consuming to train.

5.5 CONCLUSION

This study presents recent papers for visibility estimation using image-based DL techniques. This review discusses atmospheric visibility estimation systems based on TL and fine-tuning models, customized DL architectures, and transformer-based networks. Most systems have demonstrated their efficiency in the feature extraction process. The datasets used in the overall studies are also highlighted.

Based on the literature review, it is shown that DL using both CNN and transformer architectures might have significant effects on the automatic extraction of highly useful features from the images.

The major challenge of the evaluated studies on visibility estimation is the unavailability of large datasets with real weather images under a variety of weather conditions. In addition, most hybrid architectures require a long training time, necessitating the optimization of model parameters.

REFERENCES

1. D. Pomerleau, "Visibility estimation from a moving vehicle using the RALPH vision system," in *Proceedings of Conference on Intelligent Transportation Systems*, 1997.
2. R. Babari, N. Hautiere, E. Dumont, R. Bremond and N. Paparoditis, "A model-driven approach to estimate atmospheric visibility with ordinary cameras," *Atmospheric Environment*, vol. 45, pp. 5316–5324, 2011.
3. K. Ait Ouadil, S. Idbraim, T. Bouhsine, N. Carla Bouaynaya, H. Alfergani and C. Cliff Johnson, "Atmospheric visibility estimation: A review of deep learning approach," *Multimedia Tools and Applications*, vol. 83, pp. 1–26, 2023.
4. H. Chaabani, F. Kamoun, H. Bargaoui, F. Outay and A.-U.-H. Yasar, "A Neural network approach to visibility range estimation under foggy weather conditions," *Procedia Computer Science*, vol. 113, pp. 466–471, 2017.
5. R. Belaroussi and D. Gruyer, "Impact of reduced visibility from fog on traffic sign detection," in *2014 IEEE Intelligent Vehicles Symposium Proceedings*, 2014.
6. A. Palvanov and Y. I. Cho, "VisNet: Deep convolutional neural networks for forecasting atmospheric visibility," *Sensors*, vol. 19, p. 1343, 2019.
7. J. Li, W. L. Lo, H. Fu and H. S. H. Chung, "A transfer learning method for meteorological visibility estimation based on feature fusion method," *Applied Sciences*, vol. 11, p. 997, 2021.

8. Y. You, C. Lu, W. Wang and C.-K. Tang, “Relative CNN-RNN: Learning relative atmospheric visibility from images,” *IEEE Transactions on Image Processing*, vol. 28, pp. 45–55, 2018.
9. J. Liu, X. Chang, Y. Li, Y. Ji, J. Fu and J. Zhong, “STCN-Net: A novel multi-feature stream fusion visibility estimation approach,” *IEEE Access*, vol. 10, pp. 120329–120342, 2022.
10. A. Palvanov and Y. Im Cho, “DHCNN for Visibility Estimation in Foggy Weather Conditions,” in *2018 Joint 10th International Conference on Soft Computing and Intelligent Systems (SCIS) and 19th International Symposium on Advanced Intelligent Systems (ISIS)*, 2018.
11. J. Liu, J. Zhong, Y. Li, Y. Ji, J. Fu and X. Chang, “FGS-Net: A Visibility Estimation Method Based on Statistical Feature Stream in Fog Area,” 2023.
12. W. L. Lo, H. S. H. Chung and H. Fu, “Experimental evaluation of PSO based transfer learning method for meteorological visibility estimation,” *Atmosphere*, vol. 12, p. 828, 2021.
13. M. Song, X. Han, X. F. Liu and Q. Li, “Visibility estimation via deep label distribution learning in cloud environment,” *Journal of Cloud Computing*, vol. 10, pp. 1–14, 2021.
14. T. Bouhsine, S. Idbraim, N. C. Bouaynaya, H. Alfergani, K. A. Ouadil and C. C. Johnson, “Atmospheric Visibility Image-Based System for Instrument Meteorological Conditions Estimation: A Deep Learning Approach,” in *2022 9th International Conference on Wireless Networks and Mobile Communications (WINCOM)*, 2022.
15. Y. Choi, H.-G. Choe, J. Y. Choi, K. T. Kim, J.-B. Kim and N.-I. Kim, “Automatic sea fog detection and estimation of visibility distance on CCTV,” *Journal of Coastal Research*, pp. 881–885, 2018.
16. F. Outay, B. Taha, H. Chaabani, F. Kamoun, N. Werghi and A.-U.-H. Yasar, “Estimating ambient visibility in the presence of fog: A deep convolutional neural network approach,” *Personal and Ubiquitous Computing*, vol. 25, pp. 51–62, 2021.
17. H. Wang, K. Shen, P. Yu, Q. Shi and H. Ko, “Multimodal deep fusion network for visibility assessment with a small training dataset,” *IEEE Access*, vol. 8, pp. 217057–217067, 2020.
18. W. L. Lo, M. Zhu and H. Fu, “Meteorology visibility estimation by using multi-support vector regression method,” *Journal of Advances in Information Technology*, vol. 11, pp. 40–47, 2020.
19. Z. Liu, Y. Chen, X. Gu, J. K. Yeoh and Q. Zhang, “Visibility classification and influencing-factors analysis of airport: A deep learning approach,” *Atmospheric Environment*, vol. 278, p. 119085, 2022.
20. H. Qin and H. Qin, “An end-to-end traffic visibility regression algorithm,” *IEEE Access*, vol. 10, pp. 25448–25454, 2021.
21. L. Xun, H. Zhang, Q. Yan, Q. Wu and J. Zhang, “VISOR-NET: Visibility estimation based on deep ordinal relative learning under discrete-level labels,” *Sensors*, vol. 22, p. 6227, 2022.
22. J.-P. Tarel, N. Hautiere, A. Cord, D. Gruyer and H. Halmaoui, “Improved visibility of road scene images under heterogeneous fog,” in *2010 IEEE Intelligent Vehicles Symposium*, IEEE, 2010, pp. 478–485.
23. J.-P. Tarel, N. Hautiere, L. Caraffa, A. Cord, H. Halmaoui and D. Gruyer, “Vision enhancement in homogeneous and heterogeneous fog,” *IEEE Intelligent Transportation Systems Magazine*, vol. 4, pp. 6–20, 2012.
24. Li. Q, Tang, S. Peng, X and Ma Q, “A Method of visibility detection based on the transfer learning,” *Journal of Atmospheric and Oceanic Technology*, vol. 36(10), pp. 1945–1956, 2019.

Chapter 6

Advancing cloud security

Evaluating interpretable machine learning algorithms for DDoS attack detection

Mohamed Ouhssini, Karim Afdel, Mohamed Akouhar, Elhafed Agherrabi, and Abdallah Abrada

6.1 INTRODUCTION

Distributed Denial of Service (DDoS) attacks pose a significant threat to the stability and performance of cloud services, bombarding systems with overwhelming amounts of malicious traffic and rendering them inaccessible to legitimate users. The urgency of detecting and combating DDoS threats in cloud environments is underscored by the growing dependence on cloud technologies and the crucial need for effective DDoS detection strategies in these settings [1]. As organizations increasingly migrate to the cloud, the potential for cyber-attacks escalates, amplifying the risks associated with DDoS incidents on cloud-based systems [2]. Therefore, devising robust DDoS detection mechanisms is imperative for ensuring the security and resilience of cloud services. Traditional security measures like Intrusion Detection Systems (IDS), including tools such as Snort, fall short of effectively addressing the complex and evolving nature of DDoS attacks in cloud environments. This gap highlights the necessity for a DDoS detection system tailored for cloud contexts, emphasizing interpretability. The significance of this research within the artificial intelligence (AI) community lies in the development of interpretable machine learning (ML) (IML) models specifically for DDoS detection in cloud settings. Although ML methods and deep learning techniques have shown promise in detecting and mitigating DDoS threats [3, 4], the lack of interpretability in these models limits their practical application. This research seeks to fill this gap by offering interpretable models that provide insight into their decision-making processes, thereby enabling security experts to trust and understand the system's outputs. This study addresses the challenge of DDoS attack detection in cloud environments, proposing three ML algorithms – Random Forest, Decision Tree, and CatBoost – as potential solutions. These algorithms are chosen based on their demonstrated effectiveness in DDoS detection in prior research [3, 4]. The research questions guiding this study are:

- How can IML models be developed for DDoS detection in the cloud?
- How do the Random Forest, Decision Tree, and CatBoost algorithms perform in detecting DDoS attacks in the cloud?

 DOI: 10.1201/9781032664293-7

- What are the key differences between the proposed interpretable models and existing approaches?

The study of developing IML models for DDoS detection in cloud environments is important for several reasons:

- **Increasing reliance on cloud technology:** With the migration of more organizations to the cloud, there is a heightened risk and impact of cyber and DDoS attacks on cloud-based systems.
- **Evolving and sophisticated landscape of DDoS attacks:** The frequency, complexity, and sophistication of DDoS attacks are rising, highlighting the need for robust detection techniques.
- **Limitations of traditional security solutions:** Conventional security tools, such as IDS, are not fully equipped to handle the nuanced and evolving threats of DDoS attacks in cloud contexts.
- **Need for interpretability**: IML models elucidate the decision-making process, instilling confidence in security professionals regarding system outputs.
- **Contribution to AI development**: The research endeavors to enhance the efficacy and trustworthiness of AI systems for DDoS detection by focusing on interpretability.
- **Practical applications:** The outcomes of this research have tangible applications, empowering enterprises to detect and counteract DDoS threats more effectively, thus safeguarding their cloud infrastructure.

The chapter is organized as follows: Section 6.2 reviews related work on interpretable DDoS attacks, Section 6.3 describes the dataset used in the study, Section 6.4 presents the proposed system, and Section 6.5 discusses the experimental results and findings. Finally, Section 6.6 concludes the chapter and outlines future work.

6.2 RELATED WORK

In this section, we discuss the related works in the field of interpretable DDoS attack detection systems.

To enhance Vehicular Ad Hoc Networks (VANETs) security, researchers of [5] employ IDS using methods like signature, anomaly, and rule-based detection. The challenge is the opacity of ML-based IDS. The LIME toolkit clarifies model decisions. The study focuses on robust DDoS/DoS attack detection, evaluated on specific metrics, but it's based on a single dataset. Using LIME, the research offers better model understanding, strengthening VANET security.

The paper of [6] introduces an XAI-based technique for detecting DDoS attacks in IoT networks by analyzing network layer traffic and selecting key features for anomalies. It sets security policies using thresholds for different feature types and, when tested on the USB-IDS dataset, outperformed existing methods.

The paper's strengths lie in its fusion of anomaly detection via autoencoder with XAI feature explanations, a DDoS flow detection method, and a lightweight model tailored for IoT. However, its limitations include an exclusive evaluation of USB-IDS, no analysis of computational demands for IoT devices, limited discussion on advanced DDoS techniques, unexplored scalability for large IoT networks, and no consideration of adversarial evasion strategies.

The paper [7] presents an interpretable AI approach for DDoS intrusion detection. It highlights the shortcomings of current AI explanations in this field and introduces the "map, combine, and merge" (M\&M) method. This methodology converts decision tree models into boolean expressions using formal logic, with prime implicants offering clear decision justifications. While tested on real traffic data, its limitations include uncertainty about scalability and applicability to other ML models, lack of generalizability evaluation, and no mention of computational complexity.

The paper [8] presents a novel method for DDoS detection, merging a modified KNN algorithm with risk degree sorting and grid-based classification. By using a k-dimensional tree, the approach boosts efficiency and reduces query times. It creates risk profiles for DDoS traffic interpretation and filtering, and uniquely, it doesn't necessitate model retraining for new environments. With an impressive 98.4% detection accuracy and a 5-second delay, it seems promising. Yet, limitations include a lack of comparison with other methods, unexplored scalability aspects, and unspecified computational demands.

The paper [9] introduces a DDoS explainer model to offer understandable explanations for DDoS attack detection, addressing concerns over the opacity of black-box ML models. The approach applies IML techniques, specifically LIME and SHAP, to elucidate detection results. Unlike prior work that predominantly emphasized DDoS detection explanations using a single IML method, this study compares the efficacy of both LIME and SHAP, eventually choosing the best-performing one for the framework. The research uses the NSL-KDD dataset within an ensemble-supervised ML context and introduces a unique confidence score. Its effectiveness is assessed solely using the NSL-KDD dataset, calling into question its broader applicability.

The paper [10] introduces a framework combining deep neural networks (DNNs) with explainable AI techniques (SHAP, RuleFit, LIME) for IoT-based attack detection. Tested on the NSL-KDD and UNSW-NB15 datasets, it enhances IDS interpretability for cybersecurity experts. However, it overlooks computational demands.

This study [11] introduces a deep learning model targeting DoS attacks in the Internet of Vehicular Networks (IoV). Utilizing K-Means for feature ranking and an Explainable Neural Network (xNN) for classification, the model surpasses existing systems in accuracy on two datasets. It underscores the significance of data validity in IoV and the need for anomaly detection. However, the study's narrow focus on DoS attacks, absence of comprehensive comparisons, questions on scalability, and unaddressed adversarial threats present challenges.

The paper [12] presents a method for selecting universal features to enhance IoT intrusion detection. Applied to three datasets, the method identified six essential features, achieving 99.62% accuracy and reducing prediction time by 70%. SHAP was used for model explainability. Limitations include restricted dataset evaluation, a focus only on network-based detection, potential gaps in SHAP's explanations, and undiscussed scalability challenges. Further research in these areas is warranted.

The research [13] introduces an explainable deep-learning-based IDS for IoT, leveraging a short-term LSTM model combined with the SPIP framework for model interpretation. Achieving impressive detection accuracy and interpretability, it aids in understanding IoT attack behaviors. However, limitation includes evaluations using specific datasets.

The related works demonstrate diverse methods for improving the interpretability of DDoS attack detection systems in different networking environments. However, most of the papers have limitations related to dataset selection, scalability, comparisons with existing methods, and unexplored computational demands. Further research is needed to address these challenges and improve the applicability of the proposed approaches in real-world scenarios (Table 6.1).

Table 6.1 Comparative analysis of DDoS attack detection methods in network security research

Ref	*Category*	*Advantages*	*Limitations*
[5]	Machine Learning	Uses LIME for model clarity, focuses on robust DDoS/DoS attack detection, evaluates on specific metrics, strengthens VANET security.	Opacity in machine-learning-based IDS, based on a single dataset.
[6]	Deep Learning	XAI-based technique for IoT networks, fuses anomaly detection with XAI feature explanations, outperforms on USB-IDS dataset, lightweight model for IoT.	Exclusive evaluation on USB-IDS, lacks computational analysis for IoT devices, limited discussion on advanced DDoS techniques, scalability for large IoT networks unexplored, and no consideration of adversarial evasion strategies.
[7]	Machine Learning	Map, combine, and merge (M&M) method, converts decision tree models to boolean expressions, offers clear decision justifications, tested on real traffic data.	Uncertainty about scalability and applicability to other ML models, lack of generalizability evaluation, no mention of computational complexity.
[8]	Classical	Merges modified KNN with risk degree sorting, k-dimensional tree for efficiency, doesn't require model retraining, 98.4% detection accuracy.	Lack of comparison with other methods, unexplored scalability, unspecified computational demands.

(Continued)

Table 6.1 (Continued)

Ref	*Category*	*Advantages*	*Limitations*
[9]	Machine Learning	Uses IML techniques (LIME, SHAP), compares the efficacy of LIME and SHAP, introduces confidence score, uses NSL-KDD dataset.	Effectiveness was assessed only on the NSL-KDD dataset, questioning broader applicability.
[10]	Deep Learning	Combines DNNs with XAI techniques (SHAP, RuleFit, LIME), tested on NSL-KDD and UNSW-NB15 datasets, enhances IDS interpretability.	Overlooks computational demands.
[11]	Deep Learning	Targets DoS attacks in IoV, uses K-Means for feature ranking, Explainable Neural Network for classification, high accuracy on two datasets, emphasizes data validity and anomaly detection in IoV.	Narrow focus on DoS attacks, absence of comprehensive comparisons, questions on scalability, unaddressed adversarial threats.
[12]	Machine Learning	Selects universal features for IoT intrusion detection, applied to three datasets, 99.62% accuracy, reduces prediction time by 70%, uses SHAP for explainability.	Restricted dataset evaluation, focuses only on network-based detection, potential gaps in SHAP's explanations, scalability challenges undiscussed.
[13]	Deep Learning	Explainable deep-learning-based IDS for IoT, uses short-term LSTM model combined with SPIP framework, high detection accuracy and interpretability, aids in understanding IoT attack behaviors.	Evaluations using specific datasets.

6.3 DATASETS USED

6.3.1 The CICDDoS2019 dataset [14]

The CICDDoS2019 dataset is a comprehensive dataset that focuses on different types of DDoS attacks. It was generated to address the shortcomings of existing datasets and provide a valuable resource for evaluating new detection algorithms and techniques. The dataset contains both benign and malicious traffic, allowing researchers to study and develop effective detection methods. The CICDDoS2019 dataset includes various types of DDoS attacks, such as:

- SNMP-based attacks
- NetBIOS-based attacks
- LDAP-based attacks
- TFTP-based attacks

- NTP-based attacks
- SYN-based attacks
- Web DDoS attacks
- MSSQL-based attacks
- UDP-Lag attacks
- DNS-based attacks
- SSDP-based attacks
- PortScan attacks
- UDP-based attacks

In summary, the CICDDoS2019 dataset focuses on different types of DDoS attacks and provides researchers with a valuable resource for evaluating detection algorithms and techniques. It includes a variety of attack types and has been used in various research studies in the field of network security.

6.3.2 The CICIDS-2018 dataset [15]

The CICIDS-2018 dataset includes seven different attack scenarios, including DDoS attacks, brute-force attacks, heartbleed, botnet attacks, web attacks, and infiltration of the network from inside. The dataset captures network traffic and system logs of each machine, along with 80 features extracted from the captured traffic using CICFlowMeter-V3. One of the DDoS attack scenarios in the dataset is HTTP DoS, which utilizes Slowloris and LOIC as the main tools to make web servers completely inaccessible using a single attacking machine. CICIDS-2018 is a big data intrusion detection dataset that covers a wide range of attack types, including DDoS attacks, brute force attacks, and DoS attacks. The dataset contains several attack types, but these have all been labeled as "attack". The dataset has 20 independent features and 2,450,324 instances, of which roughly 2.8% typifies attack traffic.

6.3.3 The CICIDS-2017 dataset [16]

The CICIDS-2017 dataset is an intrusion detection dataset that contains different types of DDoS attacks. The dataset was generated by the Canadian Institute for Cybersecurity and comprises benign traffic and different types of DDoS attacks generated through protocols using TCP/UDP. Some of the DDoS attacks included in the dataset are GoldenEye, Hulk, and Slowloris. Some of the specific types of attacks in the dataset are:

- **DDoS**: This category includes various DDoS attacks that were simulated in the dataset, such as DoS slowloris, DoS Slowhttptest, DoS Hulk, and DoS GoldenEye.
- **Brute Force FTP**: This attack involves an attacker attempting to gain unauthorized access to an FTP server by systematically trying different username and password combinations.

- **Brute Force SSH**: Similar to Brute Force FTP, this attack aims to gain unauthorized access to an SSH server by trying different username and password combinations.
- **Heartbleed**: This attack exploits a vulnerability in the OpenSSL cryptographic software library, allowing an attacker to retrieve sensitive information from the targeted system.
- **Web Attack**: This category includes attacks targeting web applications, such as SQL injection, cross-site scripting (XSS), and remote file inclusion (RFI) attacks.
- **Infiltration**: This attack involves an attacker gaining unauthorized access to a network or system with the intention of extracting sensitive information or causing harm.
- **Botnet**: This attack involves a network of compromised computers, known as a botnet, being used to launch coordinated DDoS attacks or other malicious activities.

Table 6.2 present the features of the datasets.

6.4 THE PROPOSED APPROACH

The proposed method for DDoS attack detection in the cloud involves the following steps:

6.4.1 Data collection and preprocessing

6.4.1.1 Data collection

Our initial step involves gathering network traffic data. For this purpose, we source data from three datasets, ensuring a comprehensive and representative sample that reflects real-world scenarios.

6.4.1.2 Preprocessing

Data Cleaning: The raw data often contains inconsistencies, errors, or missing values. Through the data cleaning process, we identify and rectify these imperfections to ensure that our dataset is complete and accurate. This might involve imputing missing values, rectifying inconsistent entries, or removing duplicates.

Normalization: Different features might be recorded in different scales or units. To ensure uniformity and improve the comparability of features, we employ normalization techniques. This process scales the values such that they fall within a specified range, often between 0 and 1, making it easier for models to interpret and weigh the features correctly.

Table 6.2 Table of network features and their aliases for three datasets

Column name	*Alias*	*Column name*	*Alias*
Protocol	F1	Packet Length Max	F40
Flow Duration	F2	Packet Length Mean	F41
Total Fwd Packets	F3	Packet Length Std	F42
Total Backward Packets	F4	Packet Length Variance	F43
Fwd Packets Length Total	F5	FIN Flag Count	F44
Bwd Packets Length Total	F6	SYN Flag Count	F45
Fwd Packet Length Max	F7	RST Flag Count	F46
Fwd Packet Length Min	F8	PSH Flag Count	F47
Fwd Packet Length Mean	F9	ACK Flag Count	F48
Fwd Packet Length Std	F10	URG Flag Count	F49
Bwd Packet Length Max	F11	CWE Flag Count	F50
Bwd Packet Length Min	F12	ECE Flag Count	F51
Bwd Packet Length Mean	F13	Down/Up Ratio	F52
Bwd Packet Length Std	F14	Avg Packet Size	F53
Flow Bytes/s	F15	Avg Fwd Segment Size	F54
Flow Packets/s	F16	Avg Bwd Segment Size	F55
Flow IAT Mean	F17	Fwd Avg Bytes/Bulk	F56
Flow IAT Std	F18	Fwd Avg Packets/Bulk	F57
Flow IAT Max	F19	Fwd Avg Bulk Rate	F58
Flow IAT Min	F20	Bwd Avg Bytes/Bulk	F59
Fwd IAT Total	F21	Bwd Avg Packets/Bulk	F60
Fwd IAT Mean	F22	Bwd Avg Bulk Rate	F61
Fwd IAT Std	F23	Subflow Fwd Packets	F62
Fwd IAT Max	F24	Subflow Fwd Bytes	F63
Fwd IAT Min	F25	Subflow Bwd Packets	F64
Bwd IAT Total	F26	Subflow Bwd Bytes	F65
Bwd IAT Mean	F27	Init Fwd Win Bytes	F66
Bwd IAT Std	F28	Init Bwd Win Bytes	F67
Bwd IAT Max	F29	Fwd Act Data Packets	F68
Bwd IAT Min	F30	Fwd Seg Size Min	F69
Fwd PSH Flags	F31	Active Mean	F70
Bwd PSH Flags	F32	Active Std	F71
Fwd URG Flags	F33	Active Max	F72
Bwd URG Flags	F34	Active Min	F73
Fwd Header Length	F35	Idle Mean	F74
Bwd Header Length	F36	Idle Std	F75
Fwd Packets/s	F37	Idle Max	F76
Bwd Packets/s	F38	Idle Min	F77
Packet Length Min	F39	label	F78

6.4.2 Model training

6.4.2.1 Model selection

For our analysis, we employ three ML models:

- Random Forest
- Decision Tree
- CatBoost

Random Forest is an ensemble learning method that combines multiple decision trees to make predictions. Each decision tree is trained on a random subset of the data, and the final prediction is determined by averaging the predictions of all the trees. Random Forest is known for its ability to handle high-dimensional data, handle missing values, and reduce overfitting. It can be used for both classification and regression tasks.

Decision Tree is a simple yet powerful algorithm used for both classification and regression tasks. It partitions the data based on feature values and creates a tree-like model of decisions. Each internal node of the tree represents a feature or attribute, and each leaf node represents a class label or a predicted value. Decision Trees are interpretable, easy to understand, and can handle both categorical and numerical data. However, they are prone to overfitting and may not generalize well to unseen data.

CatBoost is a gradient-boosting algorithm that is particularly effective for categorical data. It stands for "Category Boosting" and is designed to handle categorical features without the need for extensive preprocessing. CatBoost uses a combination of ordered boosting and random permutations to improve the accuracy of predictions. It automatically handles categorical variables, missing values, and can handle large datasets efficiently. CatBoost is often used in competitions and real-world applications due to its strong performance.

Each unique model is trained with preprocessed datasets, customized to meet its specific needs. The main objective of this training is to recognize and scrutinize patterns and anomalies within the network traffic data, enabling the models to precisely classify entries as either normal or malicious.

6.4.2.2 Hyperparameter tuning

In order to achieve the highest level of model performance and precision, we engage in hyperparameter tuning for each individual model. To accomplish this, we employ the grid search method in conjunction with five-fold cross-validation. This combination allows us to thoroughly explore the parameter spaces, guaranteeing that our models are meticulously fine-tuned to deliver optimal performance when dealing with unseen data.

6.4.3 Model evaluation

To evaluate our models' performance in detecting DDoS attacks in cloud environments, we use several metrics. These include accuracy, precision, recall, F1 score, specificity, AUC, Type I error, Type II error, and detection time.

Accuracy represents the ratio of correct outcomes (comprising both true positives and true negatives) to the overall count of cases evaluated, reflecting the capabilities and potential enhancement areas of our models.

$$Accuracy = \frac{TP + TN}{TP + TN + FP + FN}$$

Precision, also known as positive predictive value, is the ratio of relevant instances to the total number of retrieved instances.

$$\text{Precision} = \frac{\text{TP}}{\text{TP} + \text{FP}}$$

Recall, also referred to as sensitivity, hit rate, or true positive rate, is the fraction of the total number of relevant instances that were successfully retrieved.

$$Recall = \frac{TP}{TP+FN}$$

The F1 score is the harmonic mean of precision and recall. The best value would be 1 and the worst value is 0.

$$F1-score = \frac{2 * \text{Precision} * \text{Recall}}{\text{Precision} + \text{Recall}}$$

Specificity (also called the true negative rate) measures the proportion of actual negatives that are correctly identified as such.

$$\text{Specificity} = \frac{TN}{TN+FP}$$

Error Type 1 (False Positive Rate): This is the proportion of positive cases in the population that are incorrectly classified as negative.

$$\text{Error Type 1} = \frac{FP}{FP+TN}$$

Error Type 2 (False Negative Rate): This is the proportion of negative cases in the population that are incorrectly classified as positive.

$$\text{Error Type 2} = \frac{FN}{FN+TP}$$

In these equations, the following notations are used:

- TP: True Positives
- TN: True Negatives
- FP: False Positives
- FN: False Negatives

The primary goal of this evaluation process is to determine how effectively our models can detect DDoS attacks in cloud environments. By using the mentioned metrics, we can acquire a holistic view of their strengths and potential areas for improvement.

6.4.4 Interpretability analysis

6.4.4.1 Decision analysis

For a more profound comprehension of our DDoS detection system, we investigate the decision-making procedures of our refined models. This exploration provides stakeholders with increased confidence in the models' predictions, fostering a deeper understanding and trust in our system.

6.4.4.2 Feature importance

A crucial element of our interpretability analysis is evaluating the rankings of feature importance. These rankings reveal the features exerting the most substantial impact on the models' decisions. Recognizing these rankings allows us to concentrate on the most pertinent features, potentially simplifying the model without forfeiting its efficacy.

6.4.4.3 Visualization

To render our analysis more understandable and user-friendly, especially for those without a technical background, we employ visualization techniques for the feature importance rankings. Methods like bar charts present a straightforward and succinct depiction, facilitating swift comparisons and insights into the model's decision-making mechanics.

6.5 EXPERIMENTS, RESULTS, AND DISCUSSION

This section provides a comprehensive analysis of the experimental trials conducted to evaluate the performance of the ML models developed using Python. The methodology employed encompasses the entire process, including the evaluation metrics utilized, the results obtained, and a meticulous examination of these outcomes. The system under investigation has been crafted employing Python, a programming language renowned for its adaptability in integrating diverse ML approaches. For data preprocessing, we have harnessed the capabilities of the NumPy and Pandas libraries, while Scikit-learn has been our choice for the remainder of the tasks. Tables 6.3–6.5 showcase the optimal hyperparameters of each model on various datasets. DT-19 refers to a decision tree model trained on the CICDDOS-2019 dataset. Similarly, such as DT-17 and DT-18, also refer to decision tree models trained on CICIDS-2018 and CICIDS-2017 datasets. The same for the others.

Table 6.6 displays the results of several ML models, specifically decision trees (DT), random forests (RF), and CatBoost model (Cat), on a dataset. The models are evaluated based on their accuracy, precision, recall, F1-score, AUC, specificity, type I error, type II error, and detection time.

The table shows that all models have high accuracy, precision, recall, and F1-score, with values close to 1.0000. The AUC values are also high, indicating good performance in distinguishing between positive and negative classes. The specificity values are also close to 1.0000, indicating that the models are good at correctly identifying negative cases. The type I error values are low, indicating that the models are good at correctly identifying positive cases. The type II error values are also low, indicating that the models are good at correctly identifying negative cases.

The detection time values are also provided, which show the time taken by each model to detect a case. The values are in seconds, and they range from 0.0084 for DT-19 to 0.3683 for RF-17.

Table 6.3 Best hyperparameters of DT model on three datasets

Model	*max_depth*	*min_samples_leaf*	*min_samples_split*
DT-19	9	1	2
DT-18	5	2	3
DT-17	9	1	2

Table 6.4 Best hyperparameters of RF model on three datasets

Model	*max_depth*	*min_samples_leaf*	*min_samples_split*	*n_estimators*
RF-19	9	2	2	50
RF-18	9	2	2	50
RF-17	9	2	2	100

Table 6.5 Best hyperparameters of Cat model on three datasets

Model	*max_depth*	*n_estimators*
Cat-19	9	150
Cat-18	7	100
Cat-17	9	75

Table 6.6 Results summary

Model	*Accuracy*	*Precision*	*Recall*	*F1-score*	*AUC*	*Specificity*	*Type I error*	*Type II error*	*Detection time (s)*
DT-19	0.9971	0.9971	0.9941	0.9956	0.9963	0.9986	0.0014	0.0059	0.0084
RF-19	0.9980	0.9987	0.9952	0.9969	0.9973	0.9994	0.0006	0.0048	0.2961
Cat-19	0.9988	0.9986	0.9979	0.9983	0.9986	0.9993	0.0007	0.0021	0.0331
DT-18	1.0000	0.9999	1.0000	0.9999	1.0000	0.9999	0.0001	0.0000	0.0214
RF-18	1.0000	1.0000	1.0000	1.0000	1.0000	1.0000	0.0000	0.0000	0.3229
Cat-18	1.0000	1.0000	1.0000	1.0000	1.0000	1.0000	0.0000	0.0000	0.0953
DT-17	0.9997	0.9999	0.9996	0.9998	0.9997	0.9998	0.0002	0.0004	0.0133
RF-17	0.9997	1.0000	0.9994	0.9997	0.9997	1.0000	0.0000	0.0006	0.3683
Cat-17	1.0000	1.0000	1.0000	1.0000	1.0000	1.0000	0.0000	0.0001	0.0436

Table 6.7 Comparison of the top ten features employed in DT, RF, and Cat

	DT-19	*RF-19*	*Cat-19*	*DT-18*	*RF-18*	*Cat-18*	*DT-17*	*RF-17*	*Cat-17*
1st Feature	F39	F8	F66	F7	F7	F3	F6	F6	F53
2nd Feature	F17	F41	F48	F37	F54	F67	F63	F8	F4
3rd Feature	F49	F39	F1	F62	F3	F55	F66	F4	F77
4th Feature	F48	F54	F39	F69	F62	F17	F12	F11	F12
5th Feature	F73	F9	F6	F31	F10	F14	F16	F53	F42
6th Feature	F75	F53	F43	F1	F5	F23	F25	F63	F66
7th Feature	F20	F49	F63	F2	F9	F29	F5	F55	F63
8th Feature	F37	F7	F38	F3	F30	F18	F19	F9	F9
9th Feature	F66	F52	F23	F4	F35	F38	F15	F2	F23
10th Feature	F64	F64	F67	F5	F67	F25	F1	F68	F35

Overall, the table suggests that all models have performed well on the datasets, with high accuracy and low error rates. The detection times vary, but they are all relatively low, indicating that the models are efficient in terms of computation time.

Table 6.7 compares the top ten features obtained in three different ML models – Decision Tree (DT), Random Forest (RF), and CatBoost (Cat) – across three different datasets (2017, 2018, 2019). Each model is presumably used for network traffic analysis.

The figures represent bar plots of feature importance scores for the three different ML models across three datasets. The features are labeled (e.g., F63, F66), with the y-axis showing the feature labels and the x-axis indicating the importance scores.

From Table 6.7 and Figures 6.1–6.9 the ten Most Frequent Features are:

- F63 (Subflow Bwd Packets) – Appears in four times: This represents the number of backward packets in a subflow. Analyzing the backward packet count can provide insights into the volume and nature of the response from a destination to a source. It might be especially useful in identifying unusual response patterns or volume which could indicate anomalies.

- F66 (Init Fwd Win Bytes) – Appears in four times: This indicates the number of bytes for the initial window size in the forward direction. Window size can provide insights into the flow control mechanism between sender and receiver. Variations or anomalies in this can signify issues or unusual patterns in network communication.
- F9 (Fwd Packet Length Mean) – Appears in four times: Represents the average size of packets in the forward direction. By understanding average sizes, it's possible to spot deviations from normal traffic patterns.
- F39 (Packet Length Min) – Appears in three times: The minimum length of the packets in a flow. This can help in identifying extremely small or "ping" packets that might be part of scanning or probing activities on the network.
- F7 (Fwd Packet Length Max) – Three occurrences: Represents the maximum length (in bytes) of the packets sent in the forward direction. It provides insight into the largest packet size in the traffic.
- F3 (Total Fwd Packets) – Three occurrences: Represents the total number of packets sent in the forward direction. It is useful for analyzing the volume of outgoing traffic.
- F6 (Bwd Packets Length Total) – Three occurrences: Represents the total length (in bytes) of the packets received in the backward direction. It is useful for analyzing the volume of incoming traffic.0

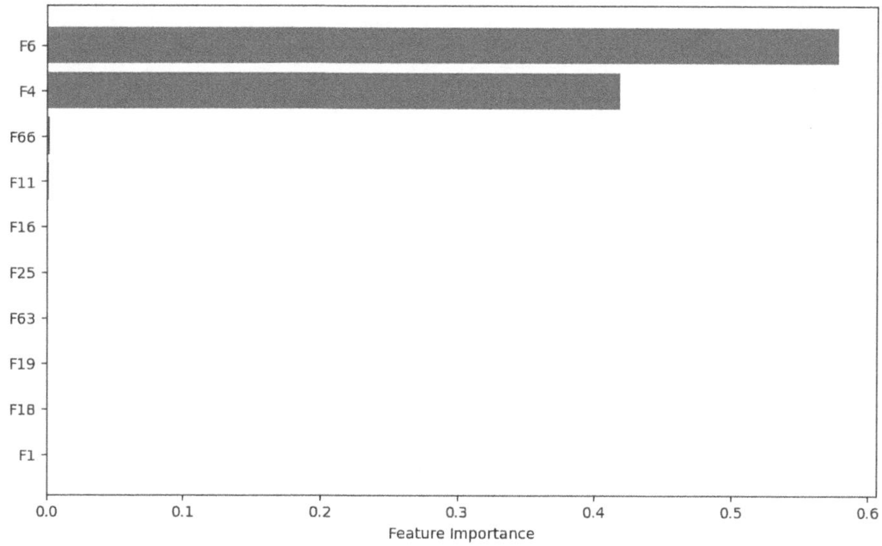

Figure 6.1 The best ten features of DT on CICIDS-17.

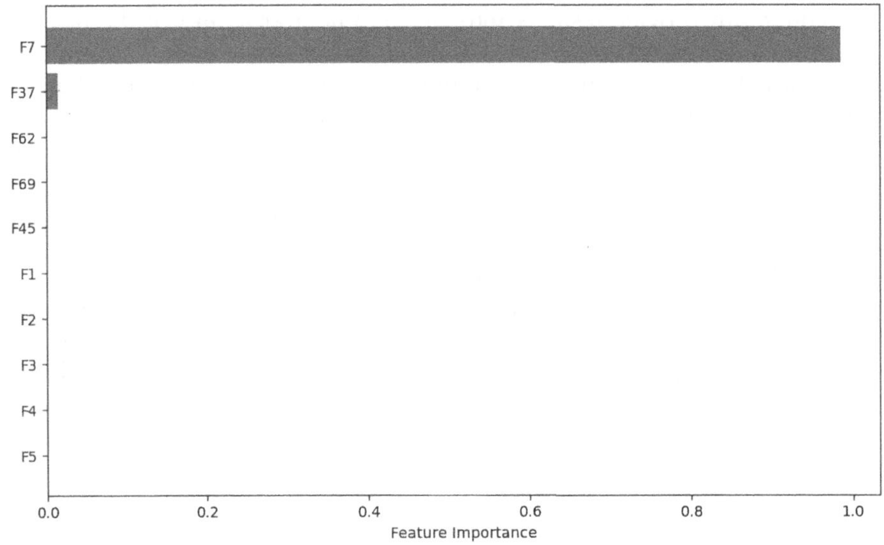

Figure 6.2 The best ten features of DT on CICIDS-18.

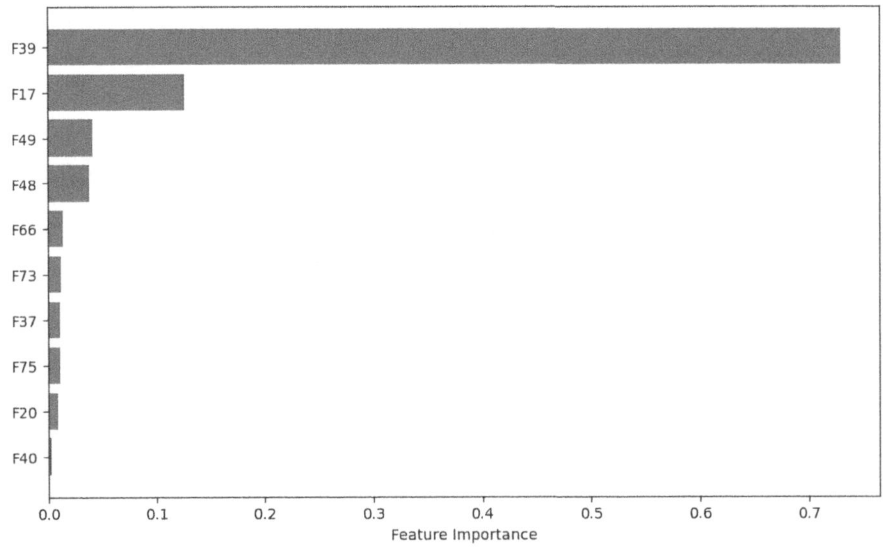

Figure 6.3 The best ten features of DT on CICDDOS-19.

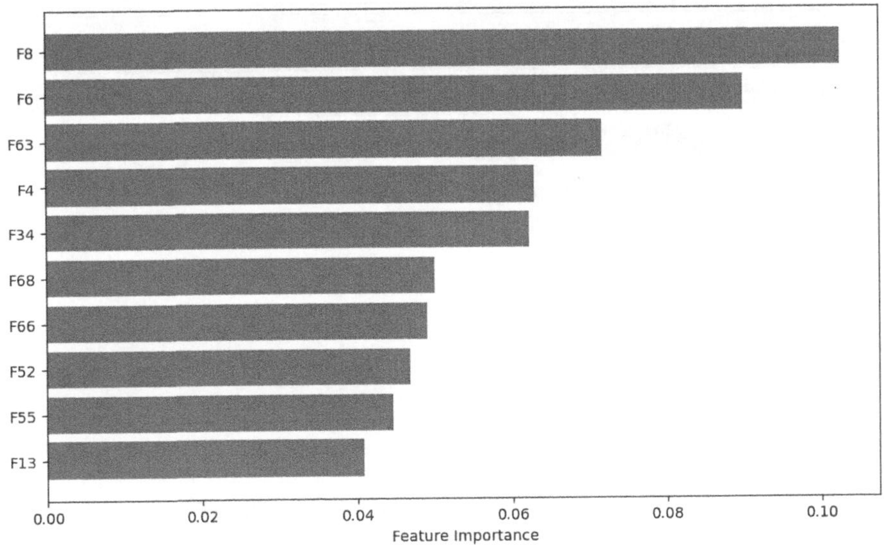

Figure 6.4 The best ten features of RF on CICIDS-17.

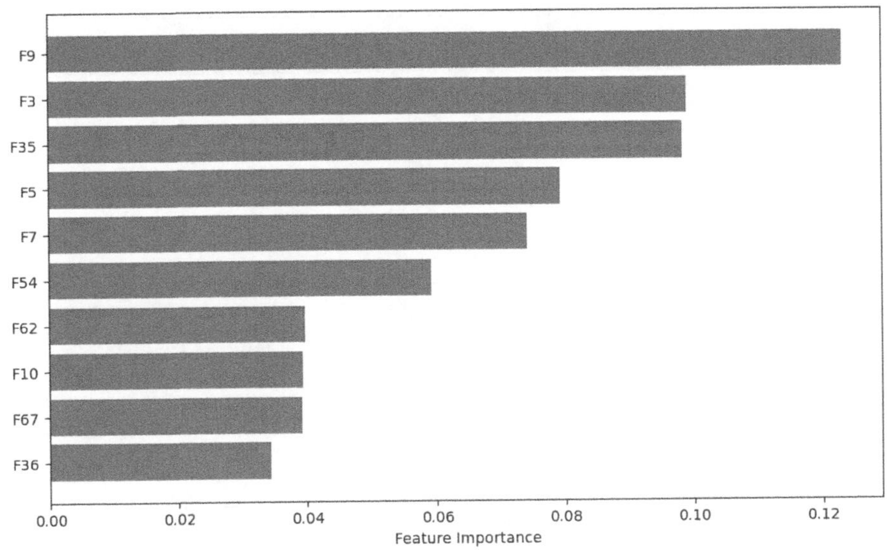

Figure 6.5 The best ten features of RF on CICIDS-18.

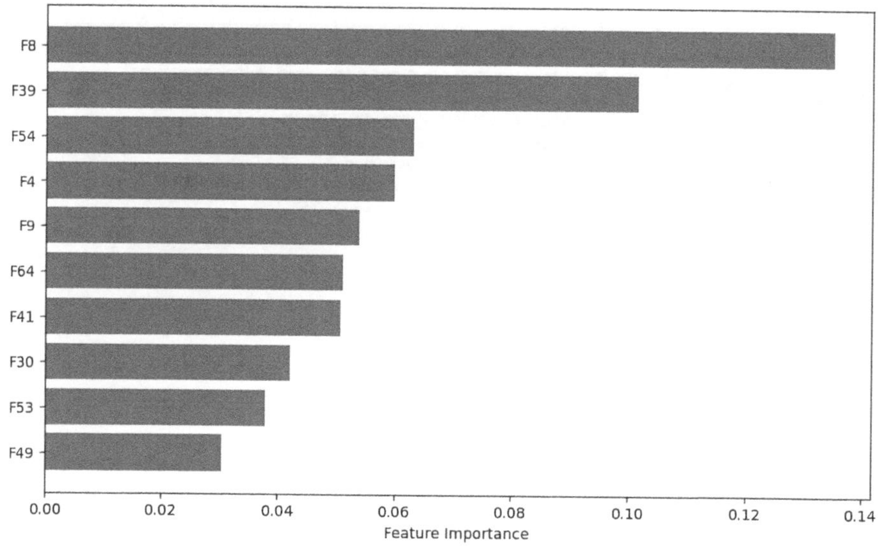

Figure 6.6 The best ten features of RF on CICDDOS-19.

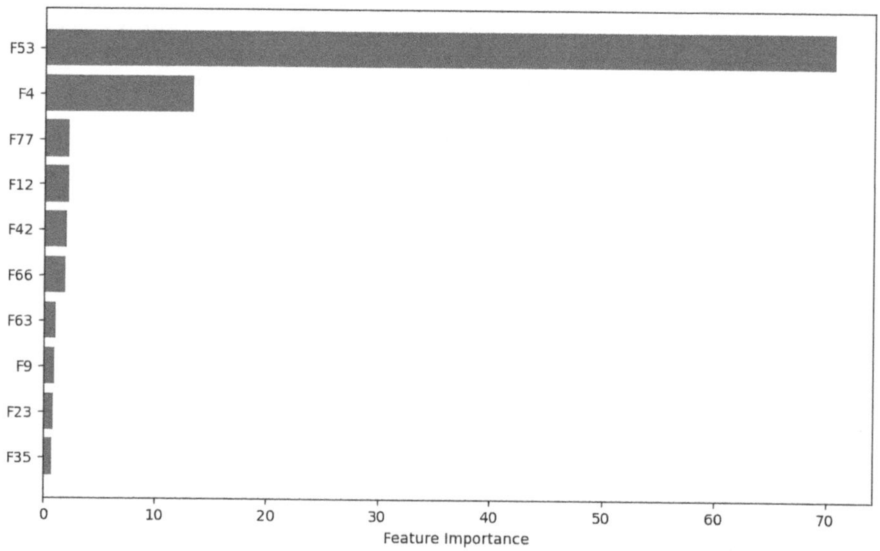

Figure 6.7 The best ten features of Cat on CICIDS-17.

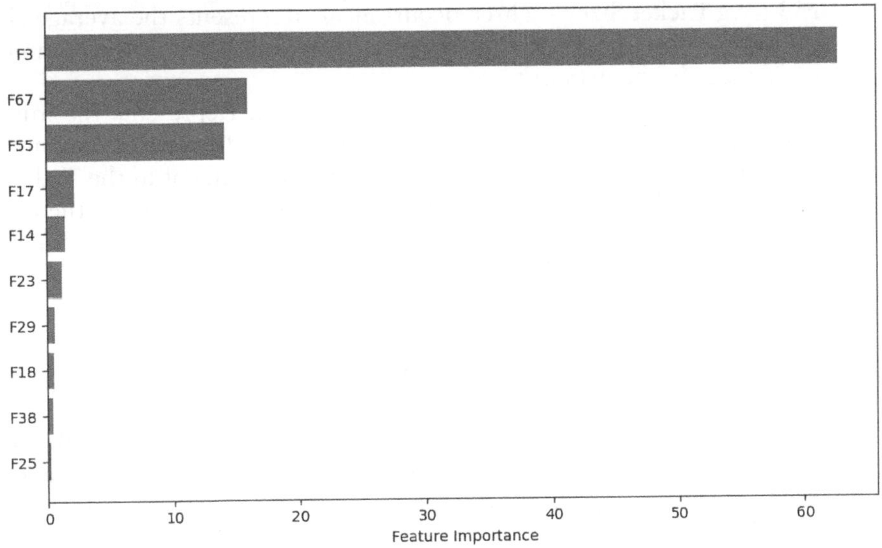

Figure 6.8 The best ten features of Cat on CICIDS-18.

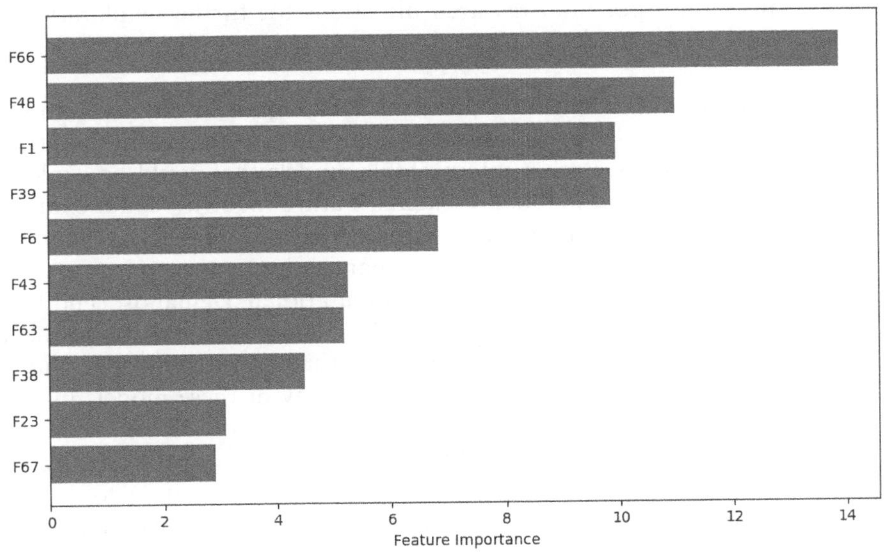

Figure 6.9 The best ten features of Cat on CICDDOS-19.

- F53 (Avg Packet Size) – Three occurrences: Represents the average size (in bytes) of the packets. It provides a general idea about the common packet size in the network traffic.
- F67 (Init Bwd Win Bytes) – Three occurrences: Represents the initial window size (in bytes) in the backward direction. It controls how much data can be received before sending an acknowledgment in the TCP.
- F4 (Total Backward Packets) – Three occurrences: Represents the total number of packets received in the backward direction. It is useful for analyzing the volume of incoming traffic.

6.6 CONCLUSION AND FUTURE WORK

In this chapter, the central issue of detecting DDoS attacks in cloud computing paradigms is thoroughly examined. The research suggests the adoption of IML frameworks, specifically the Random Forest, Decision Tree, and CatBoost algorithms, to tackle this challenge. Empirical evaluations demonstrate the effectiveness of these models in identifying DDoS attacks while providing transparency and insight into their decision-making processes. The study emphasizes the practical implications of its findings, highlighting their potential to enhance the efficiency and reliability of AI systems in DDoS detection, particularly in the dynamic environment of cloud-based services.

There are several potential research directions for further exploration in the field of DDoS attack detection in cloud environments. These include:

1. **Investigation of Alternative IML Models:** Apart from the Random Forest, Decision Tree, and CatBoost algorithms examined in this study, there are numerous other interpretable models that could be explored for DDoS detection in cloud contexts. Future inquiries could examine the efficacy and interpretability of these alternative models, comparing them with the ones discussed in this manuscript.
2. **Examination of Scalability and Computational Demands:** The scalability of the proposed models and their computational requirements in large-scale cloud frameworks need further investigation. Future research should aim to validate the feasibility of these models in real-world settings with extensive cloud infrastructures.
3. **Focus on Adversarial Evasion Tactics**: This study recognizes the importance of considering adversarial evasion strategies in DDoS attack detection. Subsequent investigations should focus on developing models that are resilient to such tactics, thereby strengthening the security and integrity of cloud-based infrastructures.
4. **Comparative Analysis with Existing Methodologies**: While this study compares the proposed interpretable models against traditional cybersecurity measures, there is an opportunity for future research to extend this comparison to include other prevalent DDoS detection strategies in

cloud environments. Such comparative analyses would provide a more comprehensive understanding of the relative strengths and limitations of different methodologies.

The dataset links are publicly accessible:

https://www.unb.ca/cic/datasets/ids-2017.html
https://www.unb.ca/cic/datasets/ids-2018.html
https://www.unb.ca/cic/datasets/ddos-2019.html

REFERENCES

[1] Madan, S., Anita, & Ali, A. (2022). DDoS attacks in cloud environment. International Journal of Health Sciences, 6(S4), 5836–5847. https://doi.org/10.53730/ijhs.v6nS4.9457

[2] Mittal, R. (2020, October). Ananlysis of DDoS Attacks In Cloud. In 2020 International Conference on Smart Technologies in Computing, Electrical and Electronics (ICSTCEE) (pp. 19–23). IEEE.

[3] Kannadhasan, S., Nagarajan, R., & Thenappan, S. (2022). Intrusion detection techniques based secured data sharing system for cloud computing using msvm. In 2022 9th International Conference on Computing for Sustainable Global Development (INDIACom) (pp. 50–56). IEEE.

[4] Kanber, B. M., Noaman, N. F., Saeed, A. M., & Malas, M. (2022). DDoS attacks detection in the application layer using three level machine learning classification architecture. International Journal of Computer Network and Information Security, 14(3), 33.

[5] Hassan, F., Yu, J., Syed, Z., Ahmed, N., Al Reshan, M. S., & Shaikh, A. (2023). Achieving model explainability for intrusion detection in VANETs with LIME. PeerJ Computer Science, 9, e1440. https://doi.org/10.7717/peerj-cs.1440

[6] Kalutharage, C. S., Papadopoulos, P., Liu, X., Chrysoulas, C., & Pitropakis, N. (2023). Explainable AI-based DDOS attack identification method for IoT networks. Computers, 12(2), 32. https://doi.org/10.3390/computers12020032

[7] Zhou, Q., Li, R., Xu, L., Nallanathan, A., Yang, J., & Fu, A. (2022). Towards Explainable Meta-Learning for DDoS Detection. Preprint submitted to Elsevier.

[8] Feng, Y., & Li, J. (2020). Toward explainable and adaptable detection and classification of distributed denial-of-service attacks. In MLHat 2020 (pp. 105–121). Springer Nature Switzerland AG. https://doi.org/10.1007/978-3-030-59621-7_6

[9] Das, S., Agarwal, N., & Shiva, S. (2021). DDoS explainer using interpretable machine learning. In Proceedings of the IEEE International Conference on Emerging Computing and Information Technology (ICECIT). https://doi.org/10.1109/IEMCON53756.2021.9623251

[10] Abou El Houda, Z., Brik, B., & Khoukhi, L. (2022). "Why should I trust your IDS?": An explainable deep learning framework for intrusion detection systems in internet of things networks. IEEE Open Journal of the Communications Society, 3, 1164–1165. https://doi.org/10.1109/OJCOMS.2022.3188750

[11] Aziz, S., Faiz, M. T., Adeniyi, A. M., Loo, K-H., Hasan, K. N., Xu, L., & Irshad, M. (2022). Anomaly detection in the internet of vehicular networks using explainable neural networks (xNN). Mathematics, 10(8), 1267. https://doi.org/10.3390/math10081267

[12] Alani, M. M., & Miri, A. (2022). Towards an explainable universal feature set for IoT intrusion detection. Sensors, 22(15), 5690. https://doi.org/10.3390/s22155690

[13] Keshk, M., Koroniotis, N., Pham, N., Moustafa, N., Turnbull, B., & Zomaya, A. Y. (2023). An explainable deep learning-enabled intrusion detection framework in IoT networks. Information Sciences, 639, 119000. https://doi.org/10.1016/j.ins.2023.119000
[14] Sharafaldin, I., Lashkari, A. H., Hakak, S., & Ghorbani, A. A. (2019). Developing realistic distributed denial of service (DDoS) attack dataset and taxonomy. In 2019 International Carnahan Conference on Security Technology (ICCST) (pp. 1–8). IEEE.
[15] Sharafaldin, I., Lashkari, A. H., & Ghorbani, A. A. (2018). Toward generating a new intrusion detection dataset and intrusion traffic characterization. ICISSP, 1, 108–116.
[16] Jazi, H. H., Gonzalez, H., Stakhanova, N., & Ghorbani, A. A. (2017). Detecting HTTP-based application layer DoS attacks on web servers in the presence of sampling. Computer Networks, 121, 25–36.

Chapter 7

Classification of gastrointestinal (GI) bleeding in WCE images based on fusing a stabilizing block with Xception

Anass Garbaz, Said Charfi, Mohamed El Ansari, and Lahcen Koutti

7.1 INTRODUCTION

Multiple gastrointestinal (GI) tract illnesses, including vascular lesions, small bowel tumors, coeliac disease, and Crohn's disease, exhibit bleeding as a relatively regular symptom [1]. Any bleeding that forms in the GI tract is referred to as GI bleeding. Although bleeding can occur anywhere along the digestive tract, it is frequently separated into upper and lower GI hemorrhages. The esophagus, stomach, and initial segment of the small intestine all exhibit upper GI bleeding. The majority of the small intestine, large intestine, rectum, and anus all experience lower GI bleeding. It can be harmful if there is significant GI bleeding. Nevertheless, any minor bleeding that persists for a long time might result in complications such as anemia or low blood levels. Numerous remedies are accessible once a bleeding site is identified to halt the bleeding or address the source. The two bleeding types indicated above are shown in Figure 7.1.

Presently, no technique could examine the small intestine's bleeding before the invention of Wireless Capsule Endoscopy (WCE) [2]. A lengthy, malleable tube fitted with a video camera is inserted via the throat or into the rectum during an endoscopy session. The GI procedure known as a capsule endoscopy, often referred to as a WCE or a video capsule endoscopy, utilizes a medication camera to capture images of the intestinal lumen. The US Food and Drug Administration legalized the use of the first capsule endoscopy in the US in 2001 after the procedure was first performed in 1999. It is a digestible technology that enhances diagnostic methods and enables visual and temporal monitoring of the whole GI tract. Its measurements are 11 mm by 26 mm, and its mass is 3.7 grams. WCE is a cutting-edge radiotelemetry camera device that has no external circuits or fiber optic containers. Within the program of 8 hours, WCE broadcasts between 50,000 and 60,000 optical frames at a typical frame rate of 2–4 frames per second. Nonetheless, WCE continues to encounter considerable obstacles in its pursuit of extensive clinical contexts. For example, it takes time and is exhausting for professionals to examine more than 50,000 frames for each patient over the course of several hours. Furthermore, specialists may overlook crucial areas due to visual

DOI: 10.1201/9781032664293-8

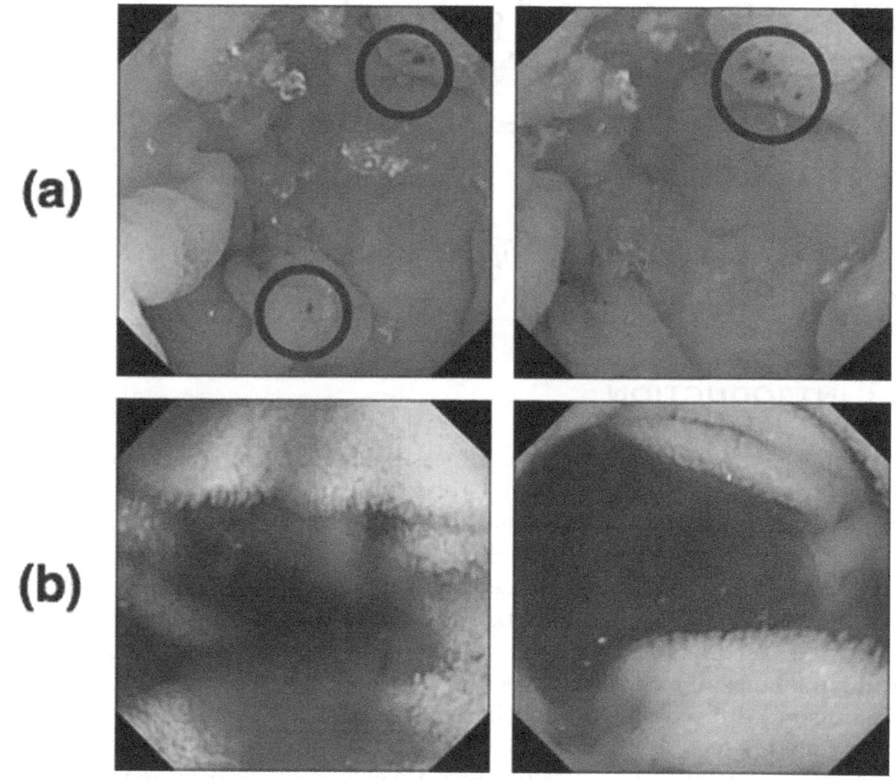

Figure 7.1 Varieties of bleeding occurrences (a) minor quantities of bleeding and (b) blood-filled frames.

tiredness or the absence of attentiveness. Different computer-aided solutions [3–5] have been developed over time to tackle these issues automatically and support medical examinations. For instance, Charfi and El Ansari [3] suggested a method for identifying polyps in colonoscopy films. It is suggested to use the concept of extracting specific regional characteristics from the image. Next, the image's descriptor is derived from the appearance graph of these characteristics. To categorize different anomalies in WCE images, a discriminative joint-feature topic model with dual constraints is presented in [4]. Using color and texture variables derived from the same image patches, they first present a joint-feature probabilistic latent semantic analysis paradigm that is jointly modeled using their dependent norms. In order to yield discriminative latent semantic themes, the joint-feature model is then simultaneously embedded with their suggested dual constraints: the relevance of visual words and the local image manifold. Initially, Liu *et al.* [5] created an approach called joint diagonalization principal component analysis, which does not use any iteration, turning, or estimation techniques. As a result, it is appropriate for GI endoscopic image dimension reduction and has low processing complexity. Afterward, by fusing the machine learning technique based on the first algorithm with the traditional feature acquisition method without learning, an innovative image feature extraction technique was created. Deep learning systems [6–9], in particular, have demonstrated outstanding classification and segmentation achievements. Vasilakakis *et al.* [6] suggested an explainable fuzzy bag-of-words feature extraction technique for the categorization of WCE frames with sparse annotations. Their model has a comparative advantage over the most recent feature extractors since it can produce an understandable classification result even when using traditional classification strategies like support vector machines. In order to address the accuracy against rapidity compromise as the foundation architecture, Souaidi and El Ansari [7] suggested a multiscale pyramidal fusion single-shot multibox detector network for identifying tiny polyp areas in WCE, colonoscopy images, or both. Following a preprocessing phase, Ellahyani *et al.* [8] collected the descriptive information from the images and sent it to an extreme learning kernel machine to carry out the classification task. In Ref. [9], in order to extract profound characteristics from WCE images, the pre-trained model ResNet50 is refined via transfer learning. We delicately combine a deep network in the Xception [10] paradigm with a proposed stabilizing block (SB) in order to address the aforementioned challenges. We extend Xception [10] with ImageNet weights [11] as the highest-level feature extractor to improve the performance of our model. We add an SB immediately following the result of the Xception to guarantee the consistency of the generated feature map.

- We present an approach for recognizing bleeding in WCE images. We adopt a deep neural network with an SB and an upper-level model based on the Xception.
- Studies on the Kvasir-Capsule dataset were conducted to ensure the efficacy of the suggested approach [12].
- It is more accurate than cutting-edge methods of identifying bleeding lesions and has excellent precision. The accuracy of the proposed architecture is 99.38%.

The demonstration of the chapter is formatted as follows: the articles deemed relevant to the current emphasis are presented in Section 7.2. The suggested approach appears in Section 7.3. The research results that verify the offered solid structure are presented and examined in Section 7.4. Section 7.5 ends with our recommendations.

7.2 RELATED WORK

It is important to notice bleeding because it may represent a sign of many diseases. How patients with cryptic bleeding are treated, specifically, has been greatly changed by WCE. In our earlier work [13], we deployed a network that combines a convolutional neural network (CNN) for the low-level model and the Inception-ResNet-V2 model for the upper level. Preprocessing, collecting features utilizing an improved CNN, and classification using gated recurrent units were the three main aspects of the method proposed by [14]. Leveraging SegNet layers with three classes, Ghosh *et al.* [15] offered a unique deep-learning-based semantic segmentation technique for bleeding region identification. Hajabdollahi *et al.* [16] looked at the issue of neural network rationalization for WCE's robotic division of bleeding areas. A multi-layer perceptron and a CNN are employed independently to perform classification, and appropriate color channels are chosen as neural network inputs. Li *et al.* [17] created concentration units and a multiple-phase architecture to assist with tiny region segmentation. The idea in [18] is built on a dual system. To define WCE images as word-based color histograms, they initially fully exploited the color information of WCE images and applied the K-means clustering approach to the pixel-represented frames to generate clustering centers. Consequently, they used the support vector machine and K-nearest neighbor algorithms to determine the condition of a WCE image. Xing *et al.* [19] suggested the use of a Saliency-aware Hybrid Network (SHNet) to automatically identify GI bleeding. The SHNet is composed of two tightly linked CNNs that are referred to as global image channels and saliency-aware streams, respectively. A short U-Net with fewer encoder-decoder groups was introduced in [20] to divide bleeding from WCE frames. Hajabdollahi *et al.* [21] suggests a simple CNN structure for detecting bleeding zones that accept a single patch as input and produce a segmented patch of identical dimensions. Lan *et al.* [22] implemented recognition operations utilizing CNNs, and other techniques were employed as well to improve WCE anomaly recognition efficiency, including transfer learning, area suggestion, and characteristics of the CNN architecture. Caroppo *et al.* [23] presented a system that uses VGG19, InceptionV3, and ResNet50, three pre-trained deep CNNs, for feature extraction. After that, several fusion rules are applied to choose and fuse the features. Amiri *et al.* [24] proposed a method that uses the expectation-maximization segmentation approach to divide the input image into potential areas of interest. Afterward, color and numerical information are retrieved from these regions. The block-based local information extraction approach proposed by [25] from each color domain produces a better depiction of features as opposed to single pixel-based characteristics. As they turned their attention to other anomalies, Jia *et al.* [26] presented computer-assisted methods for colorectal polyp detection, emphasizing the effectiveness of deep learning-based techniques in the WCE sequences. They addressed lesion zone detection, pixel-accurate segmentation, and classification as important applications of WCE polyp recognition. Wickstrøm *et al.* [27] created and assessed new developments in model interpretability and uncertainty estimation to semantically segment polyps from colonoscopy images. Additionally, they showed how readability and ambiguity may be represented for the segmentation of polyps and proposed a method for evaluating the uncertainty associated with significant characteristics in the input. Wang *et al.* [28] proposed a multi-scale context-guided deep architecture for tackling lesion localization of WCE images in the GI tract. This architecture gathers information about the global and regional surroundings to drive the evolution of the model. A selective characteristic aggregation network with border and region limits was proposed by [29]. For the purpose of forecasting polyp edges and regions, the network has two mutually limited decoders and a common encoder. Yang *et al.* [30] produced an endoscopic database for conditions related to gastritis atrophy and intestinal metaplasia. Following that, to extract essential visual elements, they proposed a machine-learning approach called local attention grouping, which was inspired by the human ocular mechanism. He *et al.* [31] introduced an in-depth hookworm recognition system for WCE images that predicts tube-like forms and visual presentations of hookworms at the same time. With just a movable pre-trained model and unlabeled data, Liu and Yuan [32] presented a consistency-based model that makes use of the source model as a proxy teacher. For automated WCE image processing, Guo and Yuan [33] proposed an adaptive, abnormal-aware attention network that incorporates an adaptive dense block with an abnormal-aware attention module. The first component is intended to assign a single attention value to each dense link in dense blocks and enhance advantageous aspects, in contrast to the second module, which aims to adaptively modify the corresponding field depending on abnormal areas and assist in drawing attention to abnormalities. Guo *et al.* [34] primarily addressed the problems in polyp segmentation by introducing ThresholdNet with a confidence-guided manifold mixup data augmentation technique.

7.3 PROPOSED METHOD

7.3.1 Xception

The CNN design dubbed Xception [10] is solely composed of depth-wise separable convolution layers [35]. It effectively asserts that cross-channel associations and spatial associations can be completely separated from one another when mapped in CNN feature maps. They called their suggested paradigm Xception, which signifies Extreme Inception because this theory is a more robust variant of the theory underpinning the Inception architecture. The feature selection core of the network in the Xception design is composed of 36 convolutional layers. Besides the initial and final components, all of the 14 components made up of the 36 convolutional layers contain linear residual connections surrounding them. The Xception system can be summed up as a linear stack of residually connected, depthwise separable convolution layers. This renders defining and changing the architecture relatively simple.

7.3.2 Proposed architecture

The hemorrhage regions' dimensions and forms are random in the capsule endoscopy images. Additionally, bleeding pixels appear in arbitrary locations. Similarly, bleeding regions are affected by brightness fluctuations over time and are not only clean red in hue but rather come in many degrees of red. In Figure 7.4, bleeding samples from the WCE are displayed. For instance, the second image in the top line has bleeding zones that are difficult to see with the unaided eye and are visible in the top-left corner. The process of detecting bleeding is more complex as a result of all these issues. Therefore, to distinguish between bleeding and regular frames, we suggested a deep learning method with various layers of filtering. First, as depicted in Figure 7.2, we improved Xception using ImageNet weights as the backbone of our training. We locked the initial layers of the pre-trained model to stop adding additional tasks to have the model learn the fundamental features, which is the same as starting from scratch and will make the process longer. The process of transfer learning [11] involves using the knowledge gained from one task to improve performance on a related one.

Additionally, the accessible data for medical imaging datasets is generally sparse. As a result, efficiency gains in deep learning applications like image classification that require a lot of resources. Transfer learning is the process of adapting a pre-trained model's pertinent components to another assignment. This will typically be the fundamental data needed for the model to work, with additional features contributed to the system to address certain problems. In our situation, the primary goal of employing it was to train the model with a small number of images and yet get acceptable results. Furthermore, the Xception model uses a technique known as depthwise separable convolution, which was first introduced in [35]. In contrast to standard convolution, which performs the channels and spatially wise computation in one operation, depthwise separable convolution performs the mathematical calculation

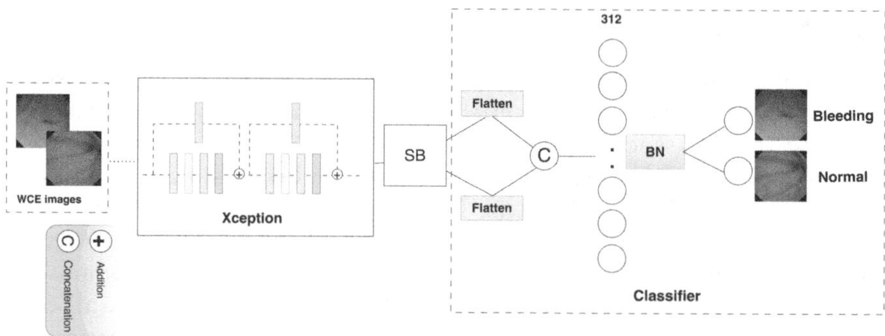

Figure 7.2 An illustration of the suggested deep network design for classifying WCE images. It is composed of a dynamic classifier, the Xception branch, and the SB module for profound feature creation.

in two actions. First, depthwise convolution applies a single convolutional filter to each input channel, and then pointwise convolution is used to create a linear arrangement of the depthwise convolution's output. Transfer learning does not come without difficulties, just like any other type of technological innovation. The issue of negative transfer remains one of the main obstacles to transfer learning. For transfer learning to be effective, the initial and target issues must be sufficiently comparable for the first training session to be meaningful. In order to balance the feature maps and introduce new features without pretrained weights, we consequently built a from-scratch module. Second, as shown in Figure 7.2, we included an extra SB to enhance the effectiveness of the proposed method. The SB is then given access to Xception's output. In Figure 7.3, two skip connections transmit the output. The first one went through a sub-block with a variety of layers to produce various feature representations. In order to preserve the common features of the Xception

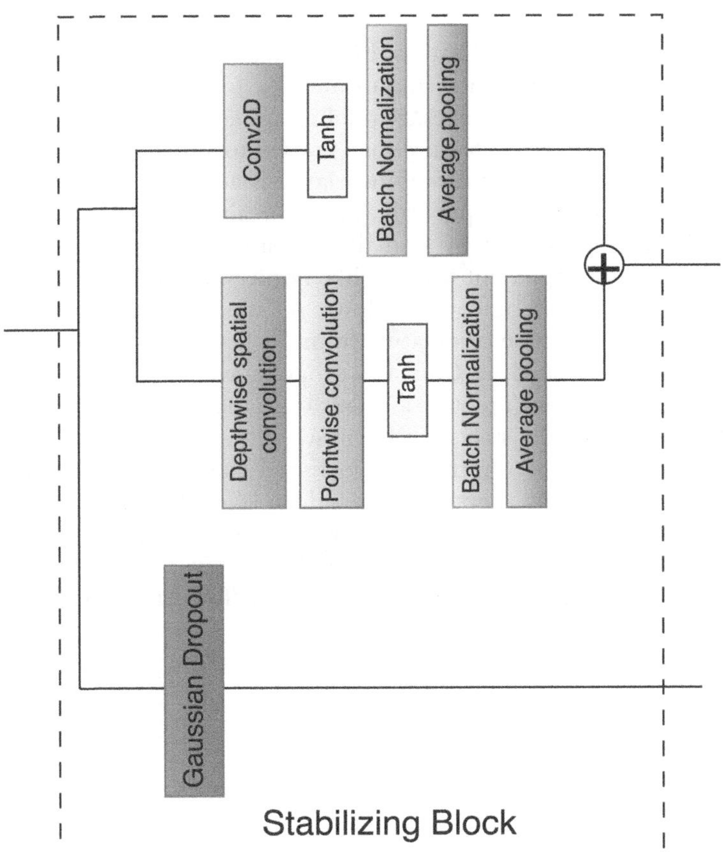

Figure 7.3 A schematic representation of the suggested stabilizing block. It is made up of many layer forms with various merging types.

model, the first skip connection underwent a regular convolution. The second is accomplished by a pointwise convolution that follows a depthwise spatial convolution and adds given filters to the data from the former. Contrary to traditional CNN, which performs convolution on all N channels at once, the depthwise operation only applies convolution to one channel at a time. Convolution is performed on the N channels in pointwise operation. The primary objective of including a depthwise separable convolution in this block is to keep the Xception's representation of transmitting features intact. Another way to put it is to avoid erasing key interest areas from the first model. 32 filters with a 1×1 kernel size are used in both levels. The two layers are then subjected to the tanh (17.1) function. The tanh function has a range of –1 to 1. The positive aspect is that the zero inputs will be traced close to zero, while the negative inputs will be highly negative in the tanh curve.

$$Tanh(x) = \frac{e^{x} - e^{-x}}{e^{x} + e^{-x}} \tag{17.1}$$

Additionally, batch normalization (BN), which employs an alteration to keep the resultant standard deviation and mean close to one and zero, is also used. Next, we included a layer for average pooling. When using average pooling, the filter's filtered region of the feature map is used to calculate the average of all the items present there. Translation invariance is added in a minor way, which means that most pooled output values are not altered significantly when the image undergoes translation by a small quantity. In contrast to max pooling, which extracts more apparent features like borders, it does so more seamlessly. A summation of the two outcomes follows. A Gaussian dropout of 30% passes the second skip connection originating from the first model. The same target is more easily accomplished by Gaussian dropout, which only uses exponential Gaussian noise and doesn't require any extra variables for the desired result. The classifying process finally starts. Different weights are added to the classifier by flattening and concatenating the sub block and Gaussian dropout findings. A 312-unit adjustment was adapted to the hidden layer. The last layer's setting of two units is due to the fact that we must predict either of the two classes, normal or bleeding. The sigmoid (17.2) function of the last layer, which outputs data in a spectrum of 0–1 as a prediction probability, is defined as follows:

$$Sigmoid(x) = \frac{1}{1 + e^{-x}} \tag{17.2}$$

7.4 NETWORK TRAINING

The process of manually adjusting hyperparameters requires attempting various hyperparameter sets. For instance, we shall carry out every experiment, using a particular set of hyperparameters. A reliable trial monitor that can monitor a wide range of parameters, including images, logs, and framework measurements, will be necessary for this technique. This means that it diverts time from crucial phases in the deep learning pipeline, such as feature engineering and result interpretation. However, a tuning method can progressively be applied to autonomously determine the appropriate hyper-parameter settings. In our case, the process was mechanized through automatic hyperparameter adjustment. This strategy, which doesn't require training, typically produces better outcomes. To accelerate the procedure and identify the most suitable set of hyperparameters to yield favorable outcomes, programmed hyperparameter adaptation was carried out. The Adam Optimizer Algorithm, a binary cross-entropy, and a batch size of 32 are used to train the suggested network over an average of 20 epochs. With customizable GPU and RAM reservations, we used Google Collaboratory. We successfully created the model using TensorFlow.

7.5 EXPERIMENTS

To enable comparisons with earlier, ongoing, and future studies as well as consistency of research experiments, the suggested architecture was examined using an open-access WCE dataset.

7.5.1 Dataset

About 47,238 frames in 14 groups that correlate to the annotated photos make up the Kvasir-Capsule Dataset. Anatomical and luminal discoveries make up the two groups. In our studies, the normal class was employed, and frames containing fresh blood and hematin were joined to produce the bleeding class. We included multiple instances from the angiodysplasia class in the test set to investigate the universality of the model by recognizing different lesions. Furthermore, 338 of the 458 bleeding frames were pivoted to balance the classes, making a total of 796. The dataset was divided into two parts: training (77.4%) and validation and testing (22.6%) (Figure 7.4).

7.5.2 Evaluation metrics

Our measurements of performance include accuracy (ACC), sensitivity (SE), precision (PRC), and specificity (SP).

- The ACC measures how closely a measurement resembles its actual value.
- The percentage of instances in which measurements taken under identical circumstances yield the same results is known as the PRC.
- The SE is the percentage of samples that pass the test and produce a good outcome.
- The SP is the percentage of samples that, when subjected to the test in issue, yield an accurate negative result.

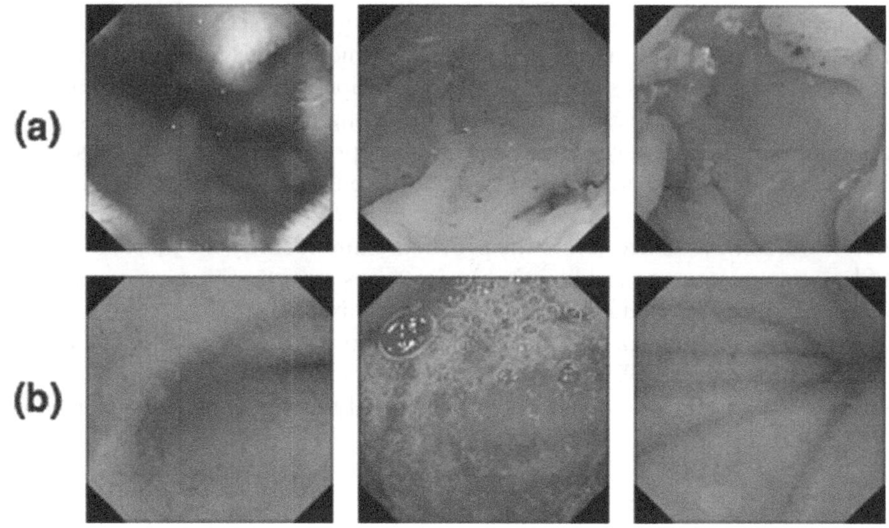

Figure 7.4 Bleeding instances in the dataset. (a) Samples with bleeding and (b) normal frames.

The following equations are listed for all measures.

$$SE = \frac{TP}{TP + FN} \tag{17.3}$$

$$PRC = \frac{TP}{TP + FP} \tag{17.4}$$

$$SP = \frac{TN}{TN + FP} \tag{17.5}$$

$$ACC = \frac{TP + TN}{TP + FP + FN + TN} \tag{17.6}$$

The true positives, true negatives, false positives, and false negatives produced by the model findings are denoted here by the letters TP, TN, FP, and FN, respectively.

7.5.3 Competing methods and discussion

An extremely thorough experimental investigation was performed in order to judge the ability of the proposed intelligent mechanism. Each subnetwork in the suggested technique, such as baseline Xception, SB, and the classifier

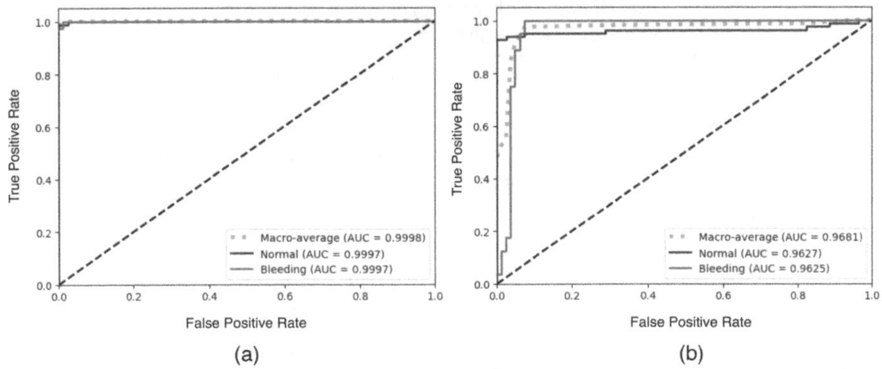

Figure 7.5 The Xception and Xception+SB ROC curves were utilized to assess the model's quality. (a) The suggested model's ROC curves and (b) the Xception model's ROC curves.

subsystem, impacts how well the recognition process functions. A receiver operating characteristic (ROC) curve was also created in Figure 7.5 by comparing predictions from the suggested model to those from the Xception model without our suggested SB. A graph that displays a classification model's effectiveness across all classification criteria is called the ROC curve. The probability curve indicates the degree to which the model can discriminate between classes. The true positive rate and the false positive rate are the two factors plotted on this curve. For every ROC, we also determined the area under the curve (AUC). The full two-dimensional region beneath the whole ROC curve is measured by AUC. It offers an overall performance indicator that spans all potential classification levels. AUC can be seen as the likelihood that a random positive example will be ranked higher by the model than a random negative example. On a comparative basis, a model's ability to discriminate between frames that have bleeding and those that don't is shown by its higher AUC. Each class's AUC is 99.97%, and their macro averages are 99.98%, indicating the model's outstanding ROC performance. For the bleeding and normal classes, the Xception model achieves 96.25 and 96.27, respectively. The proposed model successfully classifies most of the difficult lesions.

We also use the K-fold cross-validation method in Table 7.1 to determine whether the deep learning model generalizes to various inputs. Predictive evaluation of models can be done using the K-fold cross-validation method. It is a technique for assessing the extent to which a model can forecast the results of data that has not yet been seen. All of the data may still be included as a component of the testing data whenever K-fold cross-validation is employed. In this manner, the entire limited data set can be utilized for testing as well as training, improving our ability to assess the effectiveness of the suggested model. There are eight folds in the dataset. A new fold is used as the validation set on each occasion as the model is trained and assessed eight times. To calculate the rate of generalization of the framework, performance

Table 7.1 The efficiency of the suggested method using eight-fold cross-validation

Iteration	*ACC (%)*	*SE (%)*	*SP (%)*	*PRC (%)*
Iteration 1	99.38	99.39	98.78	99.38
Iteration 2	99.38	99.39	98.78	99.38
Iteration 3	99.38	99.39	98.78	99.38
Iteration 4	99.38	99.39	98.78	99.38
Iteration 5	98.77	98.78	97.56	98.78
Iteration 6	99.38	99.39	98.78	99.38
Iteration 7	99.38	99.39	98.78	99.38
Iteration 8	99.38	99.39	98.78	99.38
Average	**99.30**	**99.31**	**98.62**	**99.30**
Margin of Error	±0.07	±0.07	±0.32	±0.16
Standard Deviation	0.2	0.2	0.14	0.19

measures from each fold are averaged. This approach helps with model evaluation, selection, and hyperparameter tuning and offers a more accurate way to gauge a model's efficacy. Throughout this whole procedure, each fold training and examination would be executed precisely once. It aids in preventing overfitting. The margin of error for the accuracy is only 0.07%.

We made a variety of comparisons to demonstrate the effectiveness of the suggested approach. First, we compared our proposed method to transfer learning models (Inception-ResNet-V2, InceptionV3, ResNet50, and MobileNetV2), including Xception, which is the basis of our main model. This sort of comparison is included because our suggested method uses a transfer learning model as its basis and demonstrates the additional module enhancements over the most recent models. The suggested model performs better than the MobileNetV2 and Xception models, which use depthwise separable convolutions in various ways. The Xception model performs badly in terms of classification across the four metrics, as seen in Table 7.2. The SB integration increased the model's accuracy by 22.84%. In order to compare how well our network performed to the most effective transfer

Table 7.2 Comparisons with cutting-edge methods

Method	*ACC (%)*	*SE (%)*	*SP (%)*	*PRC (%)*
Xception	76.54	76.83	53.65	83.90
MobileNetV2	85.19	85	1	88.68
Inception-ResNet-V2	90.12	90	1	91.84
ResNet50	93.83	93.77	98.78	94.31
InceptionV3	95.06	95.12	90.24	95.45
The proposed method	**99.38**	**99.39**	**98.78**	**99.38**

Table 7.3 Effectiveness of the proposed approach employing nine-fold cross-validation

Method	*ACC (%)*	*SE (%)*	*SP (%)*	*PRC (%)*
Iakovidis *et al.* [37]	85.8	85.95	74.39	87.81
Jia and Meng [38]	79.63	79.85	62.19	83.89
Lafraxo *et al.* [36]	91.98	92.07	84	93.01
Rustam *et al.* [39]	91.98	91.87	1	93.16
The proposed method	**99.38**	**99.39**	**98.78**	**99.38**

learning techniques, we developed ResNet50, InceptionV3, and Inception-ResNet-V2 and practiced them on the dataset. The proposed approach outscored the InceptionV3 model, which came in second, by a factor of 4.32%.

We contrasted the proposed approach with several state-of-the-art WCE image classification techniques, such as those discussed in [36–39]. For an accurate comparison, we applied the authors' implementations of alternative strategies. Table 7.3's experimental findings attest to the fact that our model outperforms competing systems on almost every metric. Compared to previous deep learning-based techniques, the algorithms [24, 39] produce an accuracy that is noticeably higher. By a percentage of 7.4% in ACC, the suggested network outperforms these methods significantly. In comparison to the additional deep learning approaches [37, 38], our system achieved performance improvements of 13.58% and 19.75% in accuracy. The evaluation's findings demonstrate the benefits of the strategy utilizing our SB.

7.6 CONCLUSION

Bleeding areas can occasionally take on a variety of red hues instead of being completely red. These difficulties encourage us to create a powerful deep-learning system. In this study, a deep neural network was examined for its potential for recognizing bleeding spots in WCE images. Better classification performance is achieved by combining the Xception model with a suggested SB. Experimental testing on a public dataset verifies our framework's effectiveness. Our method outperforms other state-of-the-art methods, according to the studies, with an accuracy of 98.5%, sensitivity, specificity, and precision of 98.5%, 99%, and 98.5%, respectively. In future versions, we intend to apply our method to sizable clinical datasets and extend it to additional multi-classification issues rather than binary classification in WCE videos.

ACKNOWLEDGMENT

This work was supported by the Ministry of National Education through Vocational Training; in part by the Higher Education and Scientific Research through the Ministry of Industry, Trade, and Green and Digital Economy; in part by the Digital Development Agency (ADD); and in part by the National Center for Scientific and Technical Research (CNRST) under Project ALKHAWARIZMI/2020/20.

REFERENCES

1. Kim, B.S.M., Li, B.T., Engel, A., Samra, J.S., Clarke, S., Norton, I.D., Li, A.E.: Diagnosis of gastrointestinal bleeding: A practical guide for clinicians. World Journal of Gastrointestinal Pathophysiology **5**(4), 467 (2014).
2. Iddan, G., Meron, G., Glukhovsky, A., Swain, P.: Wireless capsule endoscopy. Nature **405**(6785), 417–417 (2000).
3. Charfi, S., El Ansari, M.: A locally based feature descriptor for abnormalities detection. Soft Computing **24**, 4469–4481 (2020).
4. Yuan, Y., Yao, X., Han, J., Guo, L., Meng, M.Q.-H.: Discriminative joint-feature topic model with dual constraints for WCE classification. IEEE Transactions on Cybernetics **48**(7), 2074–2085 (2017).
5. Liu, D.-Y., Gan, T., Rao, N.-N., Xing, Y.-W., Zheng, J., Li, S., Luo, C.-S., Zhou, Z.-J., Wan, Y.-L.: Identification of lesion images from gastrointestinal endoscope based on feature extraction of combinational methods with and without learning process. Medical Image Analysis **32**, 281–294 (2016).
6. Vasilakakis, M., Sovatzidi, G., Iakovidis, D.K.: Explainable classification of weakly annotated wireless capsule endoscopy images based on a fuzzy bag-of-colour features model and brain storm optimization. In: International Conference on Medical Image Computing and Computer-Assisted Intervention, pp. 488–498 (2021). Springer.
7. Souaidi, M., El Ansari, M.: A new automated polyp detection network MP-FSSD in WCE and colonoscopy images based fusion single shot multibox detector and transfer learning. IEEE Access **10**, 47124–47140 (2022).
8. Ellahyani, A., Jaafari, I.E., Charfi, S., Ansari, M.E.: Detection of abnormalities in wireless capsule endoscopy based on extreme learning machine. Signal, Image and Video Processing **15**, 877–884 (2021).
9. Oukdach, Y., Kerkaou, Z., El Ansari, M., Koutti, L., El Ouafdi, A.F.: Gastrointestinal diseases classification based on deep learning and transfer learning mechanism. In: 2022 9th International Conference on Wireless Networks and Mobile Communications (WINCOM), pp. 1–6 (2022). IEEE.
10. Chollet, F.: Xception: Deep learning with depthwise separable convolutions. In: Proceedings of the IEEE Conference on Computer Vision and Pattern Recognition, pp. 1251–1258 (2017).
11. Krizhevsky, A., Sutskever, I., Hinton, G.E.: Imagenet classification with deep convolutional neural networks. Advances in Neural Information Processing Systems **25**, 7 (2012).
12. Smedsrud, P.H., Thambawita, V., Hicks, S.A., Gjestang, H., Nedrejord, O.O., Næss, E., Borgli, H., Jha, D., Berstad, T.J.D., Eskeland, S.L., *et al.*: Kvasir-capsule, a video capsule endoscopy dataset. Scientific Data **8**(1), 142 (2021).
13. Garbaz, A., Lafraxo, S., Charfi, S., El Ansari, M., Koutti, L.: Bleeding classification in wireless capsule endoscopy images based on inception-resnet-v2 and cnns. In: 2022 IEEE Conference on Computational Intelligence in Bioinformatics and Computational Biology (CIBCB), pp. 1–6 (2022). IEEE.
14. Lafraxo, S., El Ansari, M., Koutti, L.: Computer-aided system for bleeding detection in wce images based on cnn-gru network. Multimedia Tools and Applications 83(7), 1–26 (2023).

15. Ghosh, T., Li, L., Chakareski, J.: Effective deep learning for semantic segmentation based bleeding zone detection in capsule endoscopy images. In: 2018 25th IEEE International Conference on Image Processing (ICIP), pp. 3034–3038 (2018). IEEE.
16. Hajabdollahi, M., Esfandiarpoor, R., Khadivi, P., Soroushmehr, S.R., Karimi, N., Najarian, K., Samavi, S.: Segmentation of bleeding regions in wireless capsule endoscopy for detection of informative frames. Biomedical Signal Processing and Control **53**, 101565 (2019).
17. Li, S., Zhang, J., Ruan, C., Zhang, Y.: Multi-stage attention-Unet for wireless capsule endoscopy image bleeding area segmentation. In: 2019 IEEE International Conference on Bioinformatics and Biomedicine (BIBM), pp. 818–825 (2019). IEEE.
18. Yuan, Y., Li, B., Meng, M.Q.-H.: Bleeding frame and region detection in the wireless capsule endoscopy video. IEEE Journal of Biomedical and Health Informatics **20**(2), 624–630 (2015).
19. Xing, X., Yuan, Y., Jia, X., Meng, M.Q.-H.: A saliency-aware hybrid dense network for bleeding detection in wireless capsule endoscopy images. In: 2019 IEEE 16th International Symposium on Biomedical Imaging (ISBI 2019), pp. 104–107 (2019). IEEE.
20. Kanakatte, A., Ghose, A.: Precise bleeding and red lesions localization from capsule endoscopy using compact u-net. In: 2021 43rd Annual International Conference of the IEEE Engineering in Medicine & Biology Society (EMBC), pp. 3089–3092 (2021). IEEE.
21. Hajabdollahi, M., Esfandiarpoor, R., Najarian, K., Karimi, N., Samavi, S., Soroushmehr, S.R.: Low complexity CNN structure for automatic bleeding zone detection in wireless capsule endoscopy imaging. In: 2019 41st Annual International Conference of the IEEE Engineering in Medicine and Biology Society (EMBC), pp. 7227–7230 (2019). IEEE.
22. Lan, L., Ye, C., Wang, C., Zhou, S.: Deep convolutional neural networks for WCE abnormality detection: Cnn architecture, region proposal and transfer learning. IEEE Access **7**, 30017–30032 (2019).
23. Caroppo, A., Leone, A., Siciliano, P.: Deep transfer learning approaches for bleeding detection in endoscopy images. Computerized Medical Imaging and Graphics **88**, 101852 (2021).
24. Amiri, Z., Hassanpour, H., Beghdadi, A.: A computer-aided method to detect bleeding frames in capsule endoscopy images. In: 2019 8th European Workshop on Visual Information Processing (EUVIP), pp. 217–221 (2019). IEEE.
25. Ghosh, T., Fattah, S.A., Wahid, K.A.: Chobs: Color histogram of block statistics for automatic bleeding detection in wireless capsule endoscopy video. IEEE Journal of Translational Engineering in Health and Medicine **6**, 1–12 (2018).
26. Jia, X., Xing, X., Yuan, Y., Xing, L., Meng, M.Q.-H.: Wireless capsule endoscopy: A new tool for cancer screening in the colon with deep-learning-based polyp recognition. Proceedings of the IEEE **108**(1), 178–197 (2019).
27. Wickstrøm, K., Kampffmeyer, M., Jenssen, R.: Uncertainty and interpretability in convolutional neural networks for semantic segmentation of colorectal polyps. Medical Image Analysis **60**, 101619 (2020).
28. Wang, S., Cong, Y., Zhu, H., Chen, X., Qu, L., Fan, H., Zhang, Q., Liu, M.: Multi-scale context-guided deep network for automated lesion segmentation with endoscopy images of gastrointestinal tract. IEEE Journal of Biomedical and Health Informatics **25**(2), 514–525 (2020).
29. Fang, Y., Chen, C., Yuan, Y., Tong, K.-Y.: Selective feature aggregation network with area boundary constraints for polyp segmentation. In: Medical Image Computing and Computer Assisted Intervention–MICCAI 2019: 22nd International Conference, Shenzhen, China, October 13–17, 2019, Proceedings, Part I 22, pp. 302–310 (2019). Springer.

30. Yang, J., Ou, Y., Chen, Z., Liao, J., Sun, W., Luo, Y., Luo, C.: A benchmark dataset of endoscopic images and novel deep learning method to detect intestinal metaplasia and gastritis atrophy. IEEE Journal of Biomedical and Health Informatics **27**(1), 7–16 (2022).
31. He, J.-Y., Wu, X., Jiang, Y.-G., Peng, Q., Jain, R.: Hookworm detection in wireless capsule endoscopy images with deep learning. IEEE Transactions on Image Processing **27**(5), 2379–2392 (2018).
32. Liu, X., Yuan, Y.: A source-free domain adaptive polyp detection framework with style diversification flow. IEEE Transactions on Medical Imaging **41**(7), 1897–1908 (2022).
33. Guo, X., Yuan, Y.: Triple ANet: Adaptive abnormal-aware attention network for WCE image classification. In: Medical Image Computing and Computer Assisted Intervention–MICCAI 2019: 22nd International Conference, Shenzhen, China, October 13–17, 2019, Proceedings, Part I 22, pp. 293–301 (2019). Springer.
34. Guo, X., Yang, C., Liu, Y., Yuan, Y.: Learn to threshold: Thresholdnet with confidence-guided manifold mixup for polyp segmentation. IEEE Transactions on Medical Imaging **40**(4), 1134–1146 (2020).
35. Sifre, L.: Rigid-motion scattering for image classification. PhD thesis (2014).
36. Lafraxo, S., Ansari, M.E., Charfi, S.: Melanet: an effective deep learning framework for melanoma detection using dermoscopic images. Multimedia Tools and Applications **81**(11), 16021–16045 (2022).
37. Iakovidis, D.K., Georgakopoulos, S.V., Vasilakakis, M., Koulaouzidis, A., Plagianakos, V.P.: Detecting and locating gastrointestinal anomalies using deep learning and iterative cluster unification. IEEE Transactions on Medical Imaging **37**(10), 2196–2210 (2018).
38. Jia, X., Meng, M.Q.-H.: A deep convolutional neural network for bleeding detection in wireless capsule endoscopy images. In: 2016 38th Annual International Conference of the IEEE Engineering in Medicine and Biology Society (EMBC), pp. 639–642 (2016). IEEE.
39. Rustam, F., Siddique, M.A., Siddiqui, H.U.R., Ullah, S., Mehmood, A., Ashraf, I., Choi, G.S.: Wireless capsule endoscopy bleeding images classification using cnn based model. IEEE Access **9**, 33675–33688 (2021).

Chapter 8

Lung tumor recognition and classification in CT scan images using CNNs, transfer learning, and ensemble learning

Amine Aaz el aarab, Oussama Smimite, and Zakaria Kerkaou

8.1 INTRODUCTION

To increase the patient's chances of survival, lung cancer identification needs to happen early; that's why doctors rely on a variety of screening techniques. These include imaging examinations like CT (computed tomography) scans, as well as chest X-rays [1]. Currently, CT scans have become paramount for lung cancer identification, especially for people who are more vulnerable to the disease such as smokers and those who have a history of lung cancer in their families [2]. CT scans provide cross-sectional images of the body, with an emphasis on the lungs, by utilizing body X-rays and computer technologies. These scans are vital for the detection of anomalies in the lungs, such as masses, nodules, and suspicious spots that all have a chance of being cancerous. As for the way that CT scans support the diagnostic process, it's by providing information about the size and shape of the anomalies as well as their location. The next task is then to decide whether the detected abnormality is benign (non-cancerous) or malignant (cancerous). Malignant masses are indicative of cancer, which needs to be treated quickly to halt its spread in the rest of the body [3]. On the other hand, benign masses are not considered cancerous and can be caused by infections, inflammation, or other conditions [4]. In the field of artificial intelligence and its applications in medical image analysis, the progress that has been made is very impressive, especially when it comes to identifying and classifying lung cancer either in lung CT scans or chest X-rays, all thanks to the growth that deep learning (DL) technology has known. Convolutional neural networks (CNNs), as well as advanced DL models, have been implemented to assist in lung cancer detection to further enhance both the accuracy and the efficiency, which has yielded, up until now, some decent results [5, 6]. The aim of this study is to categorize various abnormalities in lung CT scans into three classes: normal, benign, and malignant. The Iraq-Oncology Teaching Hospital/National Center for Cancer Diseases (IQ-OTH/NCCD) dataset is used to train a variety of pretrained models as well as a custom CNN model to find a combined approach to leverage their ensemble learning to get the best classification results. The mentioned dataset represents three classes (normal, benign, and malignant),

DOI: 10.1201/9781032664293-9

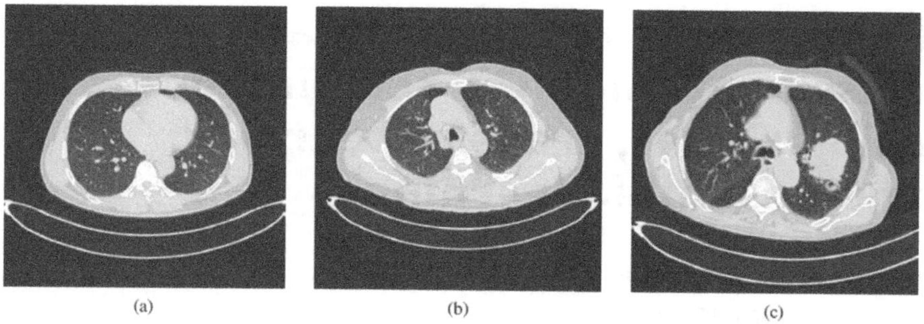

Figure 8.1 Comparison of (a) normal, (b) benign, and (c) malignant conditions.

which are illustrated in Figure 8.1. The proposed system's results are compared to some of the pre-trained known architectures. The chapter's structure is set out as follows. In Section 8.2, we discuss the various related works and state-of-the-art advancements in our research topic, and then in Section 8.3, we explain the proposed approach. Section 8.4 presents the experimental results. Section 8.5 serves as the chapter's conclusion.

8.2 RELATED WORKS

Lung cancer recognition and classification using DL techniques has been extensively studied in recent years. The capacity of CNNs to automatically extract pertinent features from raw images has shown them to be especially useful in medical image analysis. Early works by Krizhevsky et al. [7] demonstrated the potential of DL in image classification, which paved the way for its application in medical imaging. Recent studies utilized CNNs for lung cancer detection in CT scans. For instance, Shen et al. [8] employed a multi-crop CNN to improve classification accuracy by focusing on different regions of interest within the CT scans. Similarly, Hua et al. [9] presented a deep residual network (ResNet) to classify lung nodules, achieving notable performance improvements. The problem of sparse medical picture datasets has also been investigated through the use of transfer learning, which makes use of pre-trained models on big datasets. Gao et al. [10] used transfer learning with a pre-trained VGG16 model, fine-tuning it on a lung cancer dataset to enhance the classification performance. The use of transfer learning not only accelerates the training process but also helps in achieving better accuracy with fewer labeled data. Ensemble learning methods, which combine multiple models to improve robustness and accuracy, have shown promise in medical image classification. Zhou et al. [11] combined several CNN architectures using an ensemble approach to improve lung cancer detection accuracy. Their method demonstrated that ensemble learning could effectively lessen the bias and variance of individual models, leading to more reliable predictions.

Overall, the field of lung tumor recognition and classification in CT scan images has evolved greatly as a result of the integration of CNNs, transfer learning, and ensemble learning, which has established a strong framework for enhancing diagnostic accuracy and reliability.

8.3 METHODOLOGY

This section presents our proposed system for detecting and classifying pulmonary tumors, as illustrated in Figure 8.4. Data augmentation techniques, including flipping, mirroring, and random tiny rotations, are used in the initial stage. The augmented images are then utilized for training and evaluating both the DenseNet121 and our base CNN architecture separately. Finally, we employ ensemble learning by averaging the predictions from both the base CNN and the DenseNet121 to generate the final predictions.

8.3.1 Data augmentation

The transformations performed on the images in our dataset to artificially increase its size include rotations up to 10 degrees, shifts in width and height by 10%, shearing by 10%, zooming by 10%, and enabling horizontal flips. The goal of these augmentations is to introduce variance into the dataset without altering the underlying labels, thereby making the model training process more robust to variations in new, unseen images. As shown in Table 8.1, after using data augmentation techniques, the dataset's image count considerably increased. Notably, there was an increase in the "Normal" class from 416 to 1203 images, the "Benign" class from 120 to 1350 images, and the "Malignant" class from 561 to 1103 images.

8.3.2 Base CNN model

CNNs are widely used for a variety of applications, including image and video identification, image classification, and medical image analysis. This is due to their extraordinary capacity to capture spatial and temporal dependencies in data [12]. In this section, we present the structure of the baseline CNN used in our system, which is designed to efficiently process and analyze medical images for lung tumor detection and classification.

Table 8.1 Image count in the dataset before and after data augmentation

Class	*Image count before data augmentation*	*Image count after data augmentation*
Normal	416	1203
Benign	120	1305
Malignant	561	1103

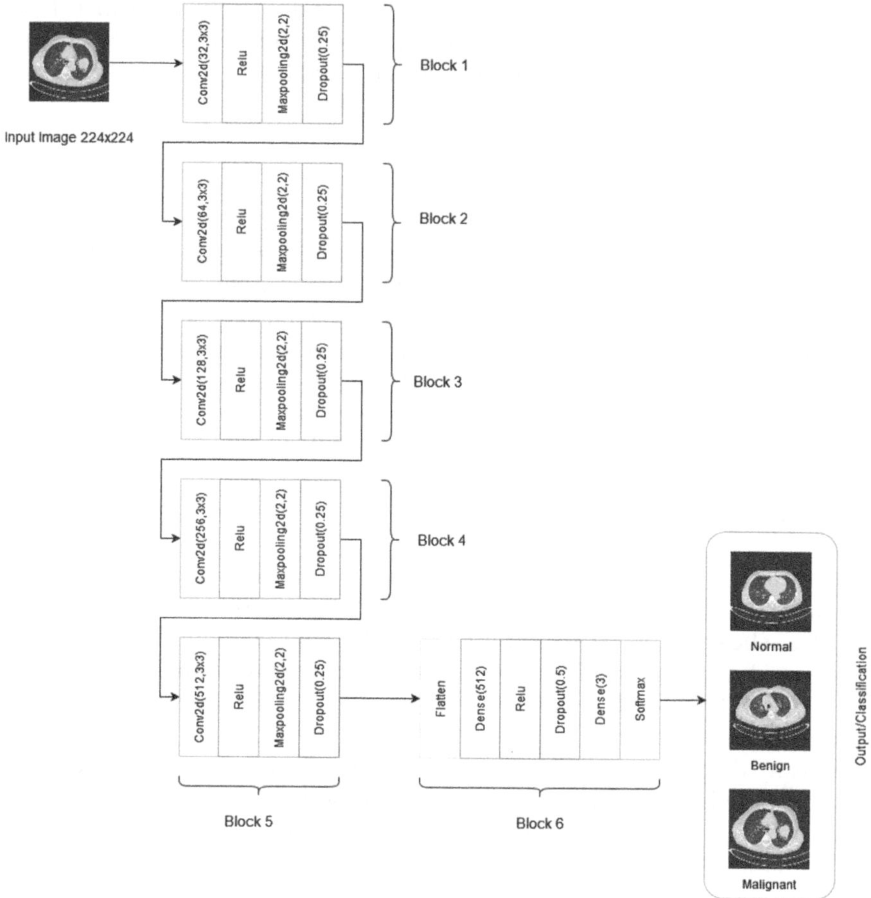

Figure 8.2 The structure of the base CNN for feature extraction from lung CT scans.

As illustrated in Figure 8.2, the baseline CNN in our proposed system is composed of six sequential blocks, each meticulously crafted to improve the model's ability to learn and generalize from the input data. The detailed structure of each block is as follows:

- Conv2D layer: The Conv2D layers, which are responsible for recognizing features in the input images, are placed as the first layer in every block. These layers have 32 filters in the first block, 64 in the second block, 128 in the third block, 256 in the fourth block, and 512 filters in the fifth block. The number of filters in these layers increases gradually from block to block. The ReLu activation function and 3 × 3 kernels are used in all the convolutional layers, giving the model non-linearity, which allows it to recognize intricate patterns and correlations in the data.

- MaxPooling2D layer: A MaxPooling2D layer is added after every Conv2D layer. This layer provides downsampling with a 2 × 2 pool size, which successfully lowers the dimensionality of the feature maps without sacrificing the most important characteristics. Pooling reduces computing complexity and reduces the chance of overfitting by concentrating on the most important features.
- Dropout layers: Dropout layers are added following each max pooling procedure in order to further alleviate the overfitting problem. A dropout rate of 0.25 is implemented in blocks one through five, randomly setting 25% of the input units to zero during each training cycle. By using this method, the network is compelled to acquire more resilient qualities that are independent of any one unit. To guarantee that the model performs well in terms of generalizing to new data, the dropout rate is raised to 0.5 in the last block.
- Flatten layer: The flatten layer handles the feature maps that the last convolutional block generated. The 3D feature maps are transformed into 1D feature vectors by this layer so that they can be used as input for the dense (fully connected) layers that come after. Because it allows the characteristics to be processed in a linear fashion by the thick layers that follow, this transformation is essential for the classification task.
- Dense layer: Following flattening, the feature vector is passed into a 512-unit dense layer, where non-linearity is introduced using ReLU activation. Moreover, L2 regularization (with a regularization parameter of 0.001) is used in this dense layer to penalize heavy weights and further prevent overfitting. Another dense layer with three units, representing the three classes in our multi-class classification problem, generates the model's final output. The SoftMax activation function is used in this output layer to translate the raw output scores into probabilities, making it easier to understand the predictions made by the model.

Overall, the design of our base CNN model emphasizes both the depth and breadth necessary to capture the intricate details present in medical images. By combining multiple convolutional layers with regularization techniques like dropout and L2 regularization, our model is capable of learning robust features while maintaining generalizability. This architecture forms a crucial component of our lung cancer detection system, providing a strong foundation for accurate and reliable classification.

8.3.3 Transfer learning

Transfer learning, an effective DL technique, entails first training a network on a sizable, well-known dataset and then altering the final layers of this pre-trained network to fit the requirements of the new task at hand. This approach leverages the pre-existing knowledge the model has acquired from the extensive initial training, allowing for efficient and effective learning

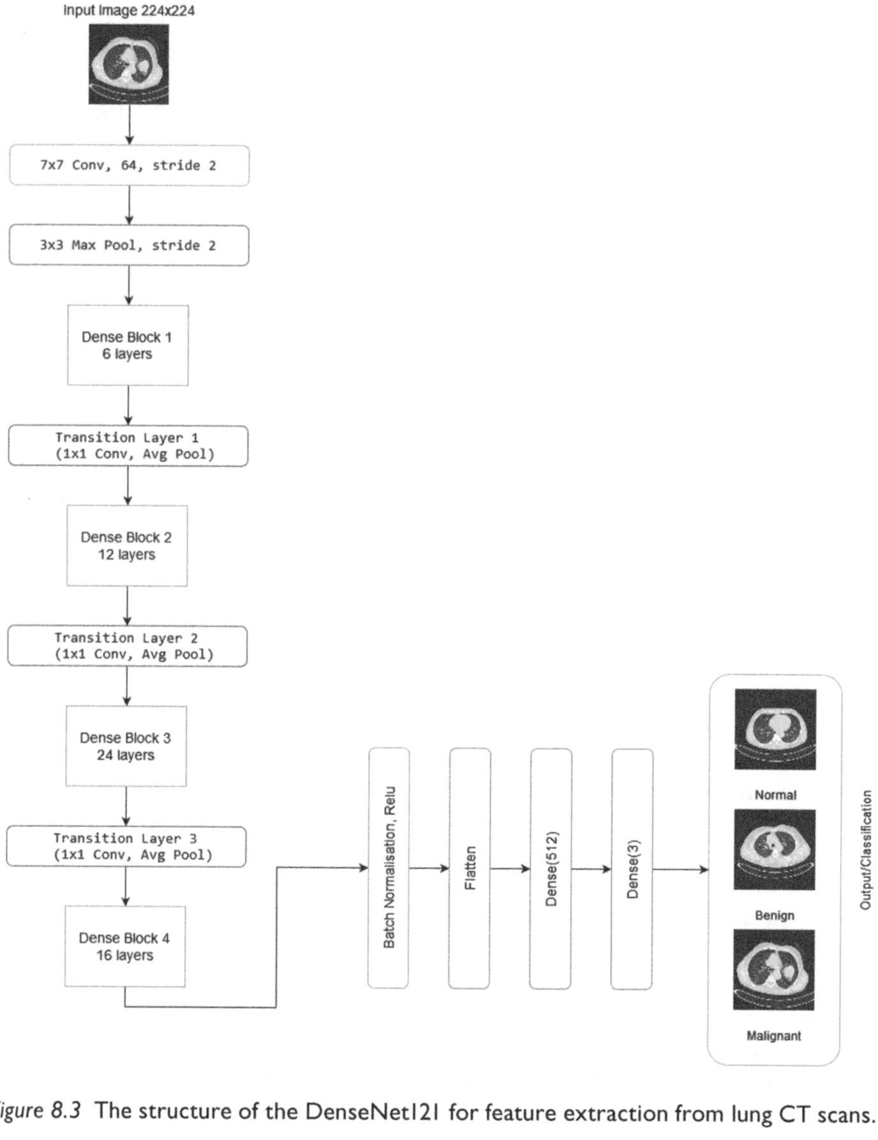

Figure 8.3 The structure of the DenseNet121 for feature extraction from lung CT scans.

on the new task, even with a relatively smaller dataset [13]. In our system, we implemented transfer learning as depicted in Figure 8.3, employing the DenseNet121 architecture. DenseNet121 is built on the concept of densely connected convolutional networks (DenseNets). This architecture stands out in particular for its unique pattern of connectivity, where every layer is feed-forwardly connected to every other layer. The vanishing-gradient problem, which is a prevalent problem in deep neural networks where gradients decrease as they are back-propagated through the layers, is lessened by this dense connectivity. More efficiently propagating gradients is how DenseNet121 makes it possible to train considerably deeper networks than with conventional architectures. Learning complex patterns in data requires strong feature reuse and improved feature propagation, which is what the DenseNet121 design accomplishes. Furthermore, the model is more efficient because of the dense connectivity, which lowers the number of parameters needed for the network.

Architecture consists of multiple dense blocks interspersed with transition layers. The dense blocks are responsible for capturing complex feature maps through a series of convolutions, while the transition layers manage the size of these feature maps via convolution and pooling operations. This combination ensures that the model not only learns rich representations of the input data but also maintains computational efficiency [14]. DenseNet121, with its 121 layers, strikes an optimal balance between depth and processing efficiency. This depth allows the network to perform extensive feature extraction, making it highly suitable for tasks such as object recognition and image classification, where detailed and nuanced feature representation is crucial. The transition layers play a vital role in controlling the model's complexity by downsampling the feature maps, thus preventing the network from becoming computationally prohibitive despite its depth [14]. In our implementation, we utilized a pre-trained DenseNet121 model, originally trained on the ImageNet dataset, a large and diverse dataset used for benchmarking image classification algorithms. By freezing the pre-trained layers during the initial stages of training on our new dataset, we preserved the valuable features learned from the ImageNet dataset. We then added new layers on top of the DenseNet121 base to adapt it to our specific classification problem. This method allowed us to benefit from the sophisticated feature extraction capabilities of DenseNet121 while fine-tuning the model to accurately classify the categories in our new dataset.

8.3.4 Ensemble learning

Ensemble learning is an approach where multiple models are trained to solve the same problem and then combined to enhance overall performance (Figure 8.4). The core concept is that by aggregating the predictions of several models, the ensemble can typically achieve better accuracy, robustness, and generalizability than any individual model alone [15]. In our ensemble training setup, two DL models, pre-trained DenseNet121 and our base CNN, are utilized to predict class probabilities from a given dataset. Each model independently predicts these probabilities, which are then averaged to mitigate model-specific biases and improve the robustness of the predictions. The averaging formula used is as follows:

$$Avg_{probs} = \frac{DenseNet121\_probs + BaseCNN_{probs}}{2} \tag{8.1}$$

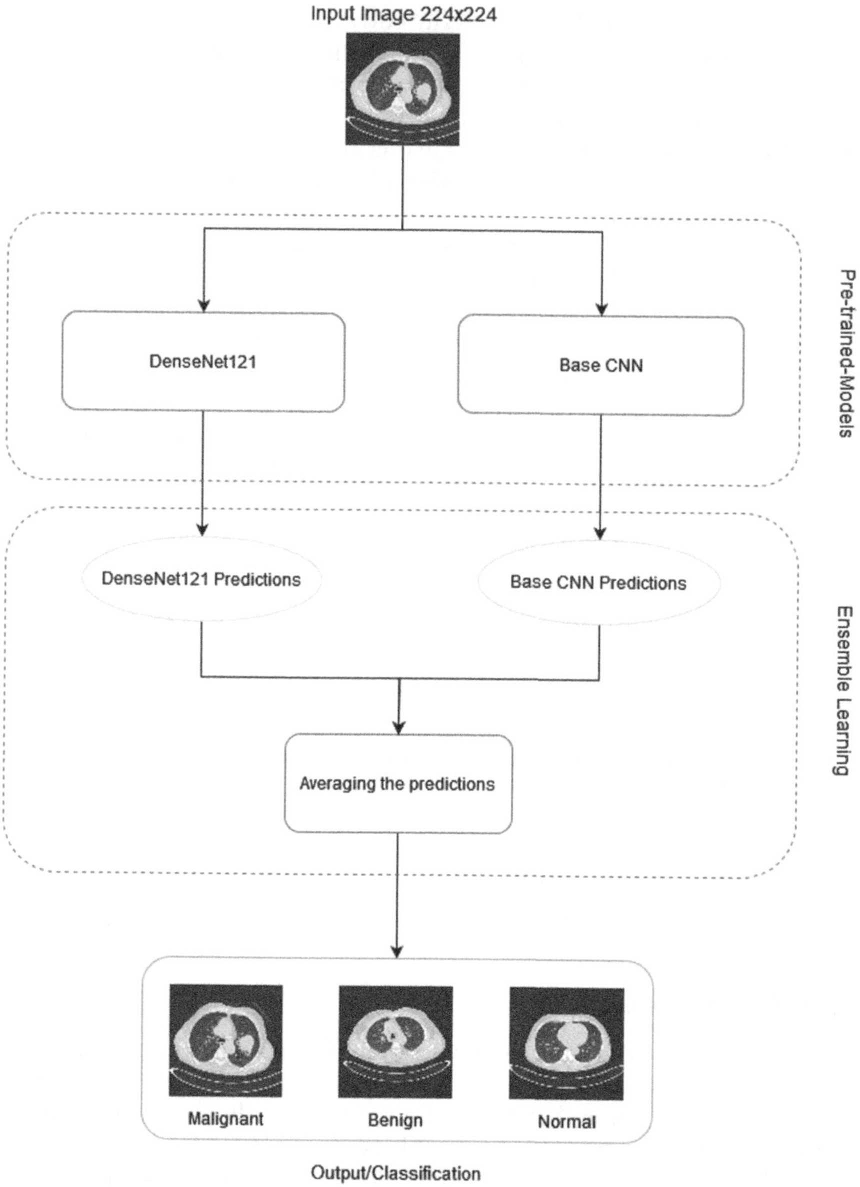

Figure 8.4 The pipeline of the suggested method for lung cancer classification.

Ensuring that each model contributes equally to the final prediction. The ensemble prediction is then determined by selecting the class with the highest probability for each instance. This method leverages the diverse strengths of both architectures to enhance prediction accuracy and stability, typical of ensemble methods that combine multiple models to reduce error likelihood.

8.4 RESULTS AND DISCUSSION

This section includes experiments that compare the performance of our suggested method to that of the base CNN as well as some of the popular transfer learning networks (DenseNet121, VGG16, VGG19, ResNet50, etc.). Information about the dataset and the implementation of the method is also included.

8.4.1 Dataset

We conducted our research using the lung cancer dataset from the IQ-OTH/NCCD. This dataset was collected over the course of three months in the fall of 2019 from various specialized hospitals. It contains CT scans of both healthy people and patients with lung cancer at different stages of the disease. Oncologists and radiologists from the two centers annotated the slides in the IQ-OTH/NCCD dataset. There are 1190 pictures or slices total from 110 CT scans in the collection. These cases fall into three categories: normal, malignant, and benign. Of the instances, 55 are normal, 15 are benign, and 40 are malignant. Every scan consists of 80–200 slices, each of which shows a different aspect and perspective of the chest. The 110 instances differ in terms of living conditions, place of residence, age, gender, and educational background [16].

8.4.2 Experimental setup

Using Google Colab's available GPUs, our suggested approach was implemented with Python and the Keras package with a TensorFlow backend. Our models underwent 50 epochs of training with a batch size of 32. We trained the models using the sparse categorical cross-entropy loss function and the "rmsprop" optimizer, with a default learning rate of 0.001. In order to improve training effectiveness and reduce the possibility of overfitting, we implemented multiple techniques. Training was stopped if the validation loss did not improve for ten consecutive epochs. This technique, known as early stopping, involved monitoring the validation loss with a patience of ten epochs. In addition, we employed Learning Rate Reduction on Plateau, which, after five epochs, automatically lowered the learning rate by a factor of 0.1 when the validation loss plateaued. By optimizing model performance, these criteria guaranteed that our training procedure was both successful and efficient.

8.4.3 Evaluation metrics

Accuracy, precision, recall, and the F1 score are the evaluation criteria that were utilized for assessing the suggested approach and the other independent models. These measures, which assess the classification's efficacy, are explained as follows:

- Accuracy: This metric evaluates the overall accuracy of the model by determining the proportion of accurate predictions across the entire dataset. It is computed by taking the sum of true positives (TPs) and true negatives (TN) and dividing it by the total number of samples:

$$Accuracy = \frac{TP + TN}{TP + TN + FP + FN} \tag{8.2}$$

- Precision: This metric assesses the percentage of TP predictions out of all positive predictions made by the model. It is determined by dividing the number of TPs by the sum of TPs and false positives (FPs):

$$Precision = \frac{TP}{TP + FP} \tag{8.3}$$

- Recall: Also referred to as sensitivity or the TP rate, it quantifies the proportion of TP predictions among all actual positive cases. It is computed by dividing the number of TPs by the sum of TPs and false negatives (FN).

$$Recall = \frac{TP}{TP + FN} \tag{8.4}$$

- F1-measure: This metric harmonizes precision and recall. It is derived as the harmonic mean of these two metrics, providing a single value that balances both precision and recall. This metric is particularly valuable when aiming for a balance between high precision and high recall, as it effectively penalizes situations where either precision or recall is extremely low:

$$F1\ Score = \frac{2TP}{2TP + FP + FN} \tag{8.5}$$

8.4.4 Experimental results

Initially, our dataset was partitioned into distinct sets for training, validation, and testing, with allocations of 70%, 15%, and 15%, respectively. The first phase of our experimentation involved the training of a base CNN model, which yielded a commendable classification accuracy of 94.01%, as detailed in Table 8.2. This initial performance served as a pivotal reference point for subsequent evaluations against more sophisticated neural network architecture. Subsequent phases of our study encompassed the refinement, training, and evaluation of various renowned DL models, including VGG16, VGG19, ResNet50, ResNet101, MobileNetV2, InceptionV3, and DenseNet121. Notably, among these models, DenseNet121 emerged as the standout performer, achieving an impressive accuracy of 97.64%, as illustrated in Table 8.2. To capitalize on the strengths of both the baseline CNN and the top-performing DenseNet121 model, we adopted an ensemble learning strategy. This innovative approach entailed averaging the predictions of the baseline

Table 8.2 Comparative results

Method	*ACC*	*PREC*	*REC*	*F1-Score*
VGG16 [17]	96.73	96.83	96.86	96.86
VGG19 [17]	96.37	96.57	96.48	96.52
ResNet50 [18]	84.75	85.27	84.90	85.06
ResNet101 [18]	86.93	87.38	87.11	87.23
MobileNetV2 [19]	96.91	96.98	97.00	96.99
InceptionV3 [20]	92.01	92.44	92.10	92.21
DenseNet169 [7]	95.10	95.12	95.29	95.17
DenseNet121 [14]	97.64	97.78	97.67	97.72
Base CNN	94.01	94.33	94.10	94.16
Proposed Method	**99.82**	**99.84**	**99.82**	**99.83**

CNN and DenseNet121 models. The ensemble method yielded a substantial performance boost, culminating in an accuracy of 99.82%, as highlighted in Table 8.2. To further solidify these findings, we delved deeper into the visual representation of the results. We present the confusion matrix of the base CNN, DenseNet121, and the proposed ensemble method in Figure 8.5.

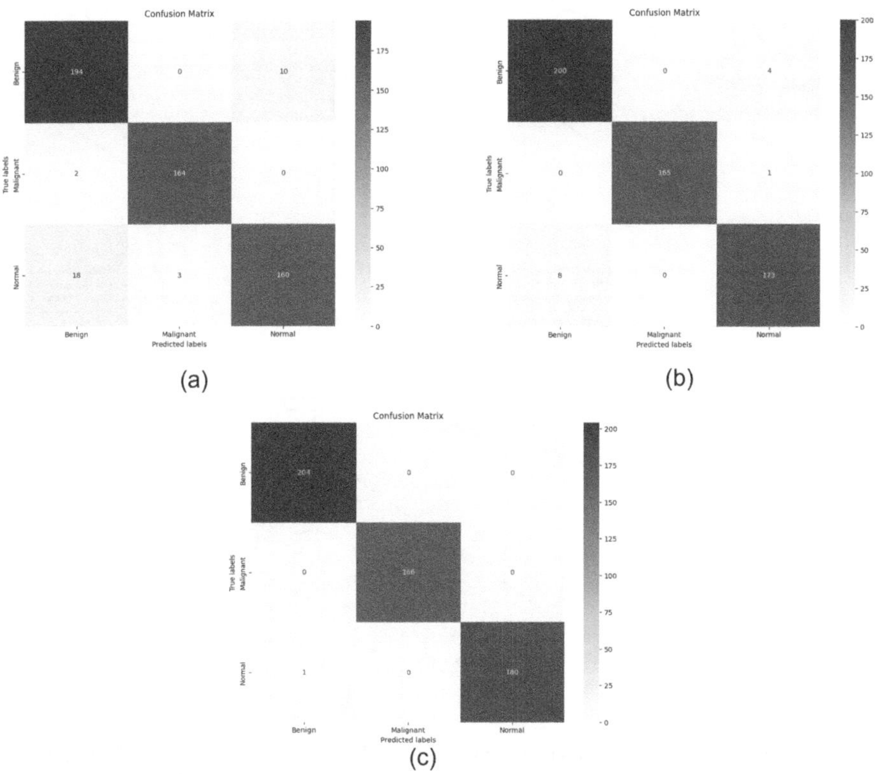

Figure 8.5 Confusion matrix for (a) base CNN, (b) DenseNet121, and (c) proposed method.

This graphic offers a detailed analysis of the model's performance across the three classes, providing an in-depth understanding of its strengths and weaknesses. By examining these confusion matrices, we can identify specific areas where each model excels or struggles, allowing for targeted improvements and a better overall grasp of their classification capabilities. In addition to the confusion matrices, Figure 8.6 showcases the multi-class ROC (Receiver-operating characteristic) curves for each of the aforementioned approaches. ROC curves are essential for evaluating a model's ability to distinguish between positive and negative classes across different thresholds. By plotting the TP rate against the FP rate, these curves provide a comprehensive assessment of each model's diagnostic performance. This is particularly important in medical applications, where accurate differentiation between normal, benign, and malignant cases is critical. The visualizations in Figures 8.5 and 8.6 demonstrate that the proposed ensemble method exhibits minimal confusion when identifying normal, benign, and malignant cases. This indicates that the ensemble method not only achieves good accuracy but also maintains a balanced performance across different categories. The ROC curves further highlight the capability of the ensemble approach to distinguish among classes with a high degree of precision. Overall, these figures provide strong

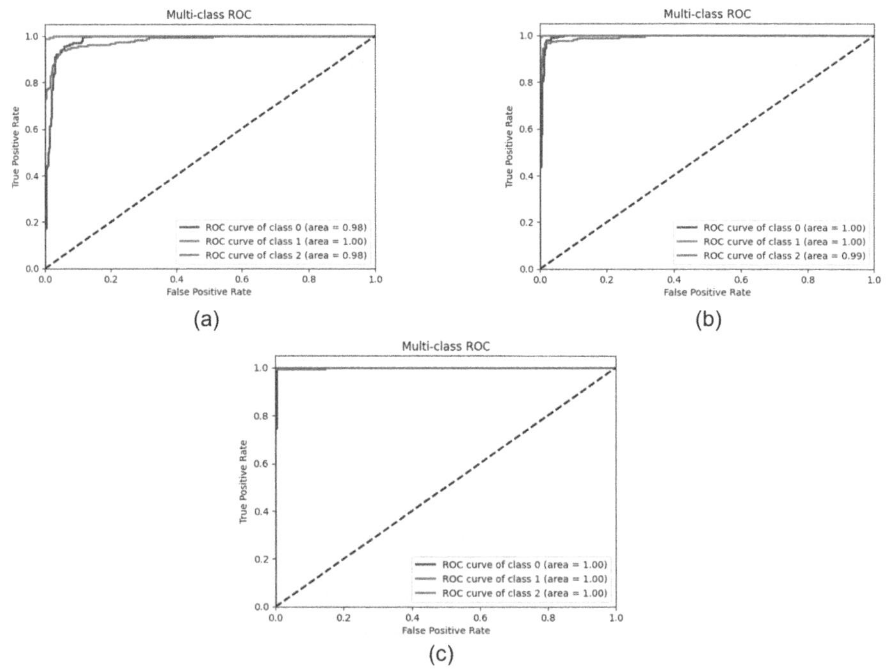

Figure 8.6 Multi-class ROC curves for (a) base CNN, (b) DenseNet121, and (c) proposed method.

evidence of the ensemble method's effectiveness in lung tumor classification. The detailed insights gained from the confusion matrices and ROC curves underscore the potential of the ensemble method to improve diagnostic accuracy and reliability in clinical settings.

8.5 CONCLUSION

In this study, we have introduced a new approach for the multi-class classification of lung CT scans, categorizing them into three classes: normal, benign, and malignant. Our proposed method involves the meticulous training of a baseline CNN in conjunction with the fine-tuning and training of DenseNet121, a well-developed DL architecture with a good performance for image recognition tasks. By employing ensemble learning techniques—specifically, by averaging the predictions from both the baseline CNN and the fine-tuned DenseNet121—we derived our final predictions. This ensemble approach harnesses the strengths of both models, leading to a synergistic effect that has outperformed other well-established networks. Notably, our method has surpassed the individual performance of each model, achieving a high accuracy rate of 99.82%. The significance of these results lies in the potential for improved diagnostic accuracy and early diagnosis of lung cancer, which is critical for effective treatment. The performance of our ensemble model demonstrates the value of integrating multiple model predictions, thus providing a more robust and reliable classification system. In future work, we aim to build upon our current research by delving deeper into the segmentation of tumors in scans identified as benign or malignant. This will involve developing and training advanced segmentation models to precisely delineate tumor boundaries, which is crucial for treatment planning and monitoring disease progression.

REFERENCES

1. National Lung Screening Trial Research Team. (2011). Reduced lung-cancer mortality with low-dose computed tomographic screening. New England Journal of Medicine, 365(5), 395–409. DOI: 10.1056/NEJMoa1102873.
2. Wood, D. E., Kazerooni, E., Baum, S. L., Dransfield, M. T., Eapen, G. A., Ettinger, D. S., ..., & Wiener, R. S. (2018). Lung cancer screening, version 3.2018, NCCN clinical practice guidelines in oncology. Journal of the National Comprehensive Cancer Network, 16(4), 412–441. DOI: 10.6004/jnccn.2018.0020.
3. Henschke, C. I., & Yankelevitz, D. F. (2015). Imaging lung cancer screening. Chest, 147(1), 14–22. DOI: 10.2214/AJR.14.12526.
4. Gould, M. K., Donington, J., Lynch, W. R., Mazzone, P. J., Midthun, D. E., Naidich, D. P., ..., & Wiener, R. S. (2013). Evaluation of individuals with pulmonary nodules: When is it lung cancer? Diagnosis and Management of Lung Cancer, 6(3), 165–176. DOI: 10.1002/14651858.CD009436.pub2.
5. Monkam, P., Qi, S., Ma, H., Gao, W., Yao, Y., & Qian, W. (2019). Detection and Classification of Pulmonary Nodules Using Convolutional Neural Networks: A Survey. IEEE Access, 7, 78075–78091.

6. Bhat, S., Shashikala, R., Kumar, S., & Gururaj, K. (2020). Convolutional Neural Network approach for the Classification and Recognition of Lung Nodules. 1310–1314. DOI: 10.1109/ICECA49313.2020.9297626.
7. Krizhevsky, A., Sutskever, I., & Hinton, G. E. (2012). ImageNet classification with deep Convolutional Neural Networks. Advances in Neural Information Processing Systems, 25, 1097–1105.
8. Shen, W., Zhou, M., Yang, F., Yu, D., Dong, D., Yang, C., & Tian, J. (2015). Multi-crop Convolutional Neural Networks for lung nodule malignancy suspiciousness classification. Pattern Recognition, 61, 663–673. DOI: 10.1016/j.patcog.2016.06.018.
9. Hua, K. L., Hsu, C. H., Hidayati, S. C., Cheng, W. H., & Chen, Y. J. (2015). Computer-aided classification of lung nodules on computed tomography images via deep learning technique. OncoTargets and Therapy, 8, 2015–2022. DOI: 10.2147/OTT.S80733.
10. Gao, M., Bagci, U., Lu, L., & Wu, A. (2017). HOLMES: Health Online Model Ensemble Serving. Proceedings of the IEEE Conference on Computer Vision and Pattern Recognition (CVPR) Workshops, 2009–2018. DOI: 10.1109/CVPRW.2017.132.
11. Zhou, Z. H., & Feng, J. (2017). Deep Forest: Towards an Alternative to Deep Neural Networks. Proceedings of the 26th International Joint Conference on Artificial Intelligence (IJCAI), 3553–3559. DOI: 10.24963/ijcai.2017/497.
12. Goodfellow, I., Bengio, Y., & Courville, A. (2016). Deep Learning. MIT Press.
13. Pan, S. J., and Q. Yang. (2010). A survey on transfer learning. IEEE Transactions on Knowledge and Data Engineering, 22(10), 1345–1359. DOI: 10.1109/TKDE.2009.191.
14. Huang, G., Liu, Z., Van Der Maaten, L., & Weinberger, K. Q. (2017). Densely Connected Convolutional Networks. In Proceedings of the IEEE Conference on Computer Vision and Pattern Recognition (pp. 4700–4708). DOI: 10.1109/CVPR.2017.243.
15. Alotaibi, Y., & Ilyas, M. (2023). Ensemble-learning framework for intrusion detection to enhance Internet of Things devices security. Sensors, 23(12), 5568. DOI: 10.3390/s23125568.
16. Alyasriy, H., & AL-Huseiny, M. (2023). The IQ-OTH/NCCD lung cancer dataset. Mendeley Data, V4, DOI: 10.17632/bhmdr45bh2.4.
17. Simonyan, K., & Zisserman, A. (2014). Very Deep Convolutional Networks for Large-Scale Image Recognition. arXiv preprint arXiv:1409.1556.
18. He, K., Zhang, X., Ren, S., & Sun, J. (2016). Deep Residual Learning for Image Recognition. Proceedings of the IEEE Conference on Computer Vision and Pattern Recognition (CVPR). DOI: 10.1109/CVPR.2016.90.
19. Sandler, M., Howard, A., Zhu, M., Zhmoginov, A., & Chen, L. C. (2018). MobileNetV2: Inverted Residuals and Linear Bottlenecks. Proceedings of the IEEE Conference on Computer Vision and Pattern Recognition (CVPR). DOI: 10.1109/CVPR.2018.00474.
20. Szegedy, C., Vanhoucke, V., Ioffe, S., Shlens, J., & Wojna, Z. (2016). Rethinking the Inception Architecture for Computer Vision. Proceedings of the IEEE Conference on Computer Vision and Pattern Recognition (CVPR). DOI: 10.1109/CVPR.2016.308.

Chapter 9

A new pedestrian detection method for intelligent surveillance systems

Redouan Lahmyed, Mohamed El Ansari, and Lahcen Koutti

9.1 INTRODUCTION

Over the last decades, pedestrian detection has attracted great attention and become one of the fastest-growing topics in computer vision. It plays a crucial role in different fields of research.

One example of these kinds of fields is video surveillance systems, which attempt to detect the presence of people and protect them and their properties from various risks including theft, riots and terrorist attacks. Different approaches have been introduced for detecting and tracking pedestrians using various sensors, such as LIDAR sensors [1, 2], thermal cameras [3, 4], and visible cameras [5–7].

Generally, the most proposed techniques use visible cameras as a main sensor due to their relatively low price and also their ability to identify objects quickly and easily. According to the published works related to pedestrian targets in images or video sequences recorded using visible sensors, pedestrian detection can be divided into two main categories. The first category aims to develop powerful learning algorithms [8–10]. JongSeok et al. [10] proposed a multiple pedestrian detection system in which the AdaBoost cascade classification is used. The works in [8] and [11] proposed experimental studies about the performance of various classifiers including support vector machine (SVM), AdaBoost and artificial neural network (ANN) in the pedestrian classification topic. Oswaldo et al. [9] introduced a novel classifier fusion scheme using learning algorithms. The adopted classifiers for combination are SVM, Fisher's Linear Discriminant Analysis (FLDA), and Gradient Descendent with momentum term and adaptive learning rate (GDX-MCI).

The second category tries to build powerful features to deal with the human body structure and height [12, 13] using different information such as texture, gradient, motion ... etc. Among these information, the majority of the proposed features for human detection prefer to use either gradient [14, 15] or texture [12, 16] due to their strongest description ability. The researchers in [15] integrated the cascade-of-rejectors approach with the histogram of oriented gradients (HOG) features to detect pedestrians. Covariance matrices are used as object descriptors together with novel learning algorithms on the Riemannian manifolds to detect pedestrians in [13]. The authors in [17]

DOI: 10.1201/9781032664293-10

introduced a feature named histogram of template (HOT) to describe the appearance of pedestrians in images. Yadong et al. [12] proposed improved local binary patterns (LBP)-based features [16] for human detection.

However, the performance of all proposed methods is not satisfying in practice due to the large variability of the human appearances and shapes caused by diversity in poses. Furthermore, the suggested methods avoid using color information or information combinations for the purpose of computations simplification.

In this chapter, we propose a novel two-stage detection method for moving pedestrians in intelligent surveillance systems. The first stage (hypotheses generation (HG)) extracts the regions of interest (ROIs) that represent suspected pedestrians. The second one (hypotheses verification (HV)) classifies the ROIs provided by HG into pedestrian or non-pedestrian classes on the basis of a new feature named gradient local binary patterns (GLBP)-COLOR. Unlike traditional feature learning proposed for pedestrian detection, which is based only on one piece of information, the creation of the proposed GLBP-COLOR is done using three kinds of information: gradient, texture, and color. The performance of the proposed feature is investigated using both AdaBoost and SVM classifiers.

The remainder of the chapter is organized as follows. The new pedestrian detection approach is detailed in Section 9.2. Experimental results and the overall performance of the proposed system are reported in Section 9.3. We conclude this chapter and present an outlook on further possible improvements in Section 9.4.

9.2 PROPOSED METHOD

This section details the proposed pedestrian detection method. It is achieved in two main stages (HG and HV). The first one is the generation of the pedestrian ROI hypothesis using motion information. The second one classifies the generated ROIs into a pedestrian and non-pedestrian, using SVM with radial basis function (RBF) kernel classifier trained on the proposed GLBP-COLOR feature representation. Here, we detail each of the two stages.

9.2.1 Hypotheses generation

The first stage of the proposed method consists of finding the ROIs (where pedestrians are likely to exist) from the images provided by the visible camera. Here, we refer to a system in which the motion information is used. Typically, the straightforward and fast approach to detecting moving objects is background subtraction. The rationale behind this approach is to detect any moving object from the difference between the current frame fc and a reference frame (background model "B_m"). In this chapter, the detection of moving objects is performed using the Gaussian Mixture Model (GMM) [18] for foreground detection. It is widely used in video surveillance systems due to its robustness for dynamic backgrounds. The main idea of GMM is to model each image pixel p_i as a mixture of Gaussians and to use an online

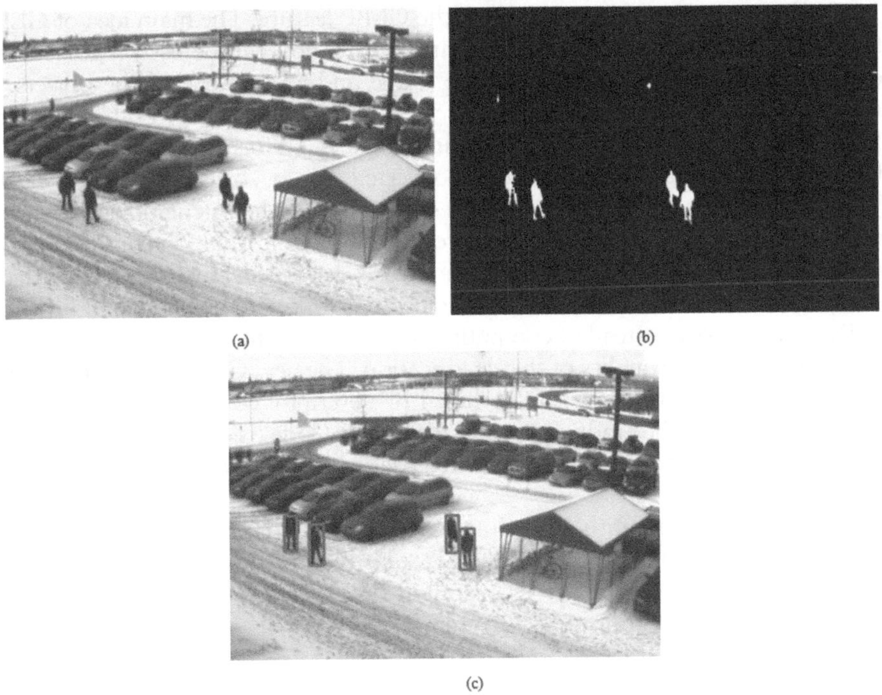

Figure 9.1 Example of hypotheses generation stage results. (a) Original image. (b) Foreground mask using GMM algorithm. (c) ROIs were obtained in the hypotheses generation step.

approximation for updating the model. In GMM, the pixels are classified either as background pixels (B_{ps}) or foreground pixels (F_{ps}) based on the persistence and the variance. The F_{ps} are grouped using 2D connected component analysis to create the foreground mask (binary image where the ROIs are presented with white spots). Once it is done, we project those spots on the visible image to generate the ROIs. Figure 9.1 depicts an example of the HG stage results obtained when the GMM technique is applied.

9.2.2 Hypotheses verification

Once the candidate ROIs are obtained from the previous stage, they are provided to the classification module in charge of identifying the pedestrians. Various experiments are performed using the INO Video Analytics dataset to choose the appropriate classifier as well as the features to be adopted by the proposed approach.

9.2.2.1 Feature extraction

Here, we present the new GLBP-COLOR, GLBP, HOG, LBP, and Gabor features involved in our experiments.

The first feature used in this work is the GLBP feature. The main idea of GLBP features is that both gradient and texture information can be used together into one descriptor to well characterize the shape and appearance of the objects. It was proposed by Jiang et al. [19] to overcome the limitations of both HOG and LBP descriptors. In [19], a preprocessing stage is performed before computing GLBP features. The input sample is converted into a corresponding grayscale. This conversion is applied with the aim of simplifying the calculations. However, there are many disadvantages to this procedure. The information captured in the red (R), green (G), and blue (B) color channels is discarded which adversely affects the quality of the descriptor. In this chapter, we propose to extend GLBP features to RGB color space instead of computing those features from grayscale images, in order to enhance the quality of the classification. The new GLBP descriptor named GLBP-COLOR has the same size as the GLBP descriptor computed from grayscale images. Except for the gradient (G) and Bit Binary Code (BBC) computation at the image pixels, the GLBP-COLOR feature is derived by following the same steps as the classical one.

For each pixel p with coordinates (i, j), we first compute three BBC (BBCR(p), BBCG(p), and BBCB(p)) derived directly from its R, G, and B components. Then, we use the three computed values to calculate the final BBC of the pixel p using the following formula:

$$\mathrm{BBC}(\mathrm{p}) = \mathrm{XOR}\big(\mathrm{BBCB}(\mathrm{p}), \mathrm{XOR}\big(\mathrm{BBCG}(\mathrm{p}), \mathrm{BBCR}(\mathrm{p})\big)\big) \tag{9.1}$$

where exclusive OR (XOR) is the exclusive disjunction logical operation.

BBC is used to generate the value of two parameters. The first parameter is the width (ω). It represents the number of occurrences of “1” in the binary code. The second parameter is angle (θ). It is the direction code of the middle pixel in the “1” area of its binary code. The two parameters used for mapping the position of the bin in the GLBP histogram. Once both ω and θ parameters are generated, we directly compute the gradient components (Gx, Gy) as follows:

$$G_x(i,j) = \max\left(G_x^R(i,j), G_x^G(i,j), G_x^B(i,j)\right) \tag{9.2}$$

and

$$G_y(i,j) = \max\left(G_y^R(i,j), G_y^G(i,j), G_y^B(i,j)\right) \tag{9.3}$$

where

$$G_x^{CC}(i,j) = CC(i+1,\ j) - CC(i-1,\ j),\ CC = B, G \text{ } or \text{ } R \tag{9.4}$$

and

$$G_y^{CC}(i,j) = CC(i,\ j+1) - CC(i,\ j-1),\ CC = B, G \text{ } or \text{ } R \tag{9.5}$$

CC is the color component which could be either R, G, or B. Both G_x and G_y are used to compute the magnitude of $p(i, j)$, which is used as a weight for voting in the histogram.

The second feature involved in this chapter is the HOG feature. It was introduced by Dalal and Triggs for pedestrian detection [14]. The basic concept of this descriptor is that both the shape and appearance of the local object can be depicted rather well by the distribution of the local intensity gradients and edge directions, even without accurate knowledge about the corresponding gradient or edge positions. The computation of the HOG feature of a given sample is simple. We first divide the sample into a set of cells. For each cell, the gradient of the edge is computed and the histogram of the gradient is accumulated in a 1D vector. Then, the obtained vector is normalized.

The third one is the LBP feature. It is the specific case of the texture spectrum model introduced by Ojala et al. [16]. The main idea of the LBP feature vector in its simplest form is almost identical to the one of the HOG. The examined sample is divided into a set of cells. For each pixel in these cells, we compare the center pixel value (I_c) and its eight neighbors and convert the result into binary numbers described by the following equation:

$$LBP\ (pc) = \sum_{i=2}^{T} s(Ii - Ic)2^i \tag{9.6}$$

$$s(Ii - Ic) = \begin{cases} 1 & Ii - Ic \geq 0 \\ 0 & Ii - Ic < 0 \end{cases} \tag{9.7}$$

where Ii is the grey value of the image and T is the total number of involved neighbors (in our case $T = 8$). Then, a histogram of the frequency of each number occurring over the cell is computed. Once all histograms are normalized, they are concatenated to create the final feature vector for the entire sample.

The last one is the Gabor feature. Gabor filters have been widely used in various signal processing and pattern recognition subjects, due to their capability to explore the local spectrum characteristics of images. A 2D Gabor filter is a band-pass spatial filter with selectivity to both orientation and spatial frequency [20], defined by the following equations:

$$Gc\ (i, j) = Be^{\frac{(i^2+j^2)}{2\sigma^2}} \cos\left(2\pi f\left(i\cos(\theta) + j\sin(\theta)\right)\right) \tag{9.8}$$

$$Gc\ (i, j) = Ce^{\frac{(i^2+j^2)}{2\sigma^2}} \sin\left(2\pi f\left(i\cos(\theta) + j\sin(\theta)\right)\right) \tag{9.9}$$

where f defines the frequency being looked for in the texture, and B and C are normalizing factors to be determined.

9.2.2.2 *Classifiers*

Once the features are computed, they are given to a classifier to perform the classification process. To select the classifier we will use, AdaBoost and SVM have been tested on various computed features.

9.2.2.2.1 SVM

SVMs are supervised learning models introduced first by Vapnik [21]. This classifier attempts to separate the negative examples (non-pedestrians) from the positive ones (pedestrians) by building a hyperplane (H). The constructed hyperplane should be described by its best generalization capability and also should guarantee that the margin between the closest positive samples and negative ones is maximal. In some situations, the data cannot be separated by a linear function. A solution is the use of a kernel function instead of a linear one. SVM is designed to solve both binary classification and multiclass problems through combinations of binary classification problems. There are two ways to achieve that: one-vs.-one or one-vs.-all.

9.2.2.2.2 AdaBoost

Adaptive Boosting (AdaBoost) is a machine learning meta-algorithm proposed by Yoav Freund and Robert Schapire [22]. The basic concept of AdaBoost is to construct a robust classifier using a summation of weak classifiers. It is well known for its simplicity, easy to construct, no-overfitting, and its generation ability that cannot be impacted by the number of iterations. It is widely used in various computer vision subjects including pedestrian detection.

9.3 EXPERIMENTAL RESULTS

This section presents the results obtained by the proposed approach. Various experiments have been performed on the INO Video Analytics dataset to assess the performances of the features as well as classifiers presented in Section 9.2.2. The hardware used in our tests is Sony Vaio Intel(R) i5 CPU 2.4 GHz running under Windows 10.

9.3.1 Dataset

As previously mentioned, the Intelligent Video Analytics dataset has been used to evaluate the performance of the proposed system. The INO dataset provides different sequences recorded in various locations and covering different weather conditions, using a Marlin F33C CCD sensor. The images size is 640 × 480 pixels and their format is 24-bit color jpg. The samples of the dataset are depicted in Figure 9.2.

(a) (b)

Figure 9.2 Samples of the INO Video Analytics dataset illustrating seasonal variations. (a) A park scene captured during snowfall in winter. (b) The same park observed in spring with clear weather conditions from a different angle.

9.3.2 Feature extraction and classifiers settings

To get optimal design parameters of each descriptor, we run some cross-validation experiments on the training set (60% of the INO dataset). By training the classifiers on the basic training set and evaluating on the validation set (40% of the INO dataset), both settings of maximum validation accuracy and lower dimensionality are chosen. Then, the classifiers are retrained one more time with selected feature extraction settings.

- To compute the GLBP-COLOR, we first normalize the detected ROIs to 64×128. Then, we divide it into 7×15 (105) blocks. Each block size is 16×16. We compute the histogram at each block using 56 bins. This results in a 5880 GLBP-COLOR vector. The same parameters are also adopted for the classical GLBP descriptor.
- For the HOG features, once the sample is normalized to 64×128, it is partitioned into 7×15 overlapping blocks. Each block is divided into 2×2 cells (8×8 pixels for each cell). At each cell, the gradient histogram is computed using 9 bins to form a final vector of 3780.
- For the LBP features, the input image is also normalized to 64×128 pixels and divided into 8×8 non-overlapping cells. By employing the uniform patterns approach, we extract 59 features per cell and finally construct the LBP feature vector.

In this work, we used an SVM with RBF kernel, $c = 4$ and $\gamma = 0.05$. The classification accuracy is impacted by the values of c and γ parameters and reaches its highest value when the parameters are equal to 4 and 0.05, respectively. Therefore, 4 and 0.05 are chosen to be used in the SVM classifier.

9.3.3 Quantitative evaluation

An example of detection results provided by the proposed approach is illustrated in Figure 9.3. First, the image is segmented based on the motion information to find out the ROIs where pedestrians are likely to exist. The detected ROIs are shown in Figure 9.3(c). However, a number of these ROIs are localized despite the fact that they do not represent pedestrians. In the aim of rejecting them, we refer to the classification stage. Figure 9.3(d) illustrates the pedestrian detection results when the proposed GLBP-COLOR and SVM are used as features and classifiers, respectively. Only two ROIs have been classified as pedestrian, while the others have been rejected. The green bounding boxes in Figure 9.3 indicate the detected pedestrian. The consuming time of HG and HV stages involved in the detection process is illustrated in Table 9.1.

In the following, we demonstrate the efficiency of the proposed method by analyzing and comparing the obtained results. Two evaluations have been considered in this chapter. The first one aims to evaluate the performance of the proposed GLBP-COLOR and compare its results to those provided by the HOG, LBP, and Gabor features. In the second evaluation, we compare

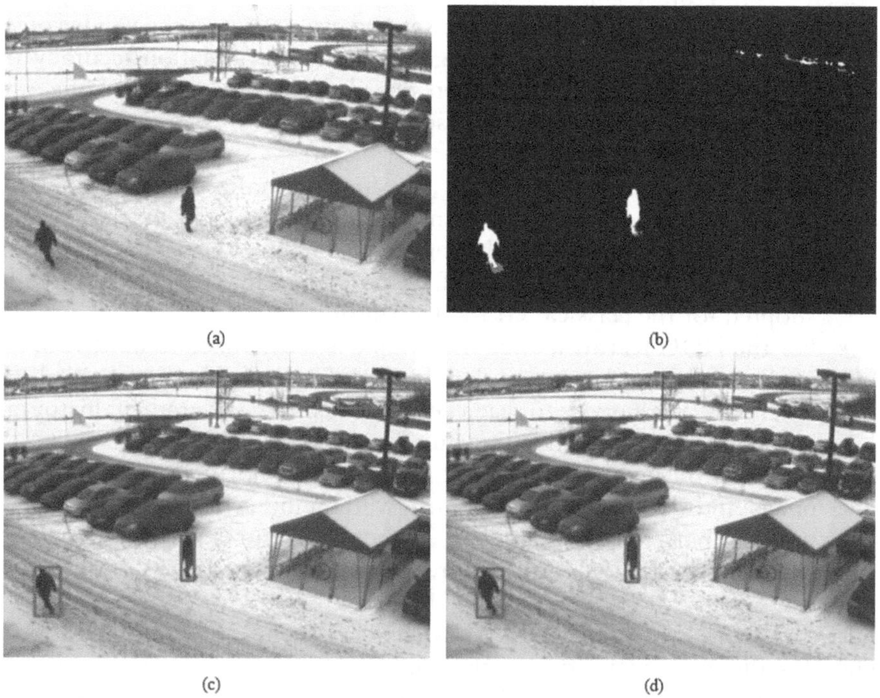

Figure 9.3 (a) Original image. (b) Its corresponding foreground mask uses the GMM technique. (c) Resulting in ROIs by mapping the obtained mask into the visible image. (d) Pedestrian detection results.

Table 9.1 Running time of each step of the proposed approach

	Hypotheses generation (HG)	*Hypotheses verification (HV)*
Run time (ms/frame)	19.07	42.34

the performances of the SVM-Kernel, SVMLinear, and AdaBoost classifiers. The sensitivity, specificity, and accuracy metrics, given below, are used for the different evaluations.

$$Specificity = \frac{Number\ of\ correct\ negative\ predictions}{Number\ of\ negatives} \tag{9.10}$$

$$Sensitivity = \frac{Number\ of\ correct\ positive\ predictions}{\text{Number of positives}} \tag{9.11}$$

$$\text{Accuracy (CCR)} = \frac{\text{Number of correct predictions}}{\text{Total samples}} \tag{9.12}$$

Table 9.2 shows the pedestrian detection results when GLBP, LBP, Gabor, HOG, and our proposed GLBP-COLOR are fed to the SVM-Kernel classifier. By reviewing the Table, it is obvious that the proposed GLBP-COLOR outperforms the other features, which confirms the importance of color information. It achieves 95.21%, 94.16%, and 94.72% in terms of specificity, sensitivity, and accuracy respectively in only 42.34 ms/f.

To justify the choice of SVM-Kernel classifier, both AdaBoost and SVM-Linear classifiers have also been tested with the same features used with the SVMKernel classifier as in Table 9.2 to decide which one will be used in this work. Table 9.3 illustrates this comparison in terms of accuracy and running time. As we can see from Table 9.3, we remark that the new GLBP-COLOR still succeeded in achieving the highest score in SVM-Linear and AdaBoost. It gives a CCR of 94.05% and 93.58%, respectively. However, their correct

Table 9.2 Performance of each feature descriptor on INO Video Analytics dataset

	Performance on INO Video Analytics dataset		
Feature	*Sensitivity (%)*	*Specificity (%)*	*Accuracy (%)*
Gabor	92.79	94.33	93.60
LBP	92.16	93.44	92.84
HOG	93.16	94.59	93.90
GLBP	93.63	94.88	94.30
GLBP-COLOR	**94.16**	**95.21**	**94.72**

Table 9.3 The accuracy and the average running time of the classifiers used in this work

	Accuracy (%) of all data set			*Running time (ms/frame)*		
Feature	*SVM-Kernel*	*SVM-Linear*	*AdaBoost*	*SVM-Kernel*	*SVM-Linear*	*AdaBoost*
Gabor	93.60	93.41	93.19	48.79	50.01	47.31
LBP	92.84	92.64	92.44	24.29	26.33	23.92
HOG	93.90	93.73	93.60	42.24	42.97	41.03
GLBP	94.30	94.15	94.00	42.39	43.00	41.18
GLBP-COLOR	**94.72**	**94.57**	**94.27**	**42.34**	**42.85**	**41.15**

classification rates (CCRs) are the lowest when compared to the ones we got by the SVM-Kernel. Therefore, we choose the SVM-Kernel algorithm as a classifier in the proposed classification method.

To assess the performances of the proposed GLBP-COLOR feature, we have also tested it on the INRIA dataset [14] using SVM-Kernel classifier. This dataset is widely used in pedestrian detection topics. It offers several normalized and centered on the person with their left-right reflections images. Figure 9.4 illustrates INRIA sample images.

Figure 9.4 Samples of the INRIA dataset. First row: Positive samples. Second row: Negative samples.

Table 9.4 Performance of each feature descriptor on INRIA dataset

	Performance on INRIA dataset		
Feature	*Sensitivity (%)*	*Specificity (%)*	*Accuracy (%)*
Gabor	92.27	96.69	93.54
LBP	91.21	95.58	92.46
HOG	93.52	97.57	94.68
GLBP	93.87	97.79	95.00
GLBP-COLOR	**95.47**	**99.12**	**96.52**

Table 9.4 shows the results of classification obtained on the INRIA dataset. We can conclude from the results depicted in the table that the best detection results are reached when the GLBP-COLOR feature is associated with the SVM-Kernel. It delivers 95.47% sensitivity, 99.12% specificity, and 96.52% accuracy.

The receiver operating characteristic (ROC) curves of the proposed GLBP-COLOR features model when applied to the INO Video Analytics and INRIA datasets are depicted in Figure 9.5(a) and (b), respectively.

Figures 9.6 and 9.7 illustrate some examples of the detection results when the proposed approach is applied to INO Video Analytics images. In Figure 9.6, all pedestrians contained in the two images are well detected. However, in Figure 9.7, several obstacles such as the presence of some objects can reduce the visibility of the pedestrian and, as a result, obstruct the detection system.

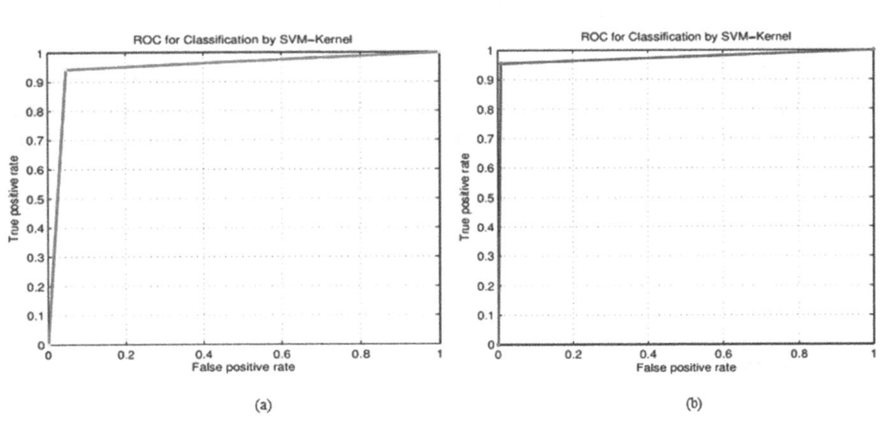

Figure 9.5 The ROC curves of the proposed approach when applied to (a) the INO Video Analytics dataset and (b) the INRIA datasets.

(a) (b)

Figure 9.6 Detection results.

(a) (b)

Figure 9.7 Examples of misdetections.

9.4 CONCLUSION AND PERSPECTIVES

In this chapter, a novel pedestrian detection approach for intelligent surveillance systems has been presented. The new approach is achieved in two main stages (HG and HV). First, extraction of ROIs based on motion information using the GMM technique. Second, the so-called GLBP-COLOR is used together with the SVM classifier to detect pedestrians from ROIs provided in the first stage. The new method has been tested on INO Video Analytics dataset data and the obtained results are promising.

REFERENCES

1. Redouan Lahmyed and Mohamed El Ansari. Multisensors-based pedestrian detection system. In *Computer Systems and Applications (AICCSA), 2016 IEEE/ACS 13th International Conference of*, pages 1–4. IEEE, 2016.

2. Mohamed El Ansari, Redouan Lahmyed, and Alain Tremeau. A hybrid pedestrian detection system based on visible images and lidar data. In *Proceedings of the 13th International Joint Conference on Computer Vision, Imaging and Computer Graphics Theory and Applications – Volume 5: VISAPP*, pages 325–334. INSTICC, SciTePress, 2018.
3. Tarek Elguebaly and Nizar Bouguila. Finite asymmetric generalized Gaussian mixture models learning for infrared object detection. *Computer Vision and Image Understanding*, 117(12):1659–1671, 2013.
4. Ichraf Lahouli, Rob Haelterman, Zied Chtourou, Geert De Cubber, and Rabah Attia. Pedestrian detection and tracking in thermal images from aerial mpeg videos. In *VISIGRAPP*, 2018.
5. Igi Ardiyanto, Teguh Bharata Adji, and Dika Akilla Asmaraman. On comprehensive analysis of learning algorithms on pedestrian detection using shape features. *Journal of Intelligent & Fuzzy Systems*, 35(4):4807–4820, 2018.
6. Ann Rija Paul and E Grace Mary Kanaga. Enhanced D-CNN architecture and centroid-based algorithm for real-time vehicle tracking and accident detection from surveillance videos. *Journal of Intelligent & Fuzzy Systems*, 46(2):1–14, 2024.
7. Daoxun Xia, Fang Guo, Haojie Liu, and Sheng Yu. Unsupervised learning of visual invariant features for person reidentification. *Journal of Intelligent & Fuzzy Systems*, 39(5):7495–7503, 2020.
8. Markus Enzweiler and Dariu M Gavrila. Monocular pedestrian detection: Survey and experiments. *IEEE Transactions on Pattern Analysis & Machine Intelligence* 31(12):2179–2195, 2008.
9. Oswaldo Ludwig Junior, David Delgado, Valter Gonçalves, and Urbano Nunes. Trainable classifier-fusion schemes: an application to pedestrian detection. In *Intelligent Transportation Systems, 2009. ITSC'09. 12th International IEEE Conference on*, pages 1–6. IEEE, 2009.
10. Jongseok Lim and WookHyun Kim. Detecting and tracking of multiple pedestrians using motion, color information and the adaboost algorithm. *Multimedia Tools and Applications*, 65(1):161–179, 2013.
11. Stefan Munder and Dariu M Gavrila. An experimental study on pedestrian classification. *IEEE Transactions on Pattern Analysis and Machine Intelligence*, 28(11):1863–1868, 2006.
12. Yadong Mu, Shuicheng Yan, Yi Liu, Thomas Huang, and Bingfeng Zhou. Discriminative local binary patterns for human detection in personal album. In *Computer Vision and Pattern Recognition, 2008. CVPR 2008. IEEE Conference on*, pages 1–8. IEEE, 2008.
13. Oncel Tuzel, Fatih Porikli, and Peter Meer. Pedestrian detection via classification on riemannian manifolds. *IEEE Transactions on Pattern Analysis & Machine Intelligence*, 30(10):1713–1727, 2008.
14. Gardezi, Navneet Dalal and Bill Triggs. Histograms of oriented gradients for human detection. In *Computer Vision and Pattern Recognition, 2005. CVPR 2005. IEEE Computer Society Conference on*, volume 1, pages 886–893. IEEE, 2005.
15. Qiang Zhu, Mei-Chen Yeh, Kwang-Ting Cheng, and Shai Avidan. Fast human detection using a cascade of histograms of oriented gradients. In *Computer Vision and Pattern Recognition, 2006 IEEE Computer Society Conference on*, volume 2, pages 1491–1498. IEEE, 2006.
16. Timo Ojala, Matti Pietikäinen, and David Harwood. A comparative study of texture measures with classification based on featured distributions. *Pattern Recognition*, 29(1):51–59, 1996.
17. Shaopeng Tang and Satoshi Goto. Histogram of template for human detection. In *Acoustics Speech and Signal Processing (ICASSP), 2010 IEEE International Conference on*, pages 2186–2189. IEEE, 2010.

18. Zoran Zivkovic. Improved adaptive Gaussian mixture model for background subtraction. In *Pattern Recognition, 2004. ICPR 2004. Proceedings of the 17th International Conference on*, volume 2, pages 28–31. IEEE, 2004.
19. Ning Jiang, Jiu Xu, Wenxin Yu, and Satoshi Goto. Gradient local binary patterns for human detection. In *Circuits and Systems (ISCAS), 2013 IEEE International Symposium on*, pages 978–981. IEEE, 2013.
20. John G Daugman. Uncertainty relation for resolution in space, spatial frequency, and orientation optimized by two-dimensional visual cortical filters. *Journal of the Optical Society of America A*, 2(7):1160–1169, 1985.
21. Vladimir Naumovich Vapnik and Vlamimir Vapnik. *Statistical learning theory*, volume 1. Wiley, New York, 1998.
22. Yoav Freund, Robert Schapire and Naoki Abe. A short introduction to boosting. *Journal-Japanese Society for Artificial Intelligence*, 14(771–780):1612, 1999.

Part II

Smart applications

Chapter 10

Compensation of harmonic currents for shunt active power filter using ADALINE neural network

Tali Mouna, Essadki Ahmed, and Nasser Tamou

10.1 INTRODUCTION

With the development of electronic components and their increasing use in industrial networks, the disturbances generated such as harmonic currents have become serious power quality problems that have great concern and attract researchers' interests. Harmonic currents must be identified and filtered so that they cannot propagate in the electrical network, nor cause harmful effects such as malfunction, poor performance, or destruction, so, several methods for identifying harmonics have already been proposed in the literature [1]. Basically, there are two kinds of compensators: passive power filters (PPFs) and active power filters (APFs), the first PPFs have been considered a simple solution and a good alternative for harmonics compensation, however, they have several disadvantages: they cannot automatically adapt to changes in loads and can also cause a resonance problem on the network [2]. That is why PPFs are gradually replaced by (APFs). APFs have the capability to overcome the above-mentioned disadvantages inherent in PPFs and nowadays, they represent an efficient and robust solution to improve power quality. The objective is to recover balanced and sinusoidal source current by injecting the compensating current at the PCC. So, the harmonic currents can be suppressed and the power factor can be compensated [3]. Generally, currents and powers are fundamental concepts that make it possible to characterize electrical systems, and in the normal case, the source current and source voltage must be pure sinusoidal waveform but the use of the nonlinear loads introduces harmonic distortion in the source current or source voltage. Hence, it is very important to overcome these undesirable features. The purpose of an APF is to generate harmonic currents having the same magnitude but in the opposite phase with harmonic current produced by the nonlinear loads. The main current obtained after compensation must be sinusoidal and in phase with the supply voltages. Several methods have been proposed and implemented for extracting the harmonic currents, these methods are classified as frequency domain techniques, Fast Fourrier Transformer (FFT), or time domain techniques. The use of a proper and adequate algorithm for current reference generation is the key to the successful implementation of APF

DOI: 10.1201/9781032664293-12

compensation. Methods suggested such as the instantaneous power theory developed and defined by Akagi [4–6] in 1984 are used to detect reactive power and to compensate harmonic current in three-phase and three-wire systems. It gives good performance for balanced sinusoidal supply conditions. However, when the voltage source is significantly unbalanced; large errors will be derived from the direct application of PQ theory. And later, it was modified and extended by Nabae to be applicable to the four wires and in unbalanced systems (called modified PQ theory or PQR theory). Other researchers use synchronous reference frequency noted dq theory. This method is ideally designed for three phases but the use of PLL is necessary, it was also modified and named modified SRFT or id-iq method without using PLL [7, 8]. For a few years, other control techniques based on Fuzzy logic, neural networks, and genetic algorithms have been used for improving the performance of the active filter. On the other hand, artificial neural networks (ANNs) have been systematically applied to electrical engineering. Nowadays, this technique is considered a new tool to design APF control circuits. In the ANN algorithms, the input-output relationships are indicated through a learning process or through an adaptive algorithm. Moreover, the parallel computing architecture increases the system speed and reliability [9–14]. In this chapter, a new design of an APF control method based on neural networks is presented. A first block with an adaptive linear neuron ADALINE has been used to estimate the reference compensation current [15–17]. These neurons, commonly used for signal processing, estimate the source current components and reference compensation currents obtained [18–22].

10.2 APF CONTROL STRATEGY BASED ON CURRENT ESTIMATOR

The principle of the APF is to use the inverter that injects reference currents at the PCC. APF approach is based on the identification of the distortion harmonics from the measures of the source current, thus the reference current is obtained and must be injected phase-opposite to cancel the unwanted harmonics. Using the ADALINE neural network, the fundamental current can be estimated to calculate the reference current.

The proposed SAPF configuration based on the ADALINE current estimator in this work is shown in Figure 10.1. The considered load to evaluate the effectiveness of the SAPF is a three-phase rectifier with an R-L load.

The control block with neural networks is developed with adaptive networks (ADALINE neurons), which allow to make an online estimation of the reference compensation current or the control. The second one is a feedforward network. After a training process, it works online as a comparator between that reference waveform and the actual compensation current.

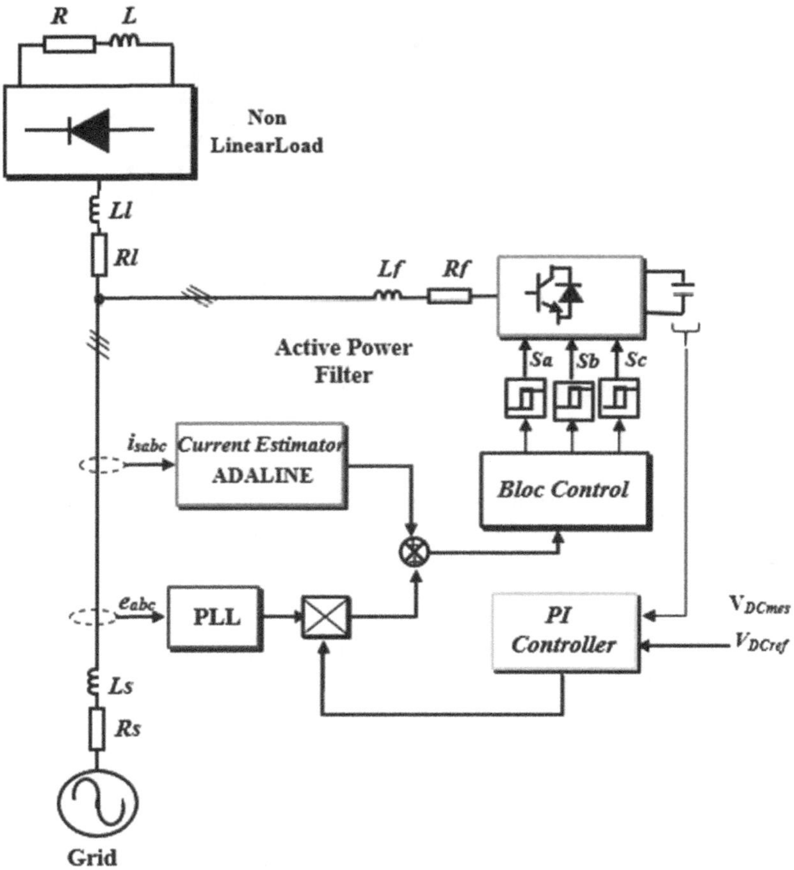

Figure 10.1 Structure of SAPF based on the current estimator.

The second block outputs are the trigger signals of IGBTs used in the Active Filter power circuit. In the control through the hysteresis band, the actual signal will stay inside a band around the reference signal.

10.3 ADALINE'S PRINCIPLE

The ADALINE algorithm is used to extract the fundamental component using the Least Mean Square (LMS) algorithm as its learning rule to update the weights W. The adaptive linear algorithm ADALINE introduced by Widrow Hoff based on the LMS learning rule now represents an efficient approach for fast prediction and estimation of signal parameters, due to its simple structure and the capability of self-adapting online. The adaptive networks can adapt to any change experimented by the load current waveform. The structure of the ADALINE is shown in Figure 10.2, where Xk is an input vector of dimension n. Wk is an adjustable weight vector of dimension n and Yk is the output which can be calculated as follows:

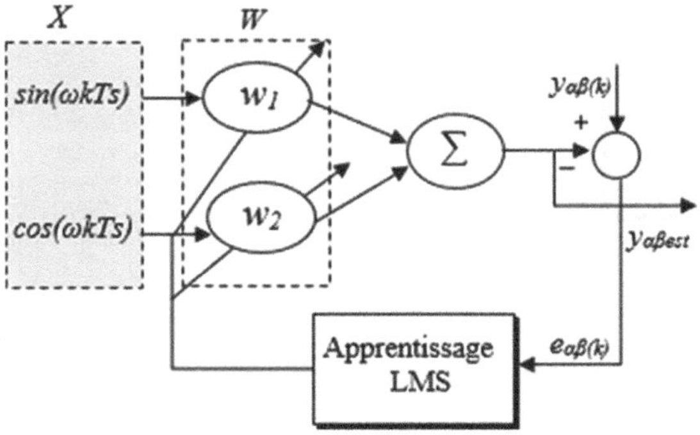

Figure 10.2 Proposed grid current estimator using ADALINE.

The designing methodology supposes that the voltage/current can be represented by the sum of a fundamental frequency and multiple of fundamental frequency components which can be written as:

$$y_\alpha(k) = n\sum_{n=1}^{N} Y_n \cos(w_n kT_S + \varphi_n)n = Y_1\cos(w_1 kT_S + \varphi_1)n + n\sum_{n=2}^{\infty} Y_n \cos(w_n kT_S + \varphi_n) \tag{10.1}$$

$$y_\beta(k) = n\sum_{v}^{N} Y_n \sin(w_n kT_S + \varphi_n)n = Y_1\sin(w_1 kT_S + \varphi_1) + n\sum_{n=2}^{\infty} Y_n \sin(w_n kT_S + \varphi_n) \tag{10.2}$$

where $e_{\alpha\beta} = y_{\alpha\beta}$, w_n, Yn, and φ_n are pulsation, amplitude, and phase angle.

The process of the line current estimator is similar to that of the adaptive filter, which extracts fundamental components and neglects all harmonic components greater than 1.

The estimated grid current using the ADALINE process can be expressed as:

$$y_\alpha(k) = nY_1\cos(w_1 kT_S + \varphi_1) \tag{10.3}$$

$$y_\beta(k) = nY_1\sin(w_1 kT_S + \varphi_1) \tag{10.4}$$

$$y_\alpha(k) = nY_1\cos(\varphi_1)\cos(w_1 kT_S) - nY_1\sin(\varphi_1)\sin(w_1 kT_S) \tag{10.5}$$

$$y_\beta(k) = nY_1\cos(\varphi_1)\sin(w_1 kT_S)n + Y_1\sin(\varphi_1)\cos(w_1 kT_S) \tag{10.6}$$

So, in vectorial notation, the estimated grid voltage using the ADALINE process can be written as:

$$y_{\alpha\beta est}(k) = W^T X(k) \tag{10.7}$$

where:

$$X(k) = \left[\cos(w_1 k T_S) \text{n} \sin(w_1 k T_S)\right]^T \tag{10.8}$$

$$W\alpha = [w_{1\alpha}\ w_{2\alpha}] = [Y_1 cos\varphi_1\ n - Y_1 sin\varphi_1] \tag{10.9}$$

$$W_\beta = [w_{1\beta}\ w_{2\beta}] = [Y_1 sin\varphi_1\ n Y_1 cos\varphi_1] \tag{10.10}$$

where $X(k)$ is the reference input vector of the ADALINE and the W represents the weight vector whose are updated by using the adaptive learning process.

The LMS algorithm is used as a learning rule in order to minimize the error $\xi(k)$, the learning rule can be expressed as:

$$W(k+1) = W(k)n + \mu\xi(k).\ X(k) \tag{10.11}$$

$$\xi(k) = Y(k) - Yest(k) \tag{10.12}$$

Thus, at each iteration, with the supervised learning process of ADALINE LMS, neural filter parameters (adaptive neural network diagram) are updated to make the above error can asymptotically converge to zero. And the estimated output $Yest(k)$ behavior becomes close to the set of desired output $Y(k)$.

An algorithm with a high rate of convergence, stability, and good tracking ability is required. The optimization method widely used for many identification applications is the gradient descent technique, but it has the drawback of having a very slow convergence rate, however, the LMS algorithm developed by Widrow-Hoff became the most used adaptive algorithm due to its simplicity of calculation and its proven robustness.

In order to enhance the precision of the proposed ADALINE voltage estimator, the weights updating are adjusted using the learning rate μ $\left(0 < \mu < 2\ [20]\right)$ ($\mu = 0,\ 2$ is chosen in order to have a faster algorithm and ensure good convergence). The stability of the proposed algorithm is proved using Lyapunov's theory [16].

A candidate Lyapunov function satisfying the learning rule (10.11) is proposed as:

$$V(k) = \xi^2(k) + \xi^2(k-1) \tag{10.13}$$

$$V(k) = \beta^k \xi^2(k) \tag{10.14}$$

$$\beta^k = \left(1 + \frac{e^2(k-1)}{\text{e}^2(\text{k})}\right) \tag{10.15}$$

The Lyapunov stability conditions are:

$$\begin{cases} V(k) > 0 \\ V(k) - V(k-1) = \Delta V(k) < 0 \end{cases} \tag{10.16}$$

$\Delta V(k)$ can be evaluated as:

$$\Delta V(k) = \beta_k \left[d(k - W^T(k)x(k) \right]^2 - \beta_k \xi^2 (k-1)$$

$$\Delta V(k) = \beta_k \left[d(k) - W^T(k-1)x(k) - \mu e^T(k) x^T(k) x(k) \right]^2 - \beta_k e^2 (k-1)$$

$$\Delta V(k) = \beta_k \left[d(k) - W^T(k-1)x(k) - \Delta W^T(k)x(k) \right]^2 - \beta_k e^2 (k-1)$$

$$y(k) = W_{op}^T (k)x(k)$$

$$\Delta V(k) = \beta_k \left[W_{op}^T (k)x(k) - W^T(k-1)x(k) - \Delta W^T(k)x(k) \right]^2 - \beta_k e^2 (k-1)$$

$$\Delta V(k) = \beta_k \left[\Delta W_{op}^T (k)x(k) - \Delta W^T (k)x(k) \right]^2 - \beta_k e^2 (k-1)$$

If $k \to \infty$ $W^T(k) \simeq W_{op}^T (k)$

then $W_{op}^T(k) - W^T(k-1) \simeq W^T(k) - W^T(k-1)$

$$\lim_{k\to\infty} \left[\Delta W_{op}^T(k)x(k) - \Delta W^T(k)x(k) \right]^2 = 0 \quad \lim_{k\to\infty} \beta^k = 2$$

$$\lim_{k\to\infty} \left(1 + \frac{e^2(k-1)}{e^2(\mathrm{k})} \right) = 2$$

$$\Delta V(k) = -\beta_k e^2 (k-1) < 0 \tag{10.17}$$

It is clear that V(k) is positive definite; therefore, the stability condition is satisfied since $V(k) < 0$ and the convergence is guaranteed.

The objective of SAPF is to compensate for harmonic currents and reactive power by injecting the compensated current into the system at the PCC as shown in Figure 10.3.

The non-linear load current iL is the sum of the source current is and the compensation current if. The objective is to get a source current without harmonic and reactive components. The suitable compensation current injected by the shunt APF corresponds to the non-active component of the load current.

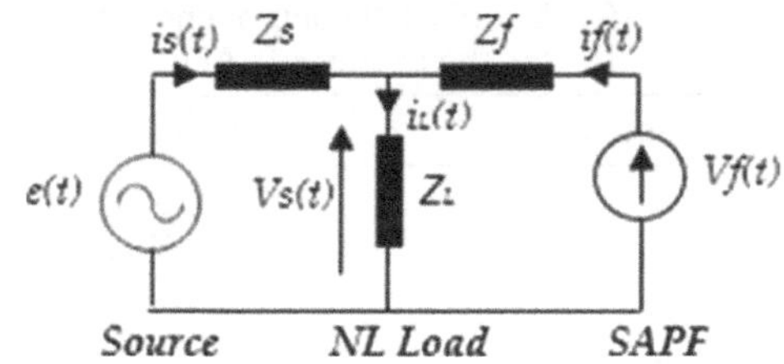

Figure 10.3 Equivalent circuit topology.

The source current in an active current at fundamental frequency is estimated and calculated by the ADALINE algorithm, so the required reference current will be obtained from the equation:

$$if = il - is \tag{10.18}$$

The amplitude of the active current in each phase is the combination of the amplitude of the fundamental active current obtained by the ADALINE neural network and the amplitude of the DC bus active current Idc. This active current occurs from the process of controlling the DC voltage equal to the reference voltage.

The DC capacitor voltage must be maintained at the reference value to compensate for the losses in the active filter. The PI controller is practically used because it carries out a good compromise between performance and cost of realization.

10.4 SIMULATION RESULTS

The performance of the proposed neural network current estimator in SAPF controller employing the ADALINE neural network method has been evaluated through simulation using MATLAB/SIMULINK. The challenge of the whole system is to achieve sinusoidal and balanced grid current under balanced and unbalanced supply conditions.

In order to ensure a good current injection into PCC, the design of the voltage source inverter is crucial, such as the selection of the reference DC bus voltage, the inductance, and the capacitor design process.

Table 10.1 summarizes the controller parameters thus obtained.

This section presents the performance of the SAPF system using the proposed method of the under-balanced and sinusoidal grid voltages, the simulation results show in Figure 10.4 that without SAPF compensation, the source currents are distorted severely with the THD around to 25.37% Figure 10.5 and at $t = 0.1s$, SAPF starts to compensate harmonics, and the estimated

Table 10.1 List of each component used for the proposed SAPF

Parameters	*Values*
Vs, fs	100 v, 50 Hz
Ls, Rs	10 μH, 0.1 Ω
LL, RL	10 mH, 30 Ω
Lf, Rf	2 mH, 0.01 Ω
Cdc, Vdc	2200 μF, 300 v
Kp, Ki	2, 0.015

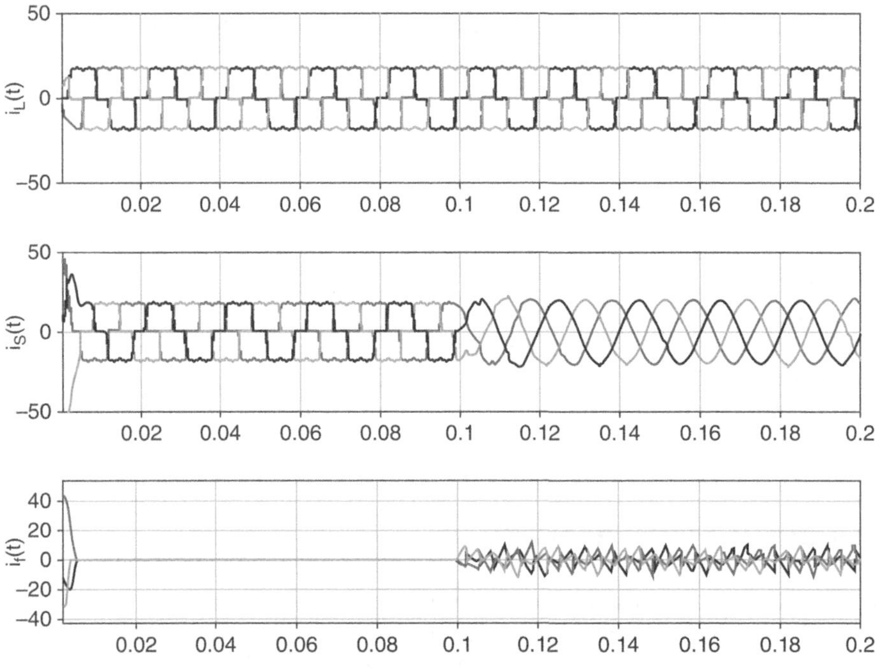

Figure 10.4 Voltage and current source waveforms before and after compensation.

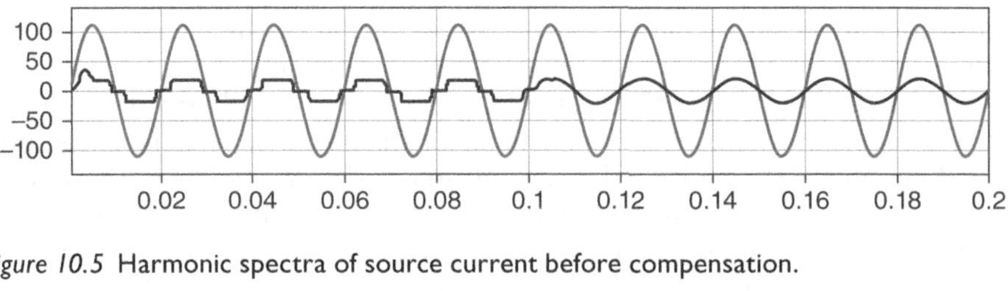

Figure 10.5 Harmonic spectra of source current before compensation.

source current using ADALINE Neural Network for each phase is sinusoidal and in phase with the grid voltage and the Total Harmonic Distortion THD current reduces to 2.86%, Figure 10.6 the compensating current injected by the SAPF at the PCC is depicted in Figure 10.7. Reduction in the THD means that the proposed controller has a better steady-state response.

The PI controller is used to control the DC bus voltage of Voltage Source Inverter VSI. Figure 10.8 shows that the DC bus voltage is maintained at the desired value.

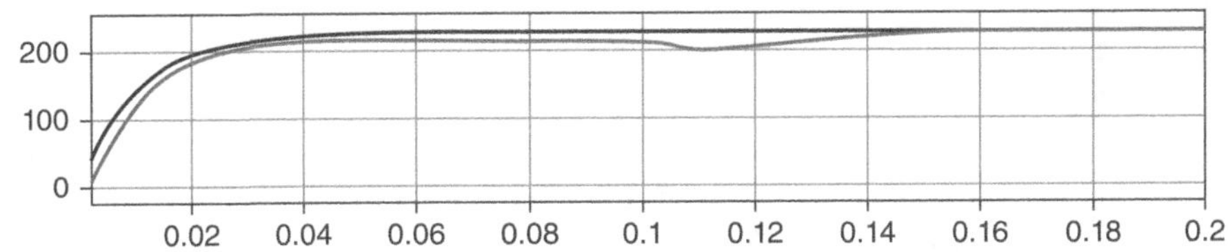

Figure 10.6 Harmonic spectrum of source current after compensation.

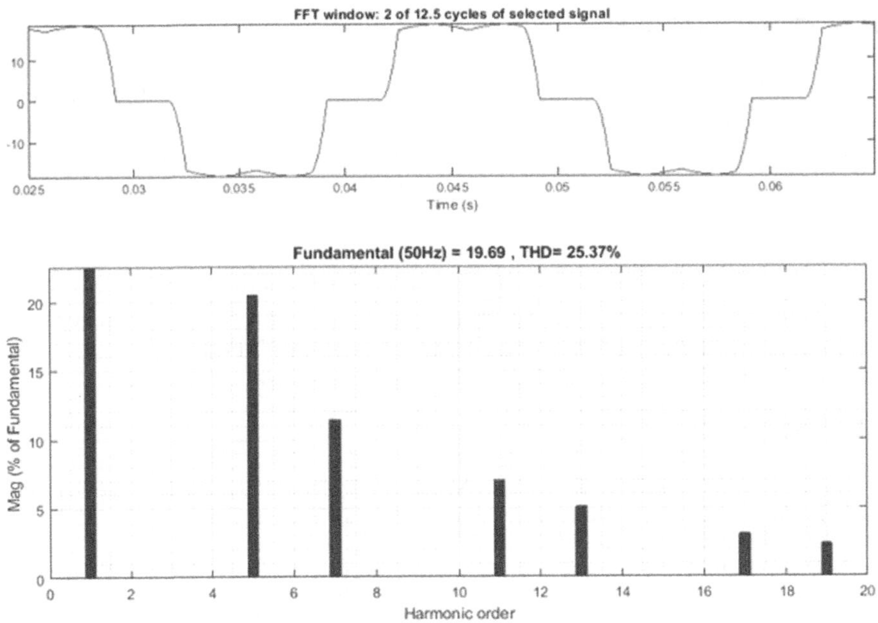

Figure 10.7 Harmonic spectra before compensation.

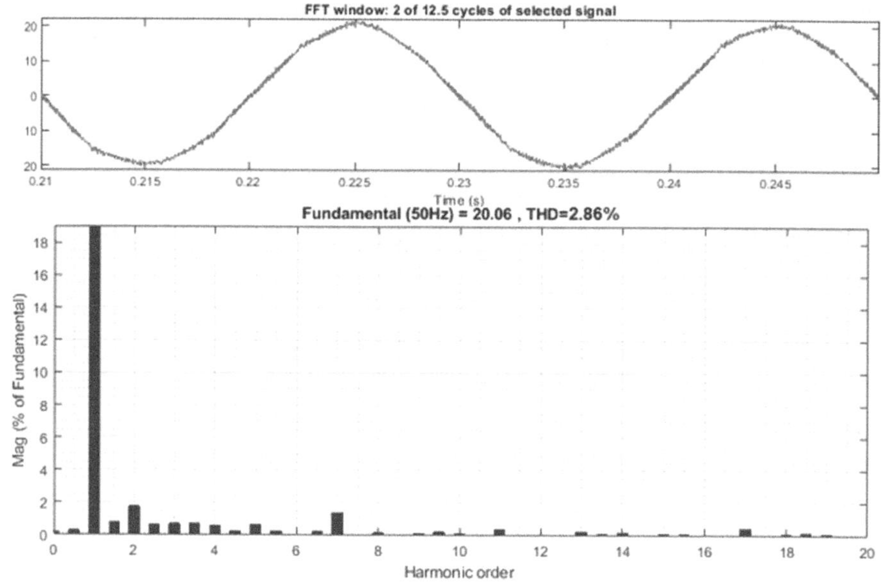

Figure 10.8 Harmonic spectra in phase (a) after compensation.

10.5 CONCLUSION

The overall aim of this chapter was to consider methods of achieving better utilization and control of APFs dealing with harmonic and reactive power compensation. An alternative and efficient method for the control three-phase shunt active power filter employing an adaptive neural network was studied. The simpler ADALINE algorithm shows fast and accurate extraction. The Lyapunov stability theorem has been exploited to ensure the convergence and stability of the proposed control performance. The simulation results confirm that the proposed ADALINE estimator algorithm represents high accuracy of harmonics mitigation with a significant reduction of THD.

REFERENCES

1. Saribulut, L., Teke, A., Meral, M., Tumay, M. (2011). Active power filter: Review of converter topologies and control strategies. Gazi University Journal of Science, 24(2): 283–289.
2. Tali, M., nObbadi, A., Elfajri, A., Errami, Y. (2014). Passive filter for harmonics mitigation in standalone PV system for nonlinear load. In 2014 International Renewable and Sustainable Energy Conference (IRSEC), Ouarzazate, Morocco, pp. 499–504. https://doi.org/10.1109/IRSEC.2014.7059834
3. Nie, X., Liu, J. (2019). Current reference control for shunt active power filters under unbalanced and distorted supply voltage conditions. IEEE Access, 7: 177048–177055. https://doi.org/10.1109/ACCESS.2019.2957946

4. Akagi, H., Kanazawa, Y., Nabae, A. (1983). Generalized theory of the instantaneous reactive power in three-phase circuits, IPEC. In Int. Power Electronics Conf., Tokyo, Japan, pp. 1375–1386.
5. Akagi, H., Kanazawa, Y., Nabae, A. (May/June 1984). Instanataneous reactive power compensator comprising switching devices without energy storage compenents. IEEE Transactions on Industry Applications, 20.
6. Tanaka, T., Akagi, H. (1995) "A new method of harmonic power detection based on the instantaneous active power in three-phase circuits." IEEE Transactions on Power Delivery, 10(4), https://doi.org/10.1109/61.473386
7. Zeng, Z., Yang, H., Guerrero, J.M., Zhao, R. (2015). Multi-functional distributed generation unit for power quality enhancement. IET Power Electronics, 8(3): 467–476. https://doi.org/10.1049/iet-pel.2013.0954
8. Chang, G.W., Chen, S.K. An a-b-c Reference Frame-Based Control Strategy for the Three-Phase Four-Wire Shunt Active Power Filter, 1, pp. 26–29, 2000.
9. Osowski, S. (March 1992). Neural network for estimation of harmonic components in a power system. IEE PROCEEDINGS-C, 139(2).
10. Salam, A.A., Ab Hadi, N.A. (2014). Fuzzy logic controller for shunt active power filter. In 2014 4th International Conference on Engineering Technology and Technopreneuship (ICE2T), Kuala Lumpur, Malaysia, pp. 256–259. https://doi.org/10.1109/ICE2T.2014.7006258
11. Abdeslam, D.O., Wira, P., Mercklé, J., Flieller, D., Chapuis, Y.A. (2007). A unified artificial neural network architecture for active power filters. IEEE Transactions 409 on Industrial Electronics, 54(1): 61–76. https://doi.org/10.1109/TIE.2006.888758
12. Bai, H., Wang, X., Blaabjerg, F. (2017). A grid-voltage sensorless resistive-active power filter with series LC filter. IEEE Transactions on Power Electronics, 33(5): 4429–4440. https://doi.org/10.1109/TPEL.2017.2717183
13. Komatsu, Y., Kawabata, T. (1997). A control method of active power filter in unsymmetrical and distorted voltage system. In Proceedings of Power Conversion Conference-PCC'97, Nagaoka, Japan, pp. 161–168. https://doi.org/10.1109/PCCON.1997.645605
14. Cichocki, A., Lobos, T. (May 1994). Artificial neural networks for real-time estimation of basic waveforms of voltages and currents. IEEE Transactions on Power Systems, 9(2).
15. Monfared, M., Sanatkar, M., Golestan, S. (2012). Direct active and reactive power control of single-phase grid-tie converters. IET Power Electronics, 5(8): 1544–1550. https://doi.org/10.1049/iet-pel.2012.0131
16. Abouelmahjoub, Y., Giri, F., Abouloifa, A., Chaoui, F.Z., Kissaoui, M. (2018). Adaptive nonlinear control of reduced-part three-phase shunt active power filters. Asian Journal of Control, 20(5): 1720–1733. https://doi.org/10.1002/asjc.1681
17. Hou, S., Fei, J., Chen, C., Chu, Y. (2019). Finite-time adaptive fuzzy-neural-network control of active power filter. IEEE Transactions on Power Electronics, 34(10): 10298–10313. https://doi.org/10.1109/TPEL.2019.2893618
18. Mohd Zainuri, M.A.A., Mohd Radzi, M.A., Che Soh, A., Mariun, N., Abd Rahim, N., Hajighorbani, S. (2016). Fundamental active current adaptive linear neural networks for photovoltaic shunt active power filters. Energies, 9(6): 397. https://doi.org/10.3390/en9060397
19. Rahoui, A., Bechouche, A., Seddiki, H., Abdeslam, D.O. (2017). Grid voltages estimation for three-phase PWM rectifiers control without AC voltage sensors. IEEE Transactions on Power Electronics, 33(1): 859–875.
20. Chang, G.W., Shee, T.-C. (2004). A novel reference compensation current strategy for shunt active power filter control. IEEE Transactions on Power Delivery, 19: 1751–1758.

21. Montero, M.I.M., Cadaval, E.R., González, F.B. (2007). Comparison of control strategies for shunt active power filters in three-phase four-wire systems. IEEE Transactions on Power Electronics, 22: 229–236.
22. Abdelkhalek, O., Benachaiba, C. (2009). Sensitivity assessment of PQ theory and synchronous detection identification methods of current harmonics under non-sinusoidal condition for shunt active power filter. Journal of Electrical & Electronics Engineering, 9: 801–807.

Chapter 11

Control of a grid-connected photovoltaic system based on MPPT and vector control

N. Ech-Cherki, Y. Errami, A. Obbadi, S. Sahnoun, and I. Nassar-Eddine

11.1 INTRODUCTION

At present, a significant proportion of the world's energy comes from fossil sources, exacerbating climate change through toxic gas emissions. Furthermore, there is an impending risk of overexploiting natural resources, potentially endangering future energy provisions. Renewable energy sources offer a promising alternative, as they are virtually inexhaustible, allowing for increasingly accessible exploitation [1]. Although certain renewable technologies may still be expensive, there is a need for continued research and development to reduce installation expenses and enhance energy efficiency in the pursuit of harnessing maximum power from these sources [2]. On the other hand, several simple and hybrid maximum power point tracking (MPPT) techniques are being used to improve the energy production potential of photovoltaic (PV) systems. In [3], researchers introduce a novel approach combining the fractional open circuit voltage FOCV and perturbation and observation (P&O) method, aimed at optimizing the duty cycle. Ref. [4] presents an alternative by proposing a current-based sliding mode MPPT algorithm that fine-tunes the performance of P&O. The study described in [5] undertakes an evaluation that scrutinizes and compares the effectiveness of six MPPT techniques rooted in artificial intelligence. In [6], a pioneering modification to the MPPT algorithm is put forth, leveraging P&O in conjunction with an adaptive duty cycle adjustment method facilitated by a Proportional Integral Derivative (PID) controller using a genetic algorithm to attain the Maximum Power Point (MPP). Conversely, current vector control (CVC) offers notable advantages, including rapid dynamic response and robustness. Consequently, CVC has found extensive application in numerous research endeavors. The authors of [7] employ CVC to transmit power from the PV system to the grid, while [8] explores the use of CVC in a two-stage cascaded inverter, the main goals being to maximize power output and to counterbalance reactive power fluctuations.

This research assesses the operational effectiveness of a grid-connected PV system. This integrated PV setup comprises a photovoltaic generator (PVG) linked to the grid through two power converters. While the PVG is

DOI: 10.1201/9781032664293-13

not optimally cost-effective [3], the use of P&O and CVC allows the PVG to operate at its MMP. The CVC method introduced stands out as a popular choice, particularly for its remarkable dynamic response capabilities.

Our document is organized as follows: The introduction is given in this section; Section 11.2 presents the modeling of the system. The control of the system using the CVC method is detailed in Section 11.3. The outcomes of the simulations are delineated in Section 11.4, while the document concludes in Section 11.5.

11.2 MODELING OF THE SYSTEM

Figure 11.1 illustrates that the system connected to the grid consists of three elements:

- PVG.
- DC-DC converter.
- Three-phase DC-AC inverter and the grid.

This work is based on two controllers: P&O for the MPPT and CVC to guarantee the unity power factor by two internal current loops [6].

11.2.1 Modeling of the PVG

We used a 100 kW PVG based on the 'SunPower SPR-305-WHT' module. Figure 11.2 presents the circuit of a cell with a single diode to analyze the characteristics of the module used.

The current and voltage delivered by the cell are given by [5] the following:

$$I_{PV} = I_{ph} - I_0 \left[\exp\left(\frac{q}{\alpha \, . \, k \, . \, T} \left(V_{pv} + I_{pv} \, . \, R_s \right) \right) - 1 \right] - \left(\frac{V_{pv} + I_{pv} \, . \, R_s}{R_{sh}} \right) \quad (11.1)$$

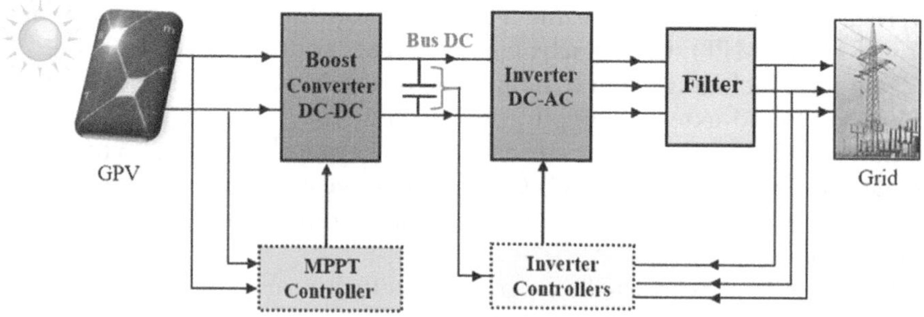

Figure 11.1 PV system topology connected to the grid.

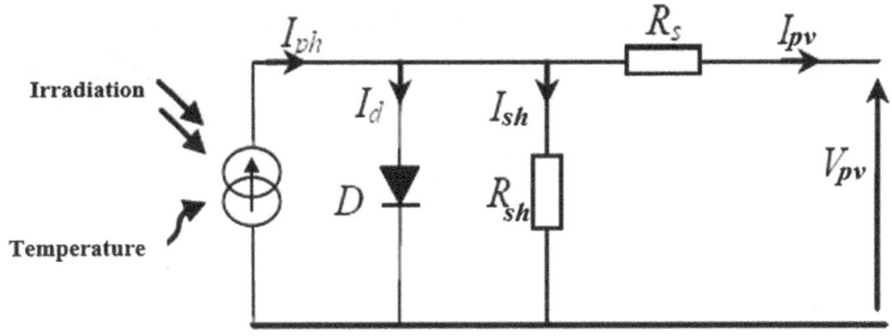

Figure 11.2 Circuit of a single-diode PV cell.

where V_{pv} and I_{pv} represent the voltage and current produced by the PV cell, respectively, I_{ph} is the photocurrent, I_0 denotes the saturation current, q signifies the elementary charge, k represents Boltzmann's constant, α represents the ideality factor of the cell, R_s and R_{sh} are the series and parallel resistances.

In order to describe a model for a PVG, we take into consideration all losses stemming from the arrangement of modules, both in series and in parallel. This consideration leads us to Equation (11.2) [6]:

$$I_{PV} = N_p . I_{ph} - N_p . I_0 \left[\exp\left(\frac{q}{\alpha . k . T} \left(\frac{V_{pv}}{N_s} + \frac{I_{pv} . R_s}{N_p} \right) \right) - 1 \right] - \frac{N_p}{R_{sh}} \left(\frac{V_{pv}}{N_s} + \frac{I_{pv} . R_s}{N_p} \right) \tag{11.2}$$

This chapter is based on the MATLAB/Simulink software environment to simulate a complete five-parameter model. The parameter identification methods outlined in [9] serve as the foundation for our model development. Figure 11.3 presents the characteristics of I(V) (Figure 11.3a) and P(V)

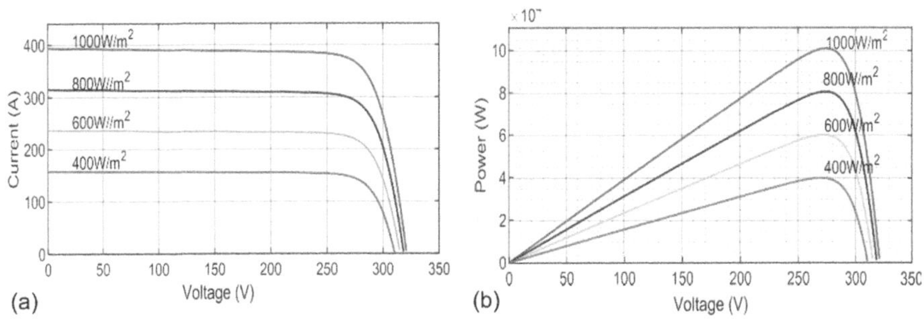

Figure 11.3 (a) I–V characteristics as a function of irradiation (b) P–V characteristics as a function of irradiation.

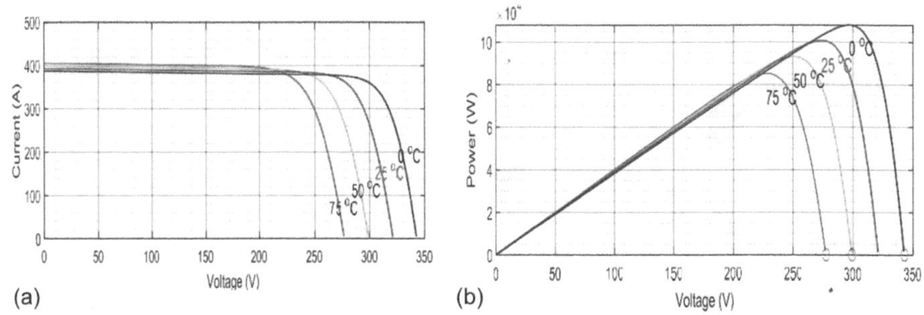

Figure 11.4 (a) I–V characteristics as a function of temperature (b) P–V characteristics as a function of temperature.

(Figure 11.3b) of PVG as a function of incident solar radiation, respectively; this figure illustrates that as solar irradiation increases, so does the output power of the PVG.

Figure 11.4 presents the curves of I_V and P_V as a function of temperature for constant irradiation. It is obvious that an increase in temperature decreases the open circuit voltage and induces a reduction in the maximum power of the PVG. From these plots, we see that the PVG characteristics are nonlinear and influenced by irradiation and temperature. To optimize its power output, an MPPT controller is employed to oversee the regulation of the DC-DC converter.

11.2.2 Modeling of DC-DC converter

The DC-DC converter serves as an intermediary between the PVG and the inverter, facilitating the extraction of maximum power. Its schematic is depicted in Figure 11.5.

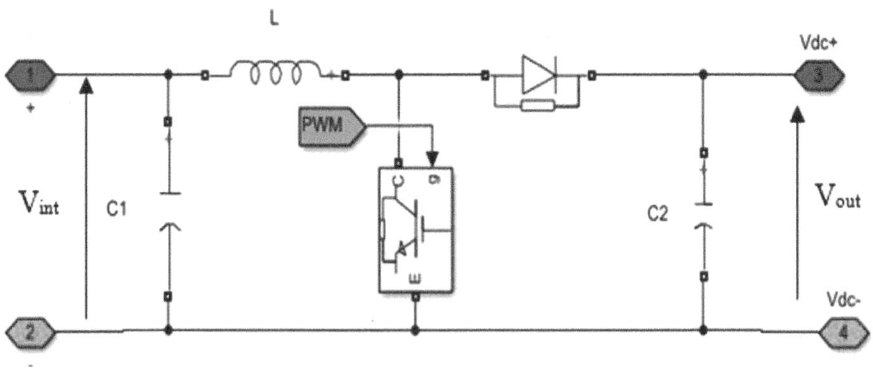

Figure 11.5 Boost converter electrical diagram.

Equation (11.3) defines the dynamic model of the DC-DC converter used [8]:

$$\begin{cases} C_1 \cdot \dfrac{dV_{pv}}{dt} = i_{pv} - i_L \\ L \cdot \dfrac{di_L}{dt} = -(1-\alpha)V_{dc} + V_{pv} \\ C_2 \cdot \dfrac{dV_{dc}}{dt} = (1-\alpha)i_L - \dfrac{V_{dc}}{R} \end{cases} \tag{11.3}$$

where V_{pv} and V_{dc} are the voltage generated by the PVG and the DC bus voltage, respectively, C_1 and C_2 are the input and continuous bus capacity, respectively, L is the inductance, and α is the duty cycle. The duty cycle is expressed as follows:

$$\alpha = 1 - \frac{V_{dc}}{V_{pv}} \tag{11.4}$$

V_{dc} is regulated by PI regulator. The Equations (11.5)–(11.7) give, respectively, the expression of the capacitances C_1 and C_2 and the inductance L of the boost [6, 7] and [8].

$$\begin{cases} C_1 = \dfrac{\alpha \cdot I_e}{\Delta V_{pv} \cdot F_s} & (11.5) \\ C_2 = \dfrac{\alpha \cdot V_{dc} \cdot I_o}{\Delta V_o \cdot F_s} & (11.6) \\ L = \dfrac{\alpha \cdot V_{pv}}{\Delta I_e \cdot F_s} & (11.7) \end{cases}$$

where F_s is the switching frequency, ΔV and ΔI are the tolerable voltage and current, and I_e and I_o are the boost converter input and output currents, respectively. The selection of the reference capacitor voltage V_{dc_ref} can be calculated using the next equation [10].

$$V_{dc\text{-}ref} = \frac{2\sqrt{2} \cdot V_{LL}}{\sqrt{3} \cdot m_a} \tag{11.8}$$

where V_{LL} is the line voltage between the lines, and m_a is the modulation coefficient.

11.2.3 Three-phase DC-AC inverter

Figure 11.6 illustrates the DC-AC converter used, comprising an inverter bridge with six controllable switches, typically insulated gate bipolar transistors (IGBTs). These switches are connected to terminals a, b, and c, which are connected to grid via inductive filter. To operate effectively when subjected to varying levels of illumination, a filtering stage becomes essential [11]. This stage serves several critical functions, including the smoothing of currents injected into the grid, the mitigation of harmonics generated by the inverter, and the restriction of voltage drop on the AC side of the inverter [12].

In Figure 11.6, L_f represents the inductance of the filter, while R_f denotes its resistance. L_f is expressed by [10] the following:

$$L_f = \frac{V_g}{6 \,.\, f_{sw} \,.\, \Delta_{ph\text{-}max}} \tag{11.9}$$

where $\Delta_{ph\text{-}max}$ is the current variation rate and 'f_{sw}' is the inverter switching frequency. The inverter's simple output voltages will be expressed as a function of the switch states and the intermediate circuit voltage V_{dc} by [9] the following:

$$\begin{bmatrix} u_a \\ u_b \\ u_c \end{bmatrix} = \frac{V_{dc}}{3} \cdot \begin{pmatrix} 2 & -1 & -1 \\ -1 & 2 & -1 \\ -1 & -1 & 2 \end{pmatrix} \cdot \begin{bmatrix} S_a \\ S_b \\ S_c \end{bmatrix} \tag{11.10}$$

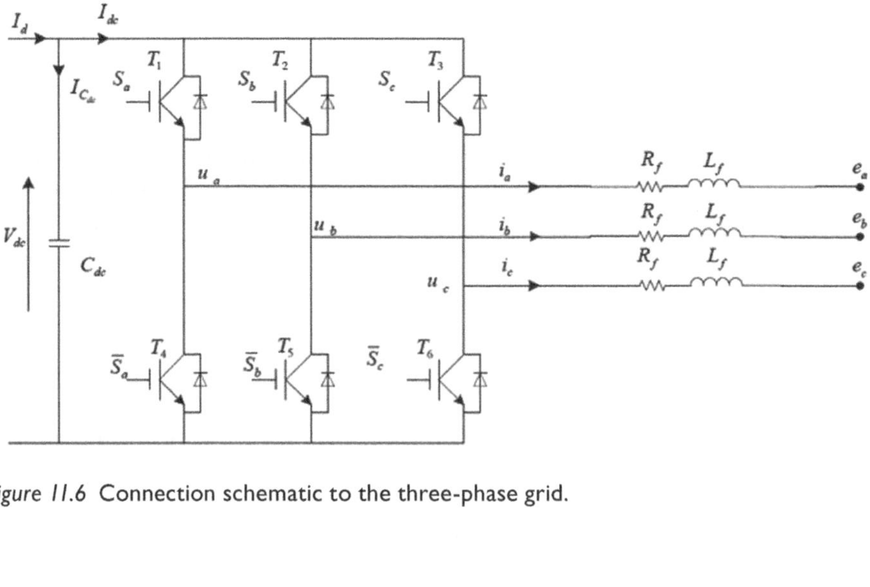

Figure 11.6 Connection schematic to the three-phase grid.

The output voltage space vectors generated by the inverter are expressed as a function of the simple voltages in Equation (11.11) and as a function of the switch states in Equation (11.12):

$$\begin{cases} V_o = \frac{2}{3}\left[V_a + \left(-\frac{1}{2} + j\frac{\sqrt{3}}{2}\right).V_b + \left(-\frac{1}{2} - j\frac{\sqrt{3}}{2}\right).V_c\right] & (11.11) \\ V_{o(i)} = \frac{2}{3}.V_{dc}\left[S_{a(i)} + \left(-\frac{1}{2} + j\frac{\sqrt{3}}{2}\right).S_{b(i)} + \left(-\frac{1}{2} - j\frac{\sqrt{3}}{2}\right).S_{c(i)}\right] & (11.12) \end{cases}$$

where 'i' is the state number {1,2…8}.

11.2.4 Grid model

Within our system, the voltage inverter establishes a connection to the three-phase grid through the inclusion of an inductive filter denoted as 'L'. The dynamic equations governing the interaction between the inverter, the inductive filter, and the network are detailed as follows [9]:

$$\begin{cases} L_f.\frac{di_a}{dt} = u_a - R_f.i_a - e_a \\ L_f.\frac{di_b}{dt} = u_b - R_f.vi_b - e_b \\ L_f.\frac{di_c}{dt} = u_c - R_f.i_c - e_c \end{cases} \quad (11.13)$$

where i_n, u_n (n = a, b, c) represent individual inverter-side voltages and currents and 'e_n' means grid-side voltages. The equations of (11.13) in the Park reference are defined by [7].

$$\begin{cases} L.\frac{di_d}{dt} = -R.i_d + Lwi_q + V_d - e_d \\ L.\frac{di_q}{dt} = -R.vi_q - Lwi_d + V_q - e_q \end{cases} \quad (11.14)$$

where ω represents the grid pulsation, i_d and i_q denote the direct and quadrature components of the current, respectively. V_d and e_d represent the direct voltage components on the inverter and grid side, while V_q and e_q refer to their respective quadrature components.

11.3 PHOTOVOLTAIC SYSTEM CONTROL

The PVG does not consistently operate at its MPP, particularly due to variations in illumination and temperature. The maximization of the electric power transfer depends on the control of the static converters used [12, 14]. By rearranging Equation (11.14), we obtain Equation (11.15):

$$\begin{cases} L.\dfrac{di_d}{dt} = U_d - R_f.i_d \\ L.\dfrac{di_q}{dt} = U_q - R_f.i_q \end{cases} \tag{11.15}$$

where

$$\begin{cases} U_d = V_d + L_f w i_q - e_d \\ U_q = V_q - L_f w i_d - e_q \end{cases} \tag{11.16}$$

The transfer function after the Laplace transformation on Equation (11.16) is given by [14] the following:

$$\begin{cases} F_d(p) = \dfrac{1}{R_f + L_f.p} \\ F_q(p) = \dfrac{1}{R_f + L_f.p} \end{cases} \tag{11.17}$$

PI controllers are used to control this PV system, whose transfer function is illustrated in Equation (11.18):

$$H(p) = k_p + \frac{k_i}{p} \tag{11.18}$$

where p represents the Laplace operator, and k_p and k_i are the parameters of the PI controller.

11.3.1 MMP control

MPPT control ensures that the PV system operates at its MPP regardless of the irradiation levels. It relies on adjusting the parameter α. Various MPPT techniques have been proposed to enhance the energy output of PV systems. In our system, P&O technique is selected due to its effectiveness compared to other methods. This algorithm adjusts α based on observing the impact of voltage variations on power [5]. The flowchart of the P&O algorithm is depicted in Figure 11.7.

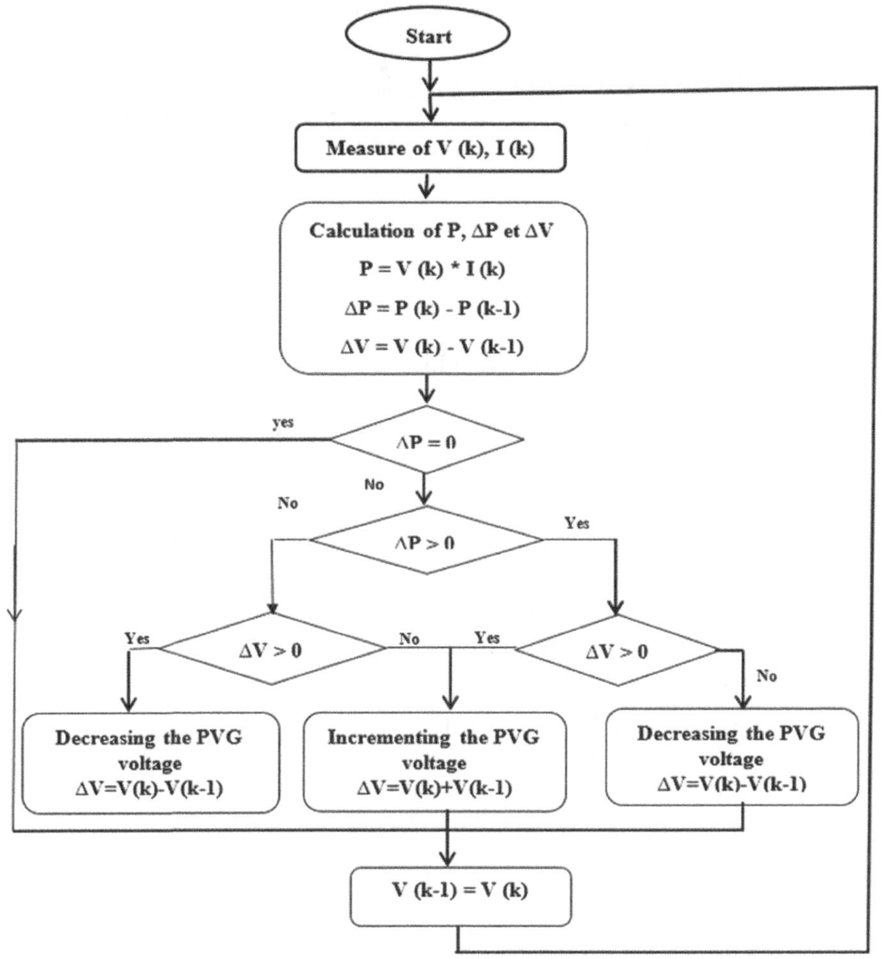

Figure 11.7 Disturbance and observation flowchart.

11.3.2 Control of DC bus voltage

The main objective of the inverter is to keep V_{dc} constant ($V_{dc} = V_{dc\text{-}ref} = 700V$). The expression for the current flowing through the capacitor is provided as follows [11]:

$$i_{dc} = C_{bus} \cdot \frac{dV_{dc}}{dt} = i_L - i_{inv} \tag{11.19}$$

So V_{dc} can be regulated by controlling i_{inv} with a PI controller whose control loop is shown in Figure 11.8.

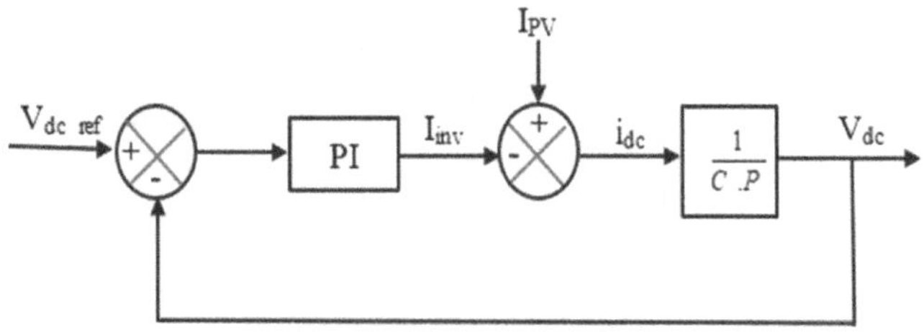

Figure 11.8 Loop for V_{dc} control.

The transfer function of the closed loop system is [14] as follows:

$$F(p) = \frac{C \,.\, k_p \,.\, p + \frac{k_i}{C}}{p^2 + k_p \,.\, p + \frac{k_i}{C}} \tag{11.20}$$

The PI controller parameters are deduced by comparing Equation (11.20) with the second-order system transfer function [10]:

$$F(p) = \frac{V_{dc\_ref}}{V_{dc}} = \frac{\left(k_p \,.\, p + k_i\right)/C}{p^2 + \frac{k_p}{C} \,.\, p + \frac{k_i}{C}} = \frac{\omega_n^2}{p^2 + 2 \,.\, \xi \,.\, \omega_n \,.\, p + \omega_n^2} \tag{11.21}$$

From Equation (11.21) the parameters of the controller are as follows:

$$\begin{cases} k_i = C \,.\, \omega_n^2 \\ k_p = 2 \,.\, \xi \,.\, C \,.\, \omega_n \end{cases}$$

11.3.3 Control of inverter on grid side

The inverter's control objective is to adjust active (P) and reactive (Q) power [1]. The power factor can be fixed at 1 by applying zero reactive power. Using Equations (11.15) and (11.16), the current control loop can be developed, as shown in Figure 11.9.

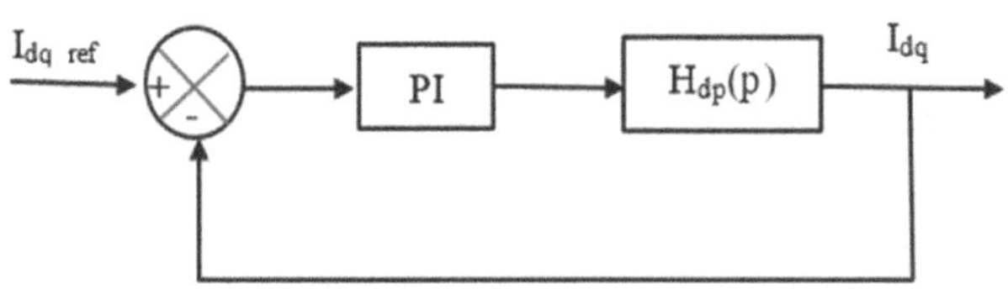

Figure 11.9 Current control cycle.

The transfer function is equal to [14] the following:

$$F(p) = \frac{(k_p \,.\, p + k_i)/L}{p^2 + \left(\frac{R + k_p}{L}\right) p + \frac{k_i}{C}} \tag{11.22}$$

The PI controller parameters are deduced by comparing Equation (11.22) with the second-order system transfer function:

$$\begin{cases} k_i = L \,.\, \omega_n^2 \\ k_p = 2 \,.\, \xi \,.\, L \,.\, \omega_n - R \end{cases}$$

In order to control P and Q separately, we must remove the link terms ($L\omega i_q$ and $L\omega i_d$) and correct the grid voltage components (e_d and e_q).

We pose the following:

$$\begin{cases} V_{d\_ref} = U_d + e_d - L\omega i_q \\ V_{q\_ref} = U_q + e_q + L\omega i_d \end{cases} \tag{11.23}$$

where

$$\begin{cases} U_d = \left(k_p + \frac{k_i}{p}\right) . \left(i_{d\_ref} - i_d\right) \\ U_q = \left(k_p + \frac{k_i}{p}\right) . \left(i_{q\_ref} - i_q\right) \end{cases} \tag{11.24}$$

The schematic for the decoupled PQ control is shown in Figure 11.10 [9].

We rearrange Equations (11.23) and (11.24) and obtain the following:

$$\begin{cases} V_{d\_ref} = \left(k_p + \frac{k_i}{p}\right) . \left(i_{d\_ref} - i_d\right) + e_d - L\omega i_q \\ V_{q\_ref} = \left(k_p + \frac{k_i}{p}\right) . \left(i_{q\_ref} - i_q\right) + e_q + L\omega i_d \end{cases} \tag{11.25}$$

To have a zero static error, we check V_{d_ref} and V_{q_ref} by regulating i_d and i_q. Then, the disturbance terms are eliminated to obtain a reference V_{d_ref} equal to the maximum amplitude of the grid voltage and a reference V_{q_ref} equal to zero ($V_{d\_ref} = \sqrt{2}.220v$ and $V_{q\_ref} = 0$) when the system is synchronized.

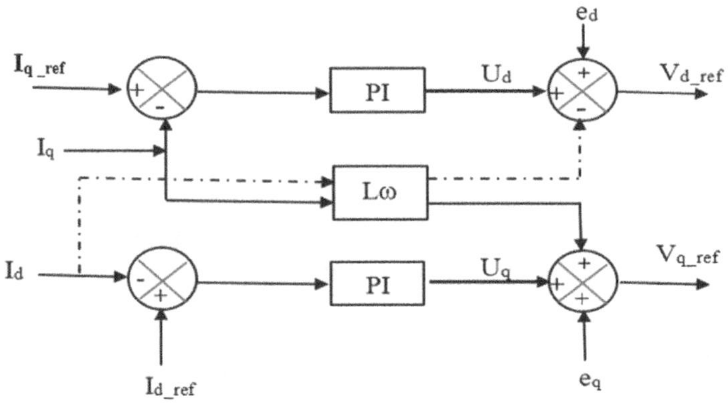

Figure 11.10 Inverter control schematic on the grid side.

The instantaneous powers in the park reference frame are given by [11] the following:

$$\begin{cases} P = \frac{3}{2}\left(V_d \cdot i_d + V_q \cdot i_q\right) \\ Q = \frac{3}{2}\left(V_q \cdot i_d - V_d \cdot i_q\right) \end{cases} \tag{11.26}$$

To achieve independent control of P and Q, a Park transformation is used such that $V_d = V$ and $V_q = 0$, so that Equation (11.26) gives the following:

$$\begin{cases} P = \frac{3}{2}\left(V_d \cdot i_d\right) \\ Q = -\frac{3}{2}\left(V_d \cdot i_q\right) \end{cases} \tag{11.27}$$

It can be observed that i_d and i_q have a linear relation with P and Q. By adjusting these currents independently, the powers will be controlled separately. Thus, from Equation (11.27), the references i_{d_ref} and i_{q_ref}, imposing the references of P and Q, are then given by the following:

$$\begin{cases} i_{d\_ref} = \frac{2}{3}\left(\frac{P_{ref}}{V_d}\right) \\ i_{q\_ref} = -\frac{2}{3}\left(\frac{Q_{ref}}{V_d}\right) \end{cases} \tag{11.28}$$

Figure 11.11 shows the general diagram of the PV system connected to the grid and the proposed controls.

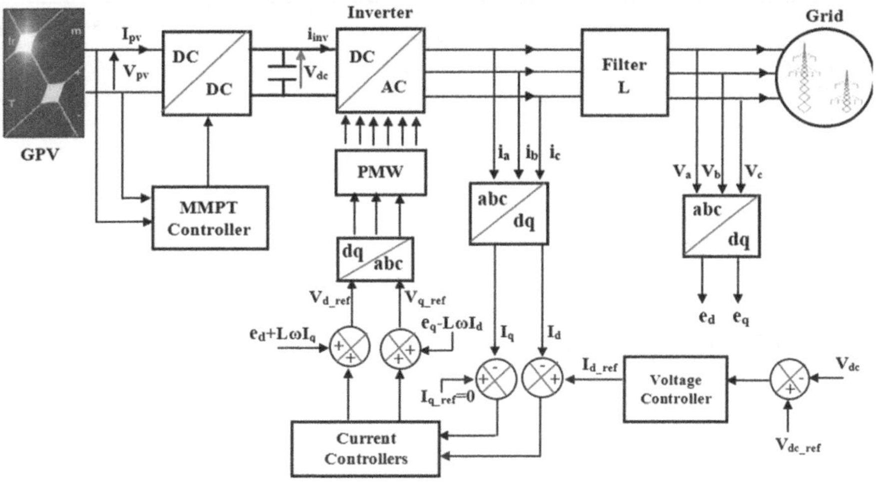

Figure 11.11 Photovoltaic system connected to the electrical grid.

11.4 RESULTS OF SIMULATIONS

The performance of the proposed algorithms has been tested under different irradiation conditions, as shown by the graph in Figure 11.12. The results for PV power are shown in Figure 11.13. PPM tracking ($P_{\_MPP}$ = 100 kW at 1 kW/m^2) is evident in this figure. Figures 11.14 and 11.15 illustrate the results for current $I_{\_PV}$ and voltage V_{dc}, respectively; $I_{\_PV}$ reaches a maximum of 368.3 A, corresponding to PVG current at 1 kW/m^2, while voltage V_{dc} remains constantly fixed at the reference value (V_{dc_ref} = 700 V). Figure 11.16 shows the inverter currents, which are effectively regulated to match their reference. As the MPPT algorithm adjusts the reference current according to the irradiation curve in Figure 11.12, the inverter output currents also adapt accordingly.

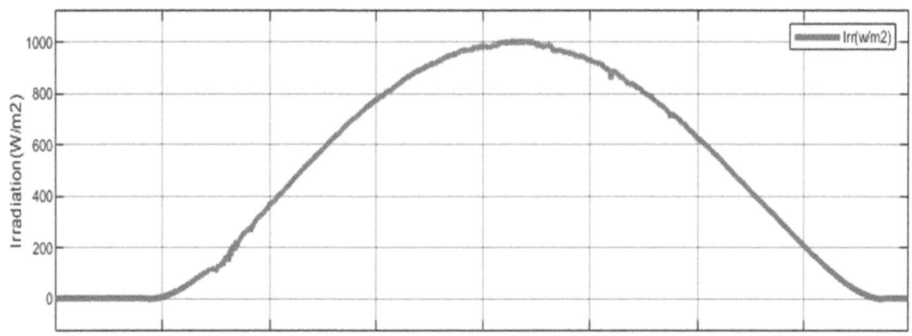

Figure 11.12 Irradiation curve.

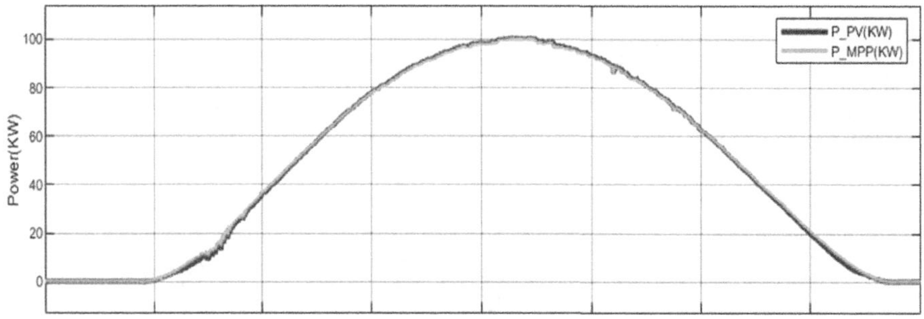

Figure 11.13 Results of photovoltaic power.

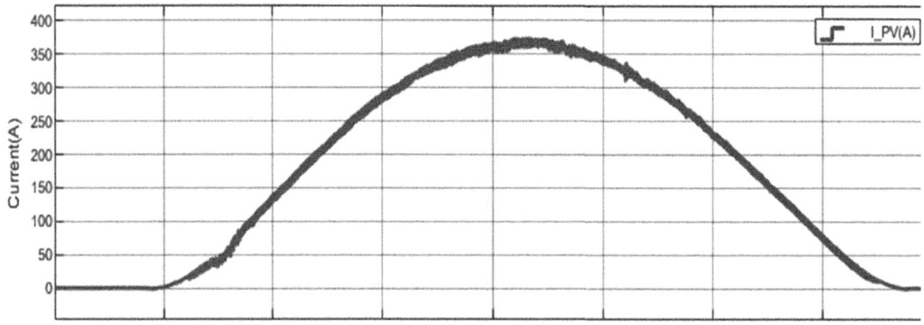

Figure 11.14 PVG current results.

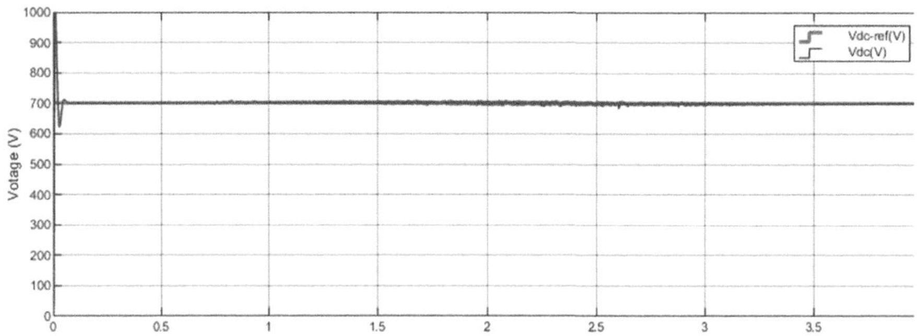

Figure 11.15 DC bus voltage results.

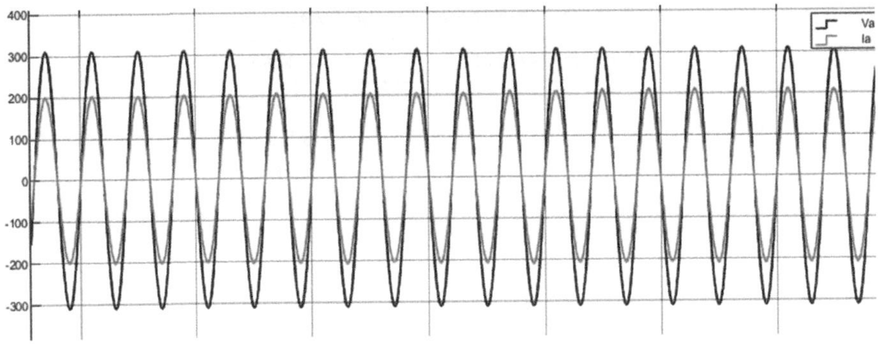

Figure 11.16 Grid voltage and current.

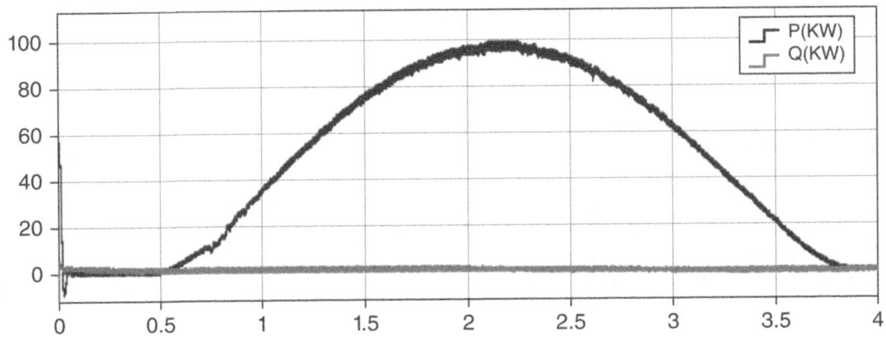

Figure 11.17 Result of active and reactive power.

The system's performance under CVC is shown visually in Figures 11.16 and 11.17, and these visualizations clearly show that, after the transient phase, the system settles into stable operating conditions. It should be noted that the voltage and current phases are synchronized, as shown in Figure 11.16. In addition, the system operates successfully at unity power factor, as shown in Figure 11.17.

11.5 CONCLUSION

This chapter explores efficient control of a PV system connected to the grid, focusing on MPPT performance tuning. It describes a P&O-based MPPT approach for the PVG and details a CVC-based control strategy for the inverter. Through simulations, the study evaluates system performance both under stationary conditions and in dynamic situations, taking into consideration

Table 11.1 Parameters of the PV system.

Parameter symbol	*Value*
C	2000 μF
L_f	3.5 mH
Fsw	5 kHz
K_i	60
K_p	20.8
K_i	75
K_p	0.51

various irradiation and temperature levels. The results show that when these conditions vary, the P&O controller efficiently follows the PPM of the PVG, ensuring stable active power production by the grid-connected PV system (Table 11.1).

REFERENCES

1. Katche, M. L., Makokha, A. B., Zachary, S. O., Adaramola, M. S. A comprehensive review of maximum power point tracking (MPPT) techniques used in solar PV systems. *Energies*, 2023, vol. 16, no 5, p. 2206.
2. Başoğlu, Mustafa Engin. Comprehensive review on distributed maximum power point tracking: Submodule level and module level MPPT strategies. *Solar Energy*, 2022, vol. 241, p. 85–108.
3. Singh, Bhuwan Pratap, Goyal, Sunil Kumar, Siddiqui, Shahbaz Ahmed. Analysis and classification of maximum power point tracking (MPPT) techniques: a review. *Intelligent Computing Techniques for Smart Energy Systems: Proceedings of ICTSES 2018*, 2020, p. 999–1008.
4. Sadick, Abubakari. Systems using perturb and observe algorithm. *Solar Radiation: Enabling Technologies, Recent Innovations, and Advancements for Energy Transition*, 2024, p. 49.
5. Baba, Ali Omar, Liu, Guangyu, Chen, Xiaohui. Classification and evaluation review of maximum power point tracking methods. *Sustainable Futures*, 2020, vol. 2, p. 100020.
6. Chao, K. H., & Rizal, M. N. (2021). A hybrid MPPT controller based on the genetic algorithm and ant colony optimization for photovoltaic systems under partially shaded conditions. *Energies*, 14(10), 2902.
7. Kharef, Fatima Zohra, Masmoudi, Ramadhan, et al. MPPT control for shading photovoltaic panels. 2021. Doctoral dissertation. Electrical controls.
8. Ibrahim, Nagwa F., Mahmoud, Karar, Lehtonen, Matti, et al. Comparative analysis of three-phase PV grid connected inverter current control schemes in unbalanced grid conditions. *IEEE Access*, 2023, vol. 11, p. 42204–42221.
9. Nassar-Eddine, I., Obbadi, A., Errami, Y., Agunaou, M. Parameter estimation of photovoltaic modules using iterative method and the Lambert W function: A comparative study. *Energy Conversion and Management*, 2016, vol. 119, p. 37–48.

10. Zhao, Yingying, An, Aimin, Xu, Yifan, *et al.* Model predictive control of grid-connected PV power generation system considering optimal MPPT control of PV modules. *Protection and Control of Modern Power Systems*, 2021, vol. 6, p. 1–12.
11. Mokhtara, Charafeddine, Negrou, Belkhir, Settou, Noureddine, *et al.* Optimal design of grid-connected rooftop PV systems: An overview and a new approach with application to educational buildings in arid climates. *Sustainable Energy Technologies and Assessments*, 2021, vol. 47, p. 101468.
12. Dhibi, Khaled, Mansouri, Majdi, Bouzrara, Kais, *et al.* An enhanced ensemble learning-based fault detection and diagnosis for grid-connected PV systems. *IEEE Access*, 2021, vol. 9, p. 155622–155633.
13. Alhejji, A., & Mosaad, M. I. (2021). Performance enhancement of grid-connected PV systems using adaptive reference PI controller. *Ain Shams Engineering Journal*, 12(1), 541-554.
14. Rehman, Haseeb Ur, Yan, Xiangwu, Abdelbaky, Mohamed Abdelkarim, *et al.* An advanced virtual synchronous generator control technique for frequency regulation of grid-connected PV system. *International Journal of Electrical Power & Energy Systems*, 2021, vol. 125, p. 106440.

Chapter 12

GeoArgania

A geolocation mapping dataset of Argania trees in the Souss region

Younes Karmoude, Taha Bouhsine, Souad Saidi, Soufian Idbraim, Manuel Arbelo, Enrique Casas-Mas, Azeddine Elhassouny, and Antoine Masse

12.1 INTRODUCTION

Argania trees (*Argania spinosa*) play a crucial role in the ecological and socio-economic landscape of the Souss region, renowned for their resilience in arid conditions and their contribution to the preservation of biodiversity [1]. These evergreen trees are not only a vital habitat for various wildlife species but also a source of sustenance and income for local communities through their valuable products, such as argan oil and timber (Figure 12.1) [2]. However, the precise distribution and abundance of Argania trees in the region have remained challenging to assess accurately [3].

Figure 12.1 Argan oil.

DOI: 10.1201/9781032664293-14

One major threat to the Argania tree is over-harvesting. The tree produces a valuable nut from which oil is extracted, and this has led to the over-exploitation of the tree for commercial gain [4]. This not only reduces the number of trees but also damages the remaining population by removing the fruit before it is ripe, which can prevent future growth and reproduction.

Another danger to the Argania tree is habitat destruction [5]. The tree is native to a small region in Morocco, and as human populations have grown and expanded, much of its natural habitat has been destroyed or degraded. This not only reduces the number of trees but also makes it more difficult for the remaining population to survive and reproduce.

Climate change is also a significant threat to the Argania tree [6]. The tree is adapted to a specific climatic range, and changes in temperature and precipitation patterns can have a significant impact on its survival.

In addition, grazing by goats is another danger to the Argania tree [7], as goats can overgraze and cause damage to the tree (Figure 12.2). To protect the Argania tree, it is important to implement sustainable harvesting practices, protect and restore its natural habitat, and address the impact of climate change. Additionally, managing the grazing of goats is also crucial.

In recent years, advancements in remote sensing technologies have opened up new opportunities for comprehensive and large-scale data collection, enabling researchers to study tree populations and monitor changes in their habitats with unprecedented accuracy [8, 9]. The Sentinel-2 satellite constellation, launched by the European Space Agency (ESA), has become a valuable

Figure 12.2 Goats overgrazing an Argania tree in the Souss Massa region.

asset for ecological research, providing high-resolution multispectral imagery with frequent revisits.

Previous work by our team involved the development of a patch-based dataset for Argania tree detection [1], where smaller image patches were extracted from Sentinel-2 imagery encompassing areas with known Argania tree presence. These patches served as labeled training data for machine learning algorithms, enabling us to develop accurate models for detecting Argania trees across the region. The success of this approach laid the foundation for the current study, where we now seek to expand our understanding by integrating a comprehensive dataset of 91,859 manually collected geolocation points for Argania trees.

Building on the success of our previous work, the present research takes a hybrid approach, combining the strengths of manual data collection and pixel precise datasets. By incorporating a vast number of geolocated Argania trees, we aim to augment the training data for machine learning models, enhancing the accuracy and reliability of our analyses. The integration of Sentinel-2 pixel values from the extensive dataset with the knowledge gained from the patch-based approach promises to deliver a more nuanced and comprehensive understanding of the Argania tree distribution in the Souss region.

This research presents a study that leverages Sentinel-2 data and Google Earth Engine (GEE) [10] to map the distribution of Argania trees in the Souss region. One of the most significant aspects of this study is the meticulous collection of geolocation data for an astounding 91,859 Argania trees, achieved through extensive field surveys and validation procedures. By combining manual geolocation and validation with advanced remote sensing techniques, we aim to obtain a comprehensive and reliable dataset that reflects the true distribution patterns of Argania trees in the region.

12.2 MATERIALS AND METHODS

12.2.1 Study site

The study focuses on the region of South Morocco, an area marked by its arid to semi-arid climate and home to a unique and diverse array of flora and fauna. The region is characterized by hot summers and mild winters, with sporadic rainfall that greatly influences the ecology of the region. The climate, coupled with unique geographical features, supports the growth of specialized vegetation adapted to these conditions.

The study area includes the Admine forest, located between Agadir and Biougra (Souss region, southwest Morocco) (Figure 12.3). It occupies an area of about 130 hectares and is delimited by the geographical coordinates 30°21′58.0″N 9°28′40.8″W and 30° 17′27.6″N 9° 19′ 04.2″W (Figure 12.4). The Admine forest is a representative area in terms of the biodiversity of the Arganeraie Biosphere Reserve declared by UNESCO in 1988 [1].

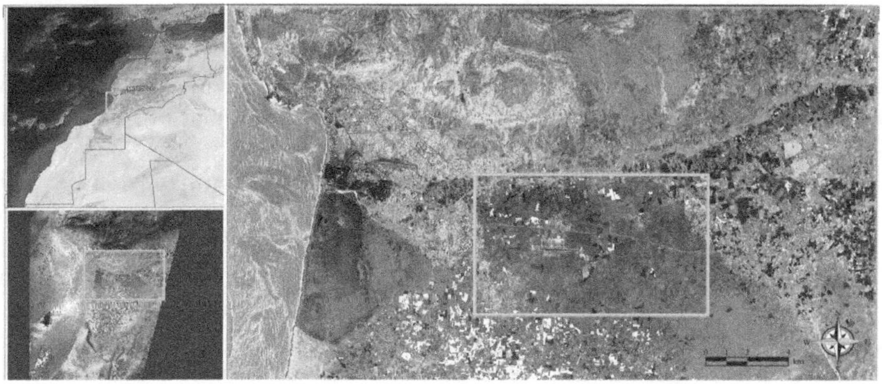

Figure 12.3 Satellite view of the study area in the Admine forest, Agadir, Souss, Morocco.

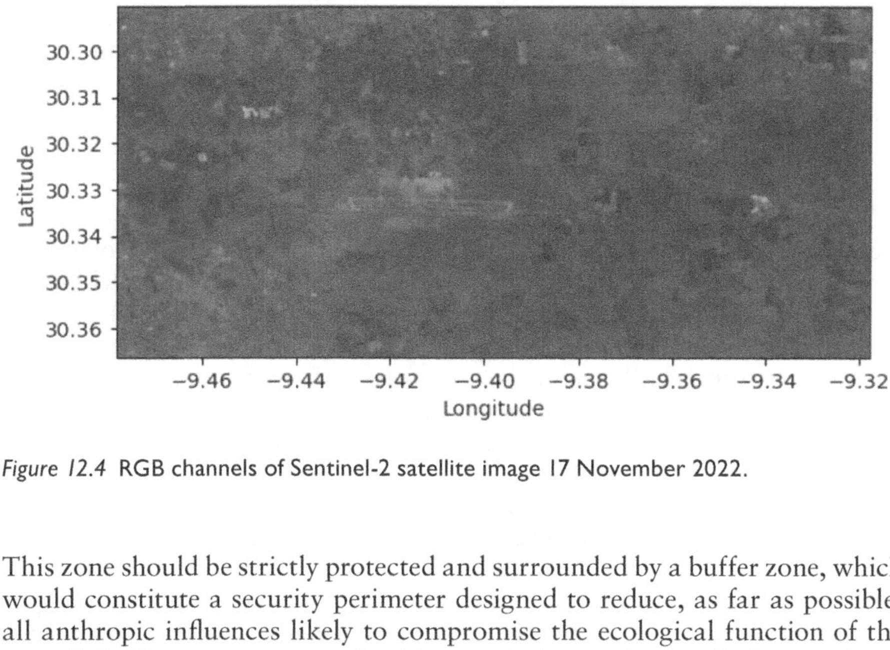

Figure 12.4 RGB channels of Sentinel-2 satellite image 17 November 2022.

This zone should be strictly protected and surrounded by a buffer zone, which would constitute a security perimeter designed to reduce, as far as possible, all anthropic influences likely to compromise the ecological function of the zone [11]. However, as a result of its proximity to the Agadir International Airport, the urban sprawl, and the proliferation of greenhouse crops in recent years, the area is threatened by deforestation [12].

From a geological point of view, it is located in the basin of the Souss river and is separated from the Sahara Desert by the Anti-Atlas mountains. The study area is characterized by a semi-arid climate with an average annual temperature ranging from 14°C to 16°C in Winter, and 19°C–25°C in Summer, with the possibility of a temperature rise to more than 40°C [13]. Average annual rainfall varies between 100 m and 443.20 mm [13].

The natural vegetation is dominated by the Argan tree (*A. spinosa*), a local endemic tree found nowhere else [14]. The argan forest provides numerous ecosystem services, including biodiversity conservation, species habitat, soil fertility, erosion prevention, food and water, climate regulation, local hydrological processes, and tourism [15].

It is estimated that Argania trees currently cover approximately 8280 sq km of land in South Morocco, demonstrating their prevalence in this landscape. These trees, typically ranging from 8 m to 10 m in height, form open woodlands and provide critical habitat for a wide variety of local fauna. Despite their significant ecological role, Argania trees are under threat, with numbers dwindling in recent years.

12.2.2 Google Earth Engine

GEE [16] is a cloud-based platform equipped with an extensive catalog of satellite imagery and geospatial datasets. GEE provides the capability for advanced processing of large-scale datasets, making it an ideal tool for environmental and ecological research [17]. In our study, GEE was employed as the primary tool for the identification and geolocation of Argania trees across the distribution range of this species.

12.2.3 Sentinel-2 satellite

Sentinel-2 is an Earth observation mission from the Copernicus Programme, operated by the ESA. This satellite provides multispectral images with 13 spectral bands at different spatial resolutions – 4 bands at 10 m, 6 bands at 20 m, and 3 bands at 60 m resolution. The imagery produced by Sentinel-2 allows for the monitoring of vegetation, soil and water cover, inland waterways, and coastal areas. Importantly for this study, it also provides crucial data for understanding the state and changes in the world's forests.

The satellite images used in this study were obtained by the Copernicus Sentinel-2 mission [18]. The mission consists of a constellation of two polar-orbiting satellites, Sentinel-2A and -2B, placed in the same sun-synchronous orbit, in phase 180°C to each other. Thanks to its open and free-data policy and high revisit frequency (10 days at the equator with one satellite and 5 days with two satellites), Sentinel-2 supports numerous services and applications, such as agricultural monitoring, emergency management, land cover classification or water quality, and it turns out to be a and an excellent and consistent source of optical images for monitoring Argania's area changes. We have acquired our dataset using GEE [2]. Each Sentinel-2 carries on board the optical Multi-Spectral Instrument (MSI) sensor. MSI samples 13 spectral bands in visible, near-infrared, and short-wave infrared wavelengths with spatial resolutions between 10 m and 60 m [19]. All bands were used in this study (Table 12.1).

Table 12.1 Spectral bands for the Sentinel-2 level 1C [20]

	Sentinel-2 level 1C		
Band number	*Central wavelength (nm)*	*Bandwidth (nm)*	*Spatial resolution (m)*
1	442.7	20	60
2	492.7	65	10
3	559.8	35	10
4	664.6	30	10
5	704.1	15	20
6	740.5	15	20
7	782.8	20	20
8	832.8	105	10
8A	864.7	21	20
9	945.1	19	60
10	1373.5	29	60
11	1613.7	90	20
12	2202.4	174	20

12.2.4 Data collection process

Using the GEE platform, we manually identified Argania trees through a detailed analysis of high-resolution satellite imagery (Figure 12.5). This identification was achieved by examining distinctive features of the Argania tree, such as crown size, color, and shadow patterns, which make them distinguishable from other tree species in the imagery. Once trees were identified, their geolocation coordinates were collected, creating a comprehensive dataset of Argania tree geolocation for 91,859 Argania trees.

In this chapter, we delve into the meticulous process of manually collecting geolocation data for 91,859 Argania trees in the Souss region in the month of January 2023. The manual data collection approach was chosen to ensure the accuracy and precision required for a comprehensive study of this magnitude. Our team of expert field researchers traversed the diverse landscapes of the region, carefully identifying and geolocating individual Argania trees.

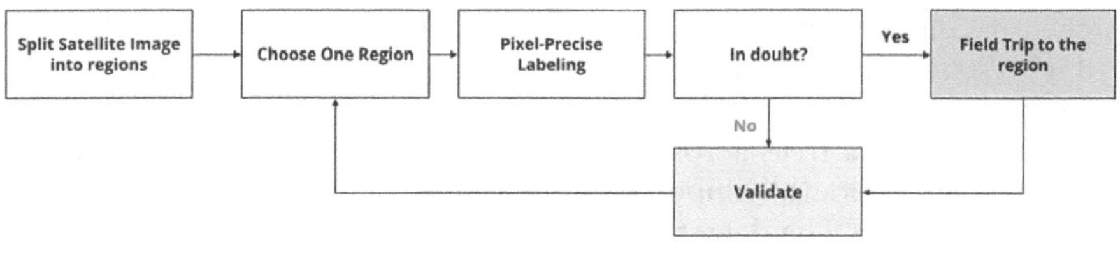

Figure 12.5 Data collection protocol used to collect Argania trees geolocation using GEE, Google Earth high-resolution satellite, and Sentinel-2 satellite.

This hands-on approach using high-resolution Google Earth RGB images provided by GEE allowed us to capture valuable information not only about the tree's location but also about its health, size, and surrounding habitat characteristics. Each geolocation point was recorded with meticulous detail, ensuring a comprehensive and representative dataset for our analysis.

The manual identification process presented some challenges. For instance, the mixed patches of Argania trees with other species complicated the identification process. To overcome this, we developed a standardized set of criteria based on the unique characteristics of Argania trees. This allowed for consistent identification, minimizing misclassification errors.

To ensure the integrity and reliability of the collected data, rigorous validation procedures were employed. First, we cross-referenced our geolocation data with existing datasets we built previously and available high-resolution satellite imagery mainly Google Earth to verify the accuracy of each point. Next, we conducted statistical analyses to detect and address any potential outliers or discrepancies on Sentinel-2 1C 13 bands and multivariate outlier detection methods on the pixels where we have Argania. Additionally, to validate the presence of Argania trees at suspicious geolocation points, we conducted field trips to confirm the tree's existence and assess its condition on the ground. These field trips provided valuable ground-truthing data and further strengthened the credibility of our dataset. Through these validation procedures, we ensured that our dataset reflects the true distribution patterns and density of Argania trees in the Souss region.

During the validation process, certain challenges were encountered, including discrepancies between the GEE dataset and Sentinel-2 spectral signatures due to factors like seasonal variations or tree health conditions. To address these issues, we made use of multiple Sentinel-2 images captured at different times to account for temporal variations in tree appearance and spectral signatures.

12.3 RESULTS AND DISCUSSION

12.3.1 Results

12.3.1.1 GeoArgania dataset

After rigorous manual geolocation and validation efforts, we compiled a comprehensive dataset (https://code.earthengine.google.com/8984e704d210c052f1d5f70f90a4f82b) consisting of identified Argania trees, complete with their respective geographic coordinates. The results showcased the extensive dispersion of Argania trees across their natural habitat (Figure 12.6), shedding light on their ecological importance as well as their potential vulnerability to various threats. The dataset serves as a reliable reference point for understanding the spatial distribution of Argania trees, providing valuable data for research and conservation initiatives. We also collected other geolocations

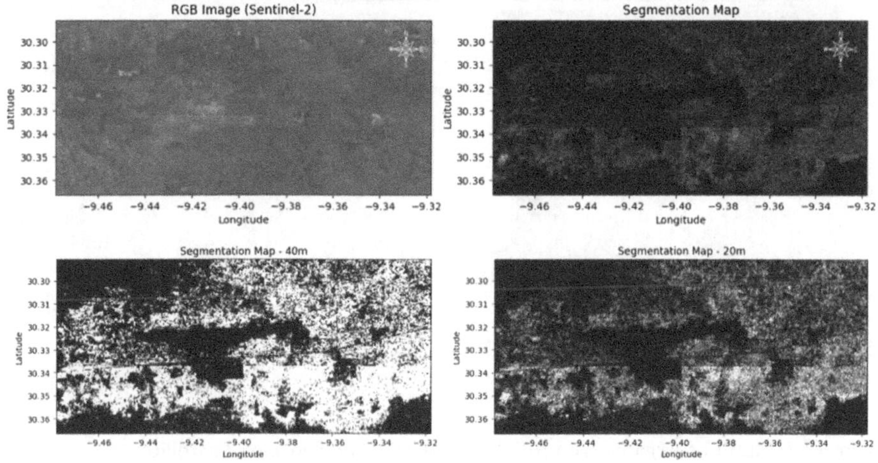

Figure 12.6 Patch-based segmentation maps built using GeoArgania dataset.

corresponding to other types of objects in the same region, e.g., buildings, roads, other types of trees ... to build the current dataset of Argania vs not Argania. Where for each geolocation we label it as Argania or not Argania.

Building on our previous work [1, 9], we integrated the insights gained from the patch-based dataset we previously built into our comprehensive study. The smaller image patches, previously used as labeled training data for machine learning algorithms, provided valuable information about spectral patterns and features characteristic of Argania trees. By leveraging this knowledge and integrating it with our extensive dataset, we aimed to enhance the accuracy and precision of our analysis. The integration of patch-based data with our manual collection allowed us to fine-tune our understanding of Argania tree distribution, particularly in areas where remote sensing data may have limitations or inconsistencies and aim for building a multispectral pixel precise machine learning model for Argania tree detection.

The combination of manual data collection, rigorous validation, and integration of patch-based datasets offers several advantages for our study. Firstly, the large and representative dataset of 91,859 geolocation points provides a comprehensive view of the Argania tree distribution across the diverse landscapes of the Souss region. This extensive dataset facilitates a more nuanced analysis, allowing us to identify high-density areas, ecological corridors, and potential threats to the population. Secondly, the ground-truthing data obtained during field trips further enhances the accuracy and reliability of our study, reducing uncertainties associated with remote sensing analyses. Lastly, by integrating patch-based datasets, we leverage machine learning insights to refine our analysis and improve the detection of Argania trees in areas with complex or heterogeneous land cover.

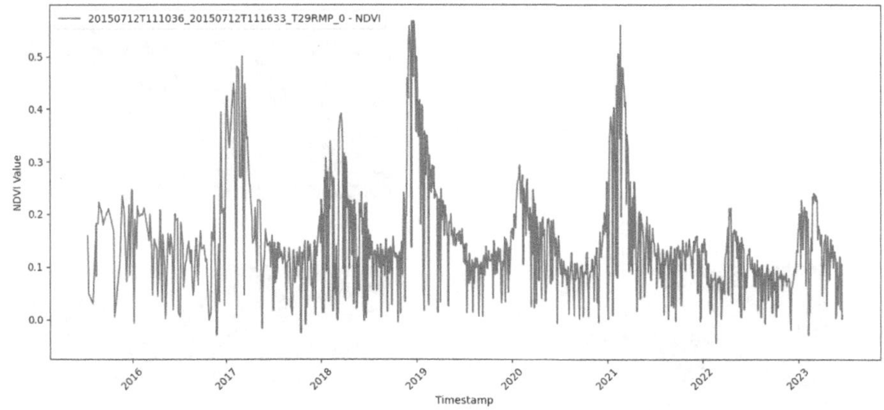

Figure 12.7 NDVI of Argania tree #ID0 calculated using Sentinel-2 satellite.

One of the key strengths of Sentinel-2 imagery is its frequent revisits to the same location, providing a time-series of data spanning from 23 June 2015 to 21 June 2023. Leveraging this temporal dimension, we extracted pixel values for each geolocation point across all available times (Figure 12.7). This allowed us to create a comprehensive dataset capturing the spectral variations of Argania trees over several years.

With the extracted pixel values from multiple time points, we conducted a temporal analysis to examine the seasonal trends and phenological changes in the Argania tree population. By analyzing the spectral profiles over time, we aimed to identify patterns related to the annual growth cycle, phenological shifts [21], and potential impacts of environmental factors on the trees' health and vitality.

12.3.1.2 Statistical analysis

The analysis of the comprehensive dataset revealed a detailed and accurate spatial distribution of Argania trees in the Admine forest (Figure 12.8). Our findings indicated that the Argania tree population was concentrated primarily in the southwestern part of the study area, covering vast stretches of semi-arid and arid lands. We observed several clusters of high tree density, indicating potential areas of ecological significance and rich biodiversity. Additionally, the spatial visualization of the dataset enabled us to identify critical corridors connecting different patches of Argania trees, highlighting the importance of preserving such ecological connections for wildlife movement and gene flow.

Through spectral analyses, we aimed to identify areas with the highest ecological significance for Argania trees and their long-term trends. By evaluating the changes in pixel values over the study period, we can infer

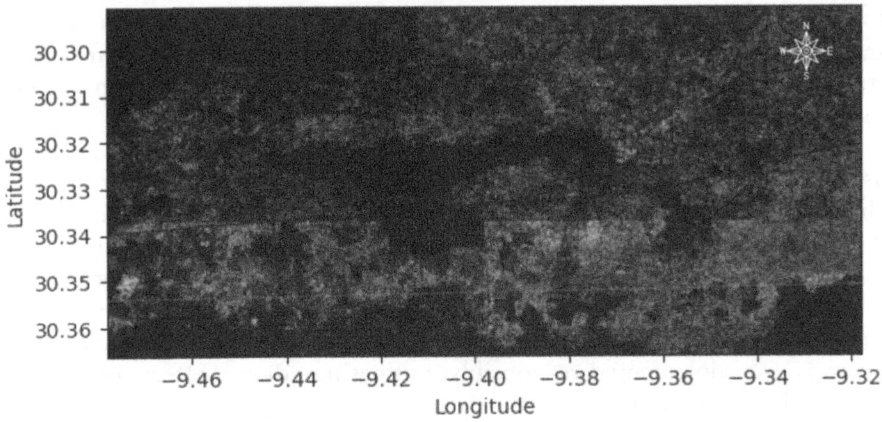

Figure 12.8 Spatial distribution of Argania trees in the Admine forest.

information about the health and stability of different Argania populations in the Souss region. Such insights are crucial for understanding the dynamics of the Argania ecosystem and guiding conservation efforts.

12.3.2 Discussion

The comprehensive dataset generated in this study, including the spatial distribution and temporal dynamics of Argania trees, can be a valuable tool for monitoring deforestation and guiding conservation strategies. By comparing current data with future surveys, change detection algorithms can be developed to assess the extent and rate of Argania deforestation over time. The identification of areas experiencing significant tree loss can serve as an early warning system, prompting swift action to protect vulnerable regions and implement measures to combat deforestation. Integrating deforestation monitoring into conservation strategies can ensure the preservation of critical Argania habitats and biodiversity.

With the ability to detect changes in Argania tree cover, the dataset can identify deforestation hotspots, areas where tree loss is occurring at an alarming rate. These hotspots can be used as focal points for conservation efforts, directing resources and interventions to combat deforestation and restore degraded areas. The timely detection of deforestation hotspots can help prevent the irreversible loss of valuable ecosystems and contribute to the protection of the Argania tree population in the Souss region.

The deforestation monitoring capability offered by the dataset can support evidence-based policy and decision-making. Government authorities and policymakers can utilize the data to assess the effectiveness of existing conservation policies and identify areas that require stronger protection measures. Moreover, the dataset can provide valuable information for designing and implementing conservation and afforestation initiatives. Informed decision-making can lead to more effective environmental policies, strengthening efforts to combat deforestation and promote sustainable land management practices.

The dataset can serve as a valuable resource for environmental impact assessment (EIA) studies. Whether for infrastructure projects or land use changes, EIAs can utilize the dataset to evaluate potential impacts on the Argania tree population and surrounding ecosystems. By understanding the potential consequences of development activities, decision-makers can make informed choices that balance socio-economic development with environmental preservation. EIAs incorporating the dataset can ensure that new projects are implemented responsibly, mitigating adverse effects on Argania trees and their habitats.

The findings of this study and the dataset it generates can contribute to global conservation efforts beyond the Souss region. The Argania tree is a unique species with ecological significance and provides valuable ecosystem services. The dataset can be shared with international conservation organizations and researchers, facilitating collaborative efforts to protect not only the Argania tree but also other endangered tree species and ecosystems worldwide. The dataset can become a valuable resource for global initiatives aimed at combatting deforestation and conserving biodiversity.

12.4 CONCLUSION

The present study represents a pioneering effort in mapping the distribution of Argania trees in the Souss region using Sentinel-2 imagery and a comprehensive dataset of 91,859 geolocated trees. Through meticulous manual data collection and rigorous validation, we obtained a robust dataset that accurately reflects the spatial distribution and ecological significance of Argania trees in this ecologically important region. Leveraging the temporal dimension of Sentinel-2 imagery, we conducted a thorough analysis of pixel values across all available times from 2015 to 2022, enabling us to explore seasonal trends and long-term dynamics of the Argania tree population.

The implications of this research are far-reaching, encompassing conservation strategies, sustainable resource utilization, climate change adaptation, and EIA. The dataset's capability for change detection in Argania deforestation can serve as a vital tool for monitoring tree loss and identifying deforestation hotspots.

Due to climate change and decreased precipitation, Argania forests face the threat of deforestation. Future efforts aim to detect changes resulting from argan deforestation. This detection will contribute to conservation planning and policy development, enabling stakeholders to make informed decisions. These decisions are crucial to protect the Argania ecosystem and promote socio-economic development in harmony with nature.

REFERENCES

1. Idbraim, S., Bouhsine, T., Dahbi, M. R., Masse, A., & Arbelo, M. (2022, May). Argania Forest Change Detection from Sentinel-2 Satellite Images Using U-Net Architectures. In *International Conference on Advanced Intelligent Systems for Sustainable Development* (pp. 174–184). Cham: Springer Nature Switzerland.
2. Lybbert, T. J., Aboudrare, A., Chaloud, D., Magnan, N., & Nash, M. (2011). Booming markets for Moroccan argan oil appear to benefit some rural households while threatening the endemic argan forest. *Proceedings of the National Academy of Sciences*, *108*(34), 13963–13968.
3. De Waroux, Y. L. P., & Lambin, E. F. (2012). Monitoring degradation in arid and semi-arid forests and woodlands: The case of the argan woodlands (Morocco). *Applied Geography*, *32*(2), 777–786.
4. Faouzi, H., & Martin, J. (2014). Soutenabilité de l'arganeraie marocaine. Entre valorisation de l'huile d'argane et non-régénération de l'arganier. *Confins. Revue franco-brésilienne de géographie/Revista franco-brasilera de geografia* (20), 20.
5. Mellado, J. (1989). SOS Souss: Argan forest destruction in Morocco. *Oryx*, *23*(2), 87–93.
6. Charrouf, Z., & Guillaume, D. (2009). Sustainable development in Northern Africa: The argan forest case. *Sustainability*, *1*(4), 1012–1022.
7. El Alaoui, N. (1999). Paysages, usages et voyages d'*Argania spinosa* (L.) Skeels (IXe-Xe siècles). *Journal d'agriculture traditionnelle et de botanique appliquée*, *41*(2), 45–79.
8. Fassnacht, F. E., Latifi, H., Stereńczak, K., Modzelewska, A., Lefsky, M., Waser, L. T., ..., & Ghosh, A. (2016). Review of studies on tree species classification from remotely sensed data. *Remote Sensing of Environment*, *186*, 64–87.
9. Idbraim, S., Mimouni, Z., Salah, M. B., & Dahbi, M. R. (2022, October). CNN Model for Change Detection of Argania Deforestation from Sentinel-2 Remote Sensing Imagery. In *The Proceedings of the International Conference on Smart City Applications* (pp. 716–725). Cham: Springer International Publishing.
10. Amani, M., Ghorbanian, A., Ahmadi, S. A., Kakooei, M., Moghimi, A., Mirmazloumi, S. M., ..., & Brisco, B. (2020). Google earth engine cloud computing platform for remote sensing big data applications: A comprehensive review. *IEEE Journal of Selected Topics in Applied Earth Observations and Remote Sensing*, *13*, 5326–5350.
11. Kirchhoff, M., Marzolff, I., Stephan, R., Seeger, M., Aït Hssaine, A., & Ries, J. B. (2022). Monitoring dryland trees with remote sensing. Part B: Combining tree cover and plant architecture data to assess degradation and recovery of *Argania spinosa* woodlands of South Morocco. *Frontiers in Environmental Science*, *10*, 896703.
12. Aouragh, M., Lacaze, B., El Mahdad, E. H., El Aboudi, A., El Mousadik, A., Fouad, M., & Defaa, C. (2013, December). Identification des changements récents de l'occupation du sol de la commune de Temsia en Souss à partir d'images de télédétection à haute résolution spatiale. In *Deuxième congrès international de l'arganier* (pp. 99–103).
13. Setti, I., Rodriguez-Castro, A., Pata, M. P., Cadarso-Suarez, C., Yacoubi, B., Bensmael, L., ..., & Martinez-Urtaza, J. (2009). Characteristics and dynamics of Salmonella contamination along the coast of Agadir, Morocco. *Applied and Environmental Microbiology*, *75*(24), 7700–7709.
14. Msanda, F., Mayad, E. H., & Furze, J. N. (2021). Floristic biodiversity, biogeographical significance, and importance of Morocco's Arganeraie Biosphere Reserve. *Environmental Science and Pollution Research*, *28*(45), 64156–64165.

15. Karmaoui, A., & Moumane, A. (2016). Changes in the environmental vulnerability of oasean system (desert oasis) pilot study in Middle Draa Valley Morocco. *Expert Opinion on Environmental Biology*, *5*(3).
16. Gorelick, N. (2013, April). Google Earth Engine. In *EGU General Assembly Conference Abstracts* (Vol. 15, p. 11997). Vienna: American Geophysical Union.
17. Mutanga, O., & Kumar, L. (2019). Google earth engine applications. *Remote sensing*, *11*(5), 591.
18. Segarra, J., Buchaillot, M. L., Araus, J. L., & Kefauver, S. C. (2020). Remote sensing for precision agriculture: Sentinel-2 improved features and applications. *Agronomy*, *10*(5), 641.
19. Main-Knorn, M., Pflug, B., Louis, J., Debaecker, V., Müller-Wilm, U., & Gascon, F. (2017, October). Sen2Cor for Sentinel-2. In *Image and Signal Processing for Remote Sensing XXIII* (Vol. 10427, pp. 37–48). SPIE.
20. Van der Werff, H., & Van der Meer, F. (2016). Sentinel-2A MSI and Landsat 8 OLI provide data continuity for geological remote sensing. *Remote Sensing*, *8*(11), 883.
21. Sparks, T., & Menzel, A. (2013). Plant Phenology Changes and Climate Change. In *Encyclopedia of Biodiversity* (pp. 103–108). Waltham: Academic Press.

Chapter 13

PSM model for NoSQL key-value databases through model programming

A Srai and F. Guerouate

13.1 INTRODUCTION

Model-driven engineering (MDE) emerged as an extension of the "all-object" approach introduced by Object-Oriented Programming (OOP) in the 1980s. OOP advocates the use of the object as an abstraction representing real-world concepts. The object serves as an abstraction encapsulating concept-specific data and processing, thus facilitating the separation of concepts. However, this approach requires an essential modeling step to the design of relevant abstractions. It is therefore necessary to put in place methods facilitating the design and exchange of these abstractions, or models. MDE offers a set of methods for modeling and managing these models. Currently, these methods are supported by the Object Management Group, a consortium developing standards related to MDE methods. On the other hand, recent years have seen the explosion of data generated and accumulated by increasingly numerous and diversified computing devices. The created databases are referred to as "Big Data" and are characterized by the so-called 3V rule. This is due to the volume of data that may exceed several terabytes and the variety of this data, which is qualified as complex. In addition, this data is often captured at very high frequency and therefore must be filtered and aggregated in real time to avoid unnecessary saturation of storage space. Conventional implantation techniques, based mainly on the relational paradigm, have limitations in managing massive DBs. Thus, new storage systems and data manipulation have been developed. Grouped under the term NoSQL, these systems are well suited to handle large volumes of data with flexible schemas. They also provide great scalability and good response time performance. There are several types of NoSQL databases: column-oriented databases, document-oriented databases like MongoDB, graph-oriented databases like Neo4j and key-value-oriented databases like Redis. In this article, we are interested in key-value-oriented databases. The main objective of the work presented in this article is the application of the model-driven approach (MDA) on NoSQL key-value platforms. We have approved by previous works cited in references the validity and applicability of the MDA approach, also known as programming by model, on NoSQL databases, in particular document-oriented.

DOI: 10.1201/9781032664293-15

Today, with this new work, we will approve and apply the MDA approach to NoSQL key-value databases. We validated our result by a study of the Redis platform. We hope that this work will be a basis and a path for researchers in the same fields.

13.2 LITERATURE REVIEW

In the literature, several research projects have been proposed as part of the integration of the MDA approach in NoSQL databases. In Chevalier et al. (2015), the authors defined a set of rules to map a star schema into two NoSQL models: column- and document-oriented. Other studies investigated the process of transforming relational databases into a NoSQL model. Li (2010) proposed an approach to transform a relational database into HBase (a column-oriented system). Gwendal et al. (2016) describe the correspondence between a conceptual UML model and graph databases via a meta-model intermediate graph. In this work, the transformation rules are specific to graph databases used as a framework for managing complex data with many connections. Generally, this type of NoSQL system is used in social networks where data is strongly connected. To our knowledge, no work has presented a global study to transform a source model (UML diagram) into a target model (graph-oriented NoSQL databases), i.e. the generation of NoSQL database key-value via an MDA approach. According to the analysis of the cited works, we found that a majority of the authors did not invoke the point of the application of the MDA approach on NoSQL platforms through a transformation of a PIM model to a platform-specific model (PSM) or through code generation through a PSM to code transformation. To the best of our knowledge, we are the first authors who have proposed a total code generation for NoSQL platforms to use models independent of all implementation platforms. No work has presented a global study to transform a source model (UML diagram) into a target model (key-value-oriented NoSQL databases), that is, a generation of a NoSQL database key-value-oriented via an MDA approach.

13.3 RESEARCH METHOD

13.3.1 Model-driven engineering

MDE improves the development of complex systems by allowing a focus on more abstract concerns than traditional programming through models. This engineering offers a methodological framework equipped to developers of embedded real-time systems, which now focuses on the development of abstract models, rather than on concepts linked to algorithms and programming.

MDE is a software development method based on the creation and use of domain-specific models. The use of these models allows you to free yourself from the choice of a specific development platform. MDE has brought about a significant change in software engineering practices. In fact, we have moved from a development practice centered on the notion of object to a practice centered on the notion of model. In this philosophy, everything is a model, and the models expressed to describe a business domain are transformed to finally obtain PSM, in this case code. The proximity of the representation to the real world rather than to the development platform allows for much better portability than with traditional programming languages. Thus, in the software specification process, the engineer can free himself from the limitations imposed by the choice of a platform. To understand the complexity of the software, while using a very high-level specification practice, IDM is based on two mechanisms: the creation of modeling languages specific to the domain and the use of transformation engines and generators. The first mechanism is the creation of domain-specific modeling languages (DSMLs). A DSML is a modeling language created to model a particular domain, to satisfy one or more objectives. The models can then be used for analysis (verification of satisfaction of temporal constraints) or code generation. A DSML generally comes with a set of tools to meet the desired objectives. In addition to creating DSML, model transformations are used. The act of model transformation describes the transition from a model to another type of model (of the same formalism or not) or to code. A transformation is defined as a set of rules or algorithms. A rule transforms a particular fragment of the source model into a fragment of the target model.

13.3.2 Model

A model is an abstract representation of a system made with a particular intention. A model can represent a system in its entirety (the structure, behavior and non-functional properties), or it can represent just one aspect of the system by obscuring these other aspects. The different stages of an application development flow require different types of models defined with appropriate precision and containing information relevant to the use made of it. In an MDE framework, a model must be interpretable (non-ambiguous) so that it can be manipulated automatically (tooling). This is only possible if the latter is expressed in a clearly structured and interpretable language, called a meta-model.

13.3.3 Meta-models

A meta-model is a model that defines a modeling language. A meta-model precisely defines the concepts of a modeling language as well as the relationships among these concepts. A meta-model is written in a language called a metalanguage. A well-trained model conforms to its meta-model. From a well-trained model, it is possible to transform it into another model or to generate code or documentation. However, verifying this compliance relationship is important before any transformation or generation. Indeed, it is necessary to ensure that a model is syntactically and semantically compliant with its meta-model before producing an application from it.

13.3.4 Model-driven architecture (MDA)

To cope with the complexity of computer applications, software specialists have turned to MDE (MDE for Model Driven Engineering), based on modeling. It is a paradigm that is characterized by an approach to generate an application from models. The basic principle of MDE is to adopt models as central elements in the development process of an application and automate the transformations among these models. With this in mind, several specifications have been proposed, such as the Model Driven Architecture (MDA) proposed by the OMG. It is an approach built on the same basic concepts as MDE, namely the model and the meta-model. The concepts manipulated in the model, as well as the relationships between these concepts, must be expressed in a well-defined modeling language. This is done through a meta-model that defines the syntax of a specific modeling language (Figure 13.1).

13.3.5 Model transformation

In MDE, a model transformation is a program that automatically generates and modifies models. Like meta-models, model transformations are a central concept in MDE. There are several model transformation standards such as Query/View/Transformation (QVT) or MOFM2T as well as numerous model transformation languages such as ATL. A transformation is the automatic generation of a target model from a source model. There are two types of model transformations: endogenous and exogenous transformations.

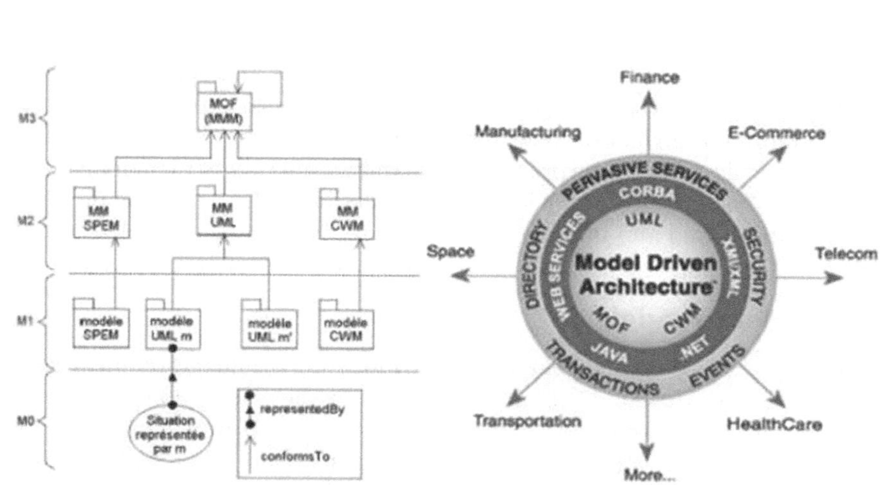

Figure 13.1 Overview of MDA principles, technologies, and architecture.

A transformation is said to be endogenous when its source model and its target model instantiate the same meta-model. For example, code refactoring or code optimization are endogenous transformations. On the contrary, a transformation is said to be exogenous when its source model and its target model do not instantiate the same meta-model. This is particularly the case in reverse engineering.

13.3.6 Model transformation

A transformation is said to be horizontal when its source model and its target model are at the same level of abstraction. For example, migrating a program from Java to C++ is an example of horizontal transformation. On the other hand, a transformation is said to be vertical when it operates at different levels of abstraction. Refinement is a vertical transformation (Figure 13.2).

13.3.7 Transformation approaches and tools

Model transformations are at the heart of the MDE approach. However, there is still no consensus on the definition and implementation of a transformation. In the literature, many approaches are proposed. To perform model

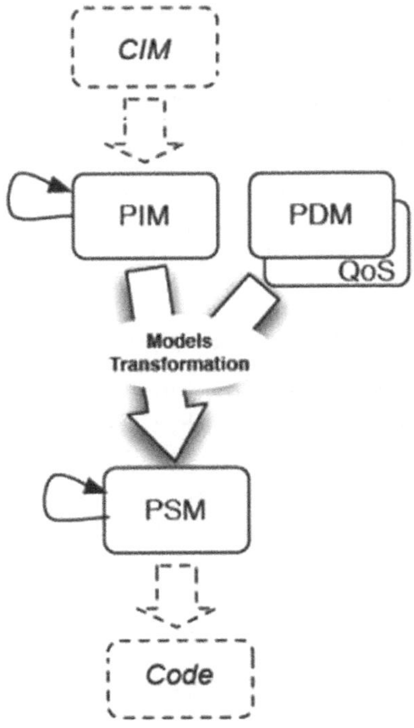

Figure 13.2 MDA: A model-driven Y-process.

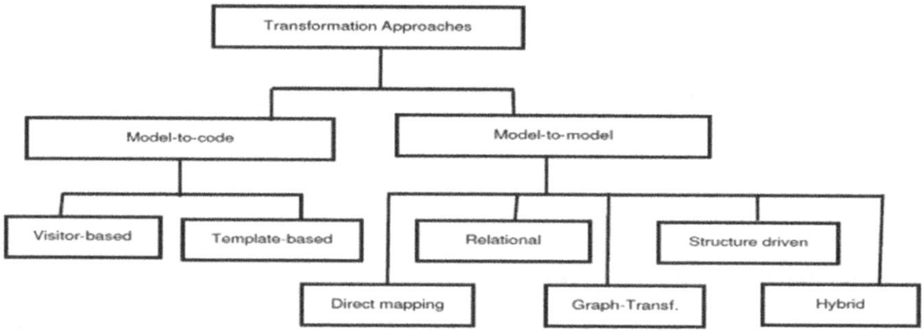

Figure 13.3 MDA: Model transformation approaches.

transformations, the latter must be expressed in a certain modeling language or meta-model. Starting from meta-model sources and targets of transformation, there are two types of transformations: endogenous and exogenous transformations. A transformation is said to be endogenous if the models involved come from the same meta-model. However, when the source models and targets are from different meta-models, the transformation is called exogenous or even translation. In view of the importance of model transformations, a particular interest has been initiated by the OMG for a standardization effort. An RFP (request for proposal) on this topic was proposed in April 2002, leading to the development of the QVT standard (Figure 13.3).

13.3.8 Syntactic and semantic transformations

A transformation is called syntactic when it only uses the syntax of the source model(s). Thus, the generation of a syntactic tree from source code by a parser is a syntactic transformation. When more complex processing takes into account the semantics of the models, the transformation is called semantic.

13.3.9 Bidirectional and unidirectional transformations

A transformation is said to be bidirectional if all the models it includes can be both source and target models. In other words, all models can be modified during a transformation execution. Otherwise, the transformation is said to be unidirectional. Bidirectional transformations form an important class of model transformations. They naturally come into play when several models must remain consistent with each other: Modifications to one model must then be reflected on the other models.

13.3.10 Transformation language: The QVT language

QVT is a standard defined by the OMG to specify transformations among models, the meta-model of which meets the MOF standard. It includes a

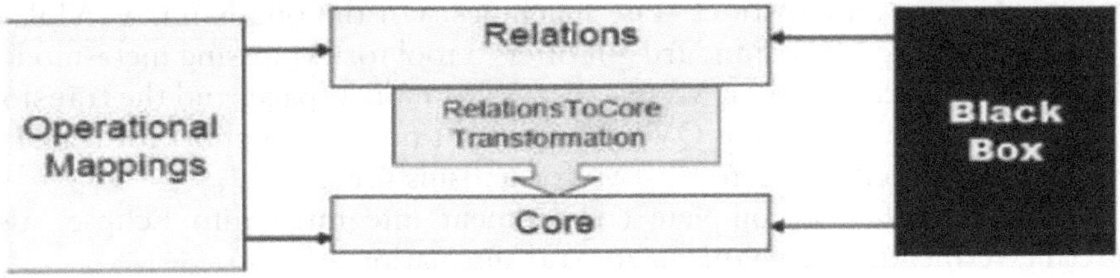

Figure 13.4 Relationships between QVT meta-models.

declarative part and an imperative part. The declarative part is made up of two parts: a part carrying out the correspondence between the two models expressed in the MOF standard called QVTr (relations) and a part that allows us to evaluate conditions on the elements of our models to make them correspond, called QVTc (Core). These two parts use OCL (Object Constraint Language) to define the correspondence rules. OCL is a formal language standardized by the OMG allowing you to specify software constraints. The imperative part, made up of QVTo (operational), makes it possible to extend the declarative language. Constructs such as for loops or if conditions are offered. QVTo also introduces the use of imperative OCL rules (Figure 13.4).

13.3.11 QVT implementations

Since the OMG proposed the QVT standard for expressing transformations, several implementations have emerged. The three main ones are ATL, operational QVT (QVTTo) and relational QVT (QVTd). There is also VIATRA2, a tool based on graph transformations, which implements certain QVT concepts. These languages are implemented in the Eclipse IDE using the Eclipse Modeling Framework (EMF) development framework. EMF is a plugin for modeling and manipulating models. EMF allows the possibility of modeling its own meta-models (complying with the MOF standard) and thus defining the source and target languages of the transformations.

QVTd implements the relational part of the language, as well as QVTCore. This plugin allows you to match elements of the source model with elements of the target model.

The rules allowing this correspondence to be conditioned are expressed using the OCL language. However, the QVTo plugin implements the imperative part of the language. This allows to define relationships when a purely declarative expression is not enough, this time relying on imperative OCL rules. These rules enrich the basic rules with more complex constructions (e.g., a while loop), while maintaining the expressiveness of the language. These two plugins offer a whole set of tools useful for their deployment on Eclipse (Java API, debugger, specific editor, etc.).

VIATRA2 implements both the declarative part and the imperative part of the language. This implementation is based on two formalisms: graph transformations and abstract state machines. On the one hand, VIATRA2 does not meet the MOF standard—it offers a tool for expressing meta-models that is more expressive than MOF called VPM model space, and the transformation language used is not QVT. However, it provides support for translating the QVT specification to the two formalisms used.

VIATRA2 offers a complete environment integrated into Eclipse, also offering verification mechanisms for transformation.

13.3.12 Processing techniques

To correctly specify and implement a transformation, one must understand the syntax and semantics of the source and target languages. There are different techniques for specifying these transformations depending on the processing carried out on the models. Transformation techniques are classified into three different categories. Direct manipulation of models: This technique uses the representation of the model in its source language and manipulates it directly through an API. The use of an intermediate representation: This time we export the model to a standardized language, which is supported by a set of tools and which allows simpler manipulation. Specifying the transformation using a language: A transformation language allows model transformations to be defined completely, independently of the source and target languages. This transformation could be generic and reusable. On the other hand, we differentiate between model-to-code and model-to-model transformations. For the implementation of model-to-model transformation, we discuss the following techniques: approaches based on direct manipulation of models—we manipulate the model directly through an API. It is in particular this method that we find in most commercial tools, which have defined their own formalism. This method nevertheless restricts the transformation to a specific language and requires advanced knowledge of the structure of the language. Relational approaches: This approach proposes defining a simple relationship between one (or more) elements of the source model and one (or more) elements of the target model. This method is the one found in the relational part of the QVT language, making it possible to carry out "pattern matching" among meta-model elements satisfying the MOF standard. Approaches based on graph transformations: represent models as graphs and apply the transformation as rules. Several works use this technique. Structure-driven approaches: They consist of two phases—first, we create the hierarchical structure of the target model from information in the source model, and then we create transformation rules for the elements of the source model that need to be modified. This technique is rarely used. One of the famous examples is OptimalJ, which has not been maintained since 2008. Hybrid approaches: a mixture of a declarative approach (relational approach) and an imperative approach allowing the elements of the source model to be extensively processed. Here we find the implementations of QVT, taking into account the relational part as well as the imperative part of the language. One of the famous tools is the ATL language.

Others: Difficult to classify, the use of the OMG CWM standard, for example, or the manipulation of XML representations with XSLT.

The use of XSLT is a mixture of the use of a transformation language (XSLT) and a representation intermediate (XML).

13.4 RESULTS AND DISCUSSIONS

13.4.1 The transformation rules M2M

We performed a model-to-model transformation, for that, we used the QVT model transformation language, and we developed a design of the source meta-model and target meta-model. The transformation of a model or the generation of models from another model through my programming by model is often a complicated task; it requires, first of all, an in-depth study of the architectures; this study makes it possible to build what we call a CIM model; this model is not discussed in this work because we want to focus our work on the development and implementation of model-based programming on NoSQL platforms.

13.4.2 UML source meta-model

Figure 13.5 illustrates the simplified UML source meta-model based on packages, including operations, associations and classes. Those classes are composed of properties with parameters.

13.4.3 NoSQL key-value target meta-model

After an in-depth study, we managed to develop a meta-model for the NoSQL key-value platform (Figure 13.6).

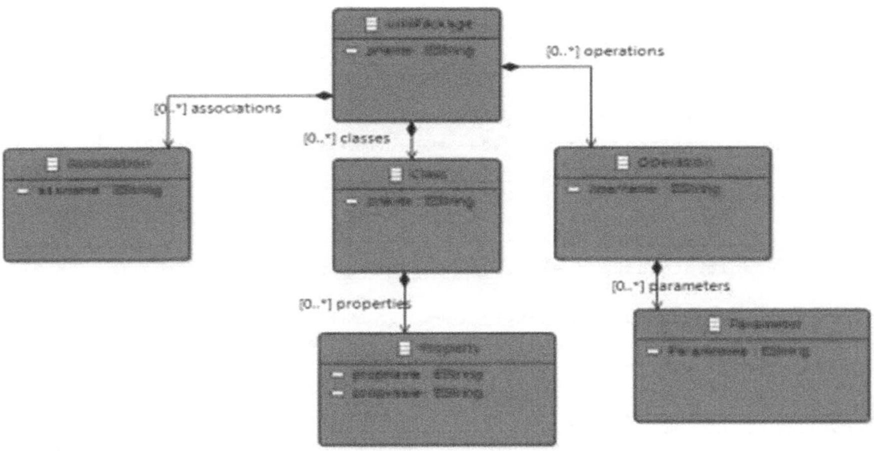

Figure 13.5 Simplified UML source meta-model.

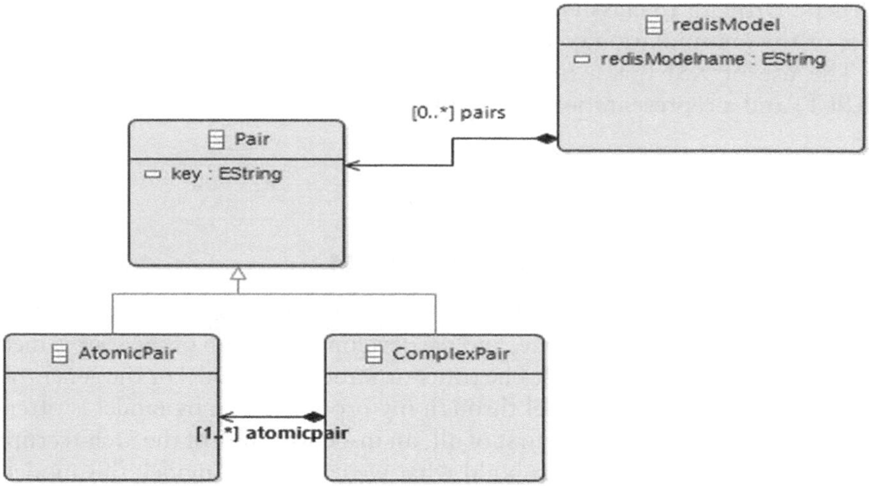

Figure 13.6 Simplified NoSQL key-value target meta-model.

The transformations use a UML-type model as input and output a NoSQL key-value target meta-model. The first transformation rule establishes the correspondence among all UML package and Redis model elements of the key-value database (Figure 13.7).

```
modeltype umlredis uses umlredis("http://umlredis.mm");
modeltype redismodel uses redismodel("http://redisModel.mm");

transformation transf1(in source:umlredis, out target:redismodel);

main() {

        source.rootObjects()[umlPackage]->map UmlRedisToRedisModel();
}
mapping umlPackage :: UmlRedisToRedisModel() : redisModel
        {
        result.redisModelname := self.pname;
        result.pairs+=self.classes->map ClassToPair();
}
mapping Class :: ClassToPair() : Pair
        {
        result.pairname := self.cname;
        result.atomicpairs+=self.properties-> map PropertyToAtomic_Pair();
}
mapping Property :: PropertyToAtomic_Pair() : AtomicPair
        {
        result.key := self.propname;
        result.value := self.propvalue;
}
```

Figure 13.7 M2M transformation with QVT from UML to NoSQL key-value Redis model.

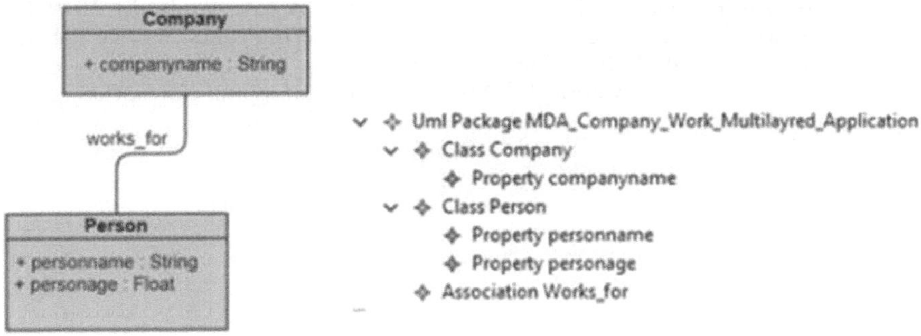

Figure 13.8 Class diagram EMF model and class diagram instance model.

```xml
<?xml version="1.0" encoding="UTF-8"?>
<redismodel:redisModel xmi:version="2.0" xmlns:xmi="http://www.omg.org/XMI" xmlns:xsi="http://www.w3.org/2001/XMLSchema-instance"
xmlns:redismodel="http://redisModel.mm" xsi:schemaLocation="http://redisModel.mm ../metamodels/Redis.ecore" redisModelname="MDARedis">
  <pairs painame="Company">
    <atomicpairs key="companyname" value="MDA and Redis"/>
  </pairs>
  <pairs painame="Person">
    <atomicpairs key="personname" value="srai"/>
  </pairs>
</redismodel:redisModel>
```

Figure 13.9 NoSQL key-value PSM target model.

To validate our transformation rules, we conducted several tests. For example, we have considered the class diagram composed of the classes Company and Person (see Figure 13.8).

After applying the transformation on the UML source model, we generated the NoSQL key-value PSM target model (see Figure 13.9).

This transformation uses a UML type model as input and a NoSQL key-value database model as output. The first transformation rule maps all elements of the UML package to the Redis type element of the key-value database.

13.5 CONCLUSION

In this article, we proposed an MDA approach to generate a PSM model for key-value NoSQL platforms. The transformation rules were developed using QVT. This work should be extended to allow the generation of other NoSQL solutions such as document- and graph-oriented. Then we can consider integrating other big data platforms like HBase, Cassandra and others. As a perspective, we want to generate an n-tier platform that integrates a NoSQL key-value database, all with model-based programming.

REFERENCES

Chevalier, M., El Malki, M., Kopliku, A., Teste, O., & Tournier, R. Implementing multidimen- sional data warehouses into NoSQL. In ICEIS (2015).

Li, C. Transforming relational database into HBase: A case study. In 2010 IEEE International Conference on Software Engineering and Service Sciences, DOI:10.1109/ICSESS.2010.5552465, July 2010.

Gwendal, D., Sunyé, G., & Cabot, J. UMLtoGraphDB: mapping conceptual schemas to graph databases, Conceptual Modeling: 35th International Conference, ER 2016, Gifu, Japan, November 14-17, 2016, Proceedings 35.

Chapter 14

Body temperature screening during COVID-19 pandemic

Addaali Bouthayna, Rachid Latif, and Amine Saddik

14.1 INTRODUCTION

During three years, the world has witnessed a significant increase in human and economic losses due to COVID-19. The COVID-19 pandemic is caused by severe acute respiratory syndrome coronavirus 2 (SARS-Cov-2), which was identified for the first time in Wuhan in December 2019 and caused millions of cases and deaths. To curb virus spread, governments have imposed screening protocols including wearing masks, disinfecting hands, keeping a distance of 2 m from others, and checking the body temperature at the entrances of public places. The fast spread of this pandemic, the lack of staffing, and the significant shortage of medical equipment, especially in developing countries have made healthcare professionals over-burdened but also put them at risk of infection.

However, vital signs monitoring of infected patients especially body temperature which is used as one of the key parameters to determine whether the person has COVID-19 is necessary [1–3]. Several works have proved the feasibility of remote monitoring systems dedicated to body temperature measurement, Somboonkaew et al. [4] have built a system to measure temperature and detect patients with fever using RGB and IR cameras at 1 m away. The system gives an alarm sound when the measured temperature is above the desired setting threshold and shows 100% sensitivity and 70% specificity. Lin et al. [5] have introduced a non-contact continuous body temperature measurement system that detects and tracks faces in thermal images using a trained model based on deep learning. A combination of the Single-Shot-Multibox Detector and MobileNet is adopted for face detection and Kernel Correlation Filter is used for face tracking. After face detection and tracking, the forehead is located as a region of interest, and then the body temperature is calculated. Ornek et al. [6] used infrared thermography and deep Convolutional Neural Networks (CNNs) together for the first time to monitor the body temperature of 38 neonates. The algorithm of this work starts with neonatal thermal image acquisition, then the data augmentation method is used to increase the number of images required for CNN training. The CNN model is created to classify the neonates as healthy and unhealthy. To evaluate the performance

DOI: 10.1201/9781032664293-16

of the classification, sensitivity, specificity, and accuracy metrics have been calculated.

This review aims to clarify the importance of body temperature as a vital sign in detecting different diseases and infections such as COVID-19 as well as to highlight the two types of body temperature (core and peripheral), impacting factors on both, how the body deals with increases or decreases in temperature and the different tools (contact and non-contact) used to determine body temperature.

This chapter is organized as follows, the first section is divided into two parts the first one presents the types of body temperature and factors causing changes in it and the second part defines the thermoregulatory system. The second section gives a brief history of thermometers and the third section is dedicated to contactless tools used for body temperature measurement and applications-based thermal cameras during the COVID-19 pandemic.

14.2 BODY TEMPERATURE: A CRUCIAL METRIC IN CLINICAL DIAGNOSIS

14.2.1 Types and impact factors on body temperature

Core body temperature (CBT) is the temperature of the inner organs such as the brain, heart, and liver, it is relatively constant (between 36.5 and 37.5°C) but can either increase or decrease due to multiple physiological variations [7]. Changes in CBT take place depending on diurnal variations usually referred to as the circadian rhythm of CBT (temperature is lowest during sleep and highest in late afternoon), sex (augmentation of female CBT during ovulation and pregnancy), age (highest for children and lowest for elderlies), prolonged exposure to a hot or cold environment, mental and inflammatory diseases, physical activity (during physical exercise muscles generate heat which raises body temperature), and emotional state (happiness, anger, and stress elevate body temperature while depression reduces it) [8–10]. The elevated value of CBT is known as hyperthermia or fever but there is a difference between the two. Hyperthermia is an uplift of the CBT above the set point due to one or more of the reasons mentioned above, in this case, the hypothalamus tries to return the body to its normal temperature while fever is a rise in CBT caused by the hypothalamus to resist diseases and infections. Hypothermia on the other hand is a decrease in body temperature below the level supported by the human body. Maintaining CBT at an appropriate level (balance between heat production and heat loss) is crucial to ensure the efficiency of the biochemical processes within the living cells (metabolism) and to protect the body from the serious repercussions of hyperthermia or hypothermia that can sometimes be deadly [11]. Invasive ways like the pulmonary artery catheter or the bladder thermistors are more accurate for CBT measurement but not favorable. The rectal thermometer gives the nearest values to the invasive measurements but is accompanied by the risk of infections [12, 13].

Skin body temperature (SBT) on the other side presents the temperature of the outermost surface of the body such as the skin, forehead, and armpit. It ranges between 33.5 and 36.9°C and is easier to measure non-invasively but has a larger margin compared to CBT. Several factors impact SBT including measurement site (normal temperature differs between the sites of measurement, normal temperature value in the forehead or armpit is between 36.4 and 36.7°C while in the ear is between 37.3 and 37.6°C), ambient temperature (the body-environment heat exchange can be via radiation through electromagnetic waves, convection e.g., when your body is warm and you take a cold shower or swim, heat transfers from your body to the water molecules, conduction which is the heat transfer resulting from the direct contact between skin and warm or cold objects and finally evaporation, when the environmental temperature raises, water on the skin evaporates which cools the body), and CBT (elevated CBT leads to a high blood flow resulting in high skin body temperature and vice versa) [14–16].

14.2.2 Thermoregulatory system

The human body is characterized by homeothermy which is provided mainly by the thermoregulatory system. This mechanism responsible for maintaining the body heat balance consists of thermoreceptors, thermoregulatory centers, and effector organs. Thermoreceptors are sensors located in the anterior hypothalamus (central thermoreceptors) as well as in different areas of the skin (peripheral thermoreceptors) that are sensitive to body temperature changes [17]. When body temperature decreases, thermoreceptors send electrical impulses to the thermoregulatory center (hypothalamus) that, in turn, sends signals to effector organs such as skin blood vessels that constrict which minimize the blood flow through the capillaries under the surface of the skin preventing heat dissipation, this operation is known as vasoconstriction [18]. Piloerection is another mechanism also responsible for increasing internal body temperature. In case of body exposure to cold, arrector pili muscles (small smooth muscles attached to hair follicles) contract causing the erection of skin hairs which constitute an insulating layer of air around the skin to reduce heat loss since the air is a bad conductor of heat [19]. If vasoconstriction and piloerection are insufficient, the body starts producing heat by shivering which is an involuntary muscle contraction and relaxation occurs in response to the stimulation of the center of shivering in the posterior hypothalamus [20].

Besides the response to the decrease in body temperature, the thermoregulatory system restores the temperature to its normal level if it increases through the vasodilation mechanism, the augmentation of body temperature gives rise to the dilation of blood vessels which increases the blood flow in the capillaries resulting in elevated heat loss across the epidermis [21]. The sweating mechanism also helps the body to get cold through sweat glands that produce sweat, sweat evaporation requires energy which releases heat from the body [22].

14.3 TRADITIONAL MEASUREMENT TOOLS OF BODY TEMPERATURE

Thermoscopes were the first instruments used for temperature measurement invented by Galileo Galilei based on studies that have proved that gases, notably air, are affected by temperature changes. The thermoscope consists of a tube with its upper part attached to a bulb and its lower part submerged in a liquid such as water, oil, alcohol, and mercury. Fluctuations in temperature lead to air expansion and contraction which introduce liquid level changes [23].

The origin of the word thermometer is the Greek words thermos "hot" and metron "measure". The difference between a thermometer and a thermoscope is that the thermometer possesses a scale, while the thermoscope does not. Several scales were suggested by scientists but the first universal scale was proposed by Daniel Gabriel Fahrenheit in 1724 and the second by Anders Celsius in 1742 [24]. The thermometer has boomed in medical services since the 19th century after the invention of a portable clinical thermometer by Thomas Clifford Allbutt in 1866 and its development over the years, especially in terms of length and response time [25]. The first medical thermometer was a liquid-in-glass thermometer, containing mercury most of the time and colored alcohol sometimes. Mercury in glass is the most accurate liquid in glass thermometers, it has been used for decades to monitor body temperature in medical services but should always be protected from breakage since mercury is toxic, multiple countries have banned the use of this tool to avoid damaging environment and health which increased the rate of use of other technologies such as electronic digital thermometers [26, 27]. To obtain an appropriate value of temperature, a mercury thermometer should be positioned in the measurement site (mouth, rectum, or armpit) and left for a sufficient time. The invention of the commercially viable thermistor by Samuel Ruben in 1930 led to the development of the electronic digital thermometer, this final consists of a thermistor sensitive to temperature changes where the output is linked to an analog-to-digital converter (ADC), the obtained digital signal is processed by a microcontroller and sent to an LCD to read the measured value [28].

14.4 CONTACTLESS BODY TEMPERATURE SCREENING

14.4.1 Infrared thermometers

As we explained beforehand, principal body-environment interactions are radiation, convection, conduction, and evaporation, during radiation heat is transferred from the body to the environment in the form of electromagnetic waves, infrared thermometer detects these waves and converts them to a numerical value that reflects skin body temperature. Alone or in a complex system, infrared thermometers were widely used during the COVID-19 pandemic to overcome the virus outbreak. Attention has focused on this tool to develop its accuracy. Reference [29] assessed the reliability of body temperature measurements obtained with infrared point thermometers for COVID-19 detection and studied the impact of working distance, angle of inclination, and light conditions on temperature measurements. The experiments affirm that those factors can strongly impact values obtained which could invalidate the results. Ramelan et al. [30] have proposed a low-cost and portable contactless body temperature measurement system based on the Internet of Things (IoT). The algorithm used consists of the perception layer, where the body temperature data is collected in real-time using a GY-906 Infrared Thermometer sensor and then processed through an Arduino Uno microcontroller, the transport layer where the data obtained is uploaded via the internet network (Wi-Fi) using the Hypertext Transfer Protocol (HTTP) Post protocol, and application layer where the body temperature data is displayed in tabular or graphic form on the mobile application and the web application.

14.4.2 Thermal camera

Thermal cameras have the same working principle as infrared thermometers, they were largely used during the COVID-19 pandemic to measure body temperature both in clinical and non-clinical applications. Several studies have developed models for body temperature screening to prevent the spread of the virus. Paramasivam et al. [31] have designed an (IOT) system for temperature reading and disinfection. The system is a pathway chamber that contains an IR thermal camera used to estimate the CBT by reading the skin surface temperature, a UV disinfection system for viruses' elimination, and a control station that opens and closes the automated barrier gate after temperature measurement and hand sanitization. The system also uses blockchain technology for storing and managing data. Ulleri et al. [32] have built a contactless employee management system to mark the presence of employees, make sure they wearing masks, and measure their body temperature during the COVID-19 pandemic. A CMOS camera was used for obtaining the facial data of the person and a thermal camera was used for measuring infrared radiations emitted by the person's face. Several machine learning algorithms were implemented on a Raspberry Pi to ensure face Recognition, mask detection, and body temperature measurement. Figure 14.1 shows a flowchart of the proposed system in work [32].

Zeng et al. [33] have proposed a system based on the temperature difference method, the system combines visible and infrared thermal imaging for face recognition and real-time temperature screening. The proposed method aims to detect patients with COVID-19 fever symptoms in public places and eliminate environmental interference and equipment errors that have a huge impact on the obtained values. Reference [34] presents a free-flow fever screening system based on deep learning algorithms that can measure body temperature without contact, in real-time, and for multiple persons simultaneously. Body temperature was measured with a big accuracy using visible and thermal cameras even when the face was partially covered or the acquired

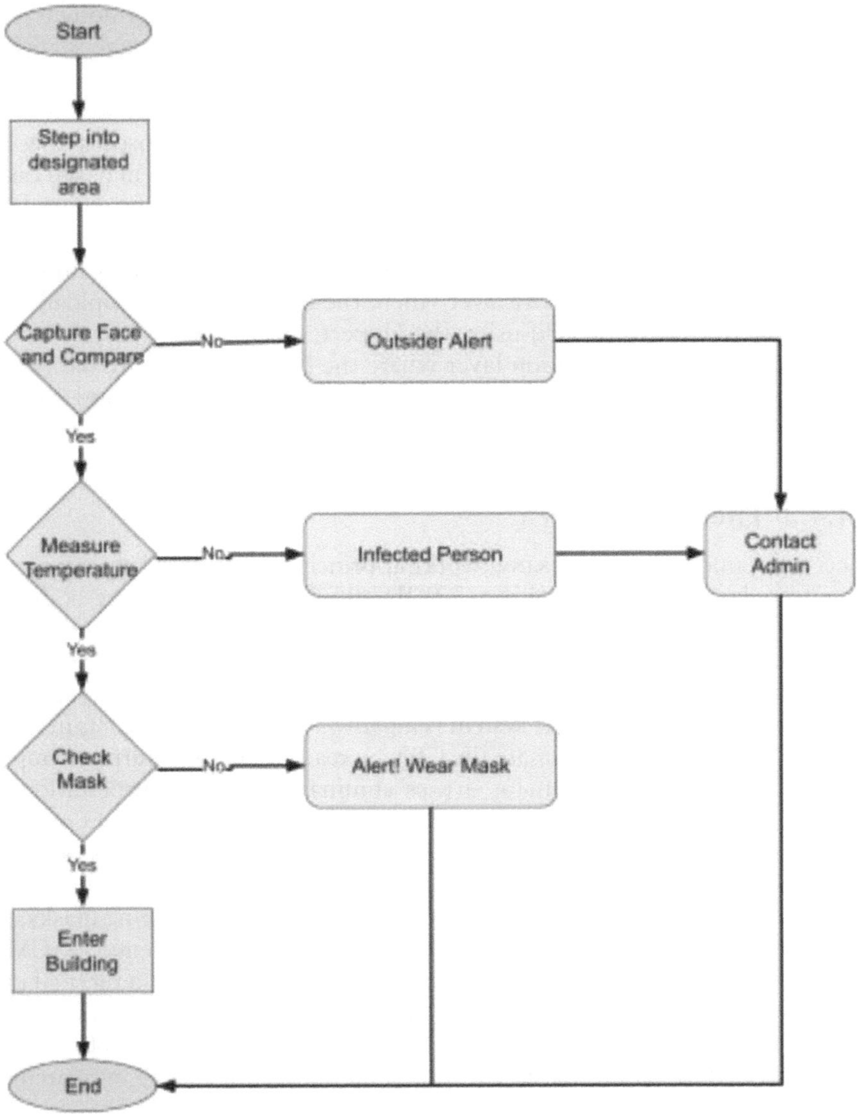

Figure 14.1 Flowchart of the proposed system in work [32].

images were blurred. Netinant et al. [35] developed a smart infrared thermal camera system using a Long-Wave Infrared (LWIR) micro thermal camera and a Raspberry PI 4 and exploiting IoT technologies to improve the accuracy of measurement, different temperatures between 35 and 39°C and two distances (100 m and 150 m) were used in the experience to study the impact of these two factors on body temperature measurement and finally, the obtained

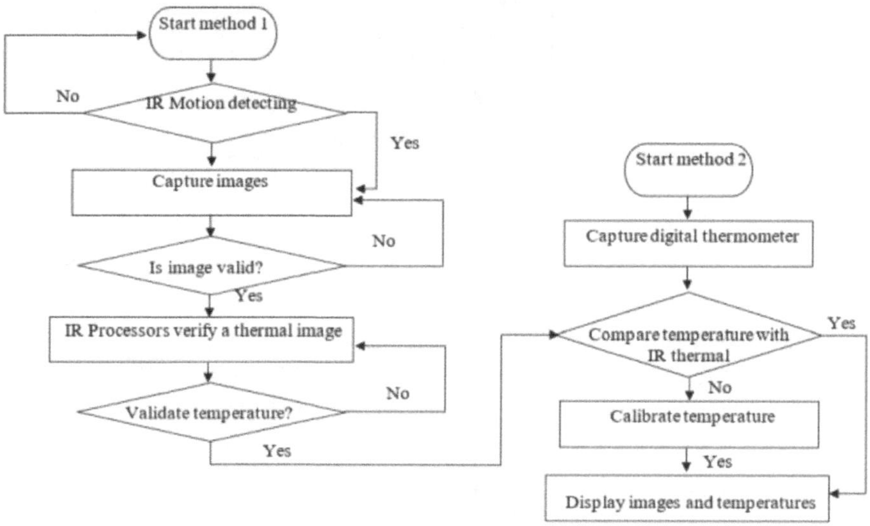

Figure 14.2 Body temperature screening algorithm in work [35].

values were compared to those obtained with Infrared Thermometer to calibrate the values. Figure 14.2 presents the algorithm used in work [35] for body temperature measurement using an Infrared camera.

Abdulrazaq et al. [36] designed a real-time body temperature measurement system using a thermal camera integrated into a smart helmet and combined with IoT technology. The data is collected using a thermal camera and processed to know if the body temperature is normal or not, if there is a high temperature detected; the optical camera captures the face, specifies the GPS location of the concerned person, and sends a notification to health officer with the captured face and GPS location. Figure 14.3 shows the workflow of the proposed system by Abdulrazaq et al.

In [37] authors have also used an artificial-intelligence-enabled IoT-based system implemented into a UAV for COVID-19 scanning. The algorithm starts with face detection using deep learning algorithms and facial features are extracted to determine the person's name based on his registration in the created web portal. After detecting the face, the body temperature is measured from the person's forehead to cheek if there is a fever. In case of high-temperature detection, a notification will be sent to the concerned person. Figure 14.4 presents the algorithm of work [37].

Peddinti et al. [38] have proposed a framework for real-time detection of COVID-19 cases in public places. The process starts with capturing people with thermal cameras operating as surveillance cameras at the airport, then separating the foreground from the background to remove any effect of background on measured values. An architecture of CNN is applied to people's thermal images to detect infected cases and a decision of quarantining the

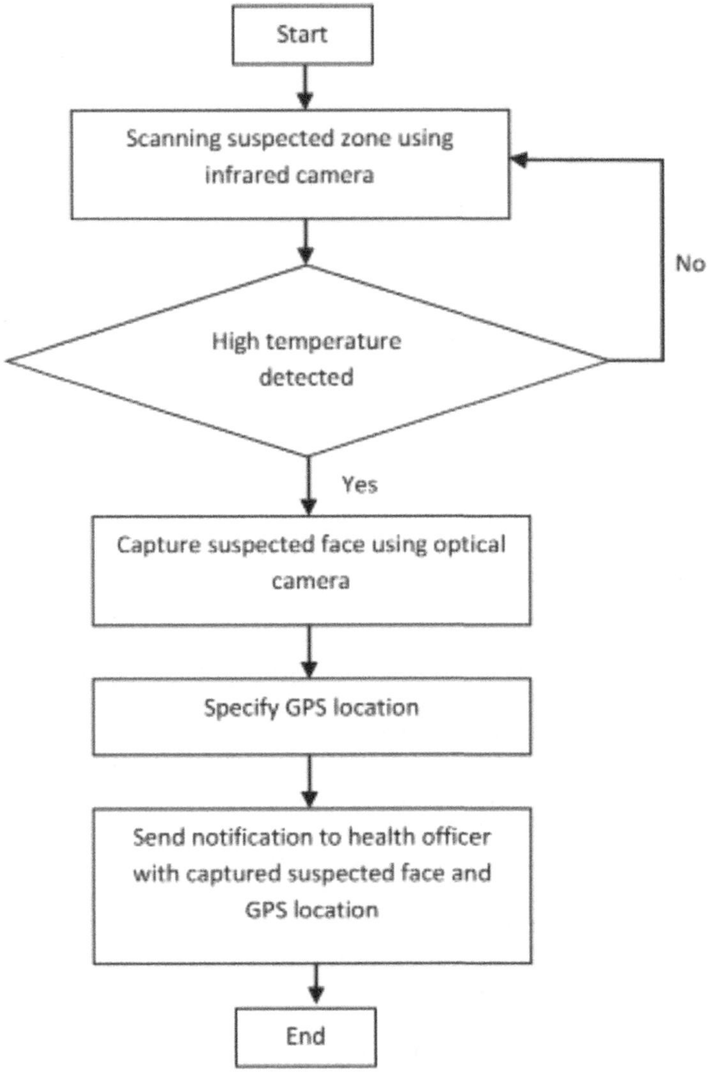

Figure 14.3 Proposed system by Abdulrazaq et al.

person or not is made by the algorithm and sent to the authority. Figure 14.5 presents the proposed algorithm for real-time detection of infected people by COVID-19.

Chin et al. [39] evaluated the influence of subject-sensor distance on remote temperature screening and proposed a real-time temperature screening system that used a thermal image, corresponding face box coordinates, sensor-subject distance, and ambient temperature reading as inputs into the thermal compensation model to estimate the corrected temperature values.

Algorithm 1: Proposed Body Temperature Identification algorithm

Input:
$V_{th} \leftarrow$ video recorded from thermal camera,
$V_n \leftarrow$ video recorded through normal camera
$L \leftarrow$ length of video in seconds
Output: A list of potential COVID-19 patients
for *i in L* **do**
  $v^i_{th} \leftarrow$ captured 30 frames per second by thermal camera
  $v^i_n \leftarrow$ captured 30 frames per second by normal camera
  // taking 5 frames per second only
  for *j in [1, 7, 13, 19, 25]* **do**
    $I_{th} \leftarrow j^{th}$ frame of thermal camera
    $I_n \leftarrow j^{th}$ frame of normal camera
    // extracting region of interest (ROI)
    ROI $\leftarrow$ YOLOv3(I_n)
    // extracting face's overlay coordinates and the person name/ID using face recognition system (FRS)
    (face_overlay, IDs) $\leftarrow$ FRS(I_n, ROI)
    // finding febrile and afebrile list of people using TIS
    (febrile, afebrile) $\leftarrow$ TIS(I_{th}, face_overlay, IDs)
    // sending notification to people in febrile list
    message(febrile)
  end
end

Figure 14.4 Algorithm proposed by Barnawi et al.

Those values are evaluated based on an alarm threshold and depicted as green for normal values and red for elevated values. Face box coordinates are obtained based on upper body key point information, and whole-body key points are detected using OpenPose real-time multi-person 2D pose estimation. Figure 14.6 shows the schematic diagram of the temperature screening system used in the work [39].

Zhou et al. [40] evaluated the performance of Infrared thermographs used for fever detection during epidemics. Multiple facial locations including the forehead, canthi, mouth, and entire face have been extracted from face images of 596 subjects and compared with oral thermometer measurements that have been used as reference. Results show that the accuracy of measurement depends on the Region of Interest (ROI). In inner canthi and full-face regions, fever detection is more accurate than the other facial locations. Figure 14.7 presents the flowchart of the temperature measurement procedure used in the work [40].

Table 14.1 presents some references that have developed systems for contactless body temperature measurement during COVID-19 and its specifications.

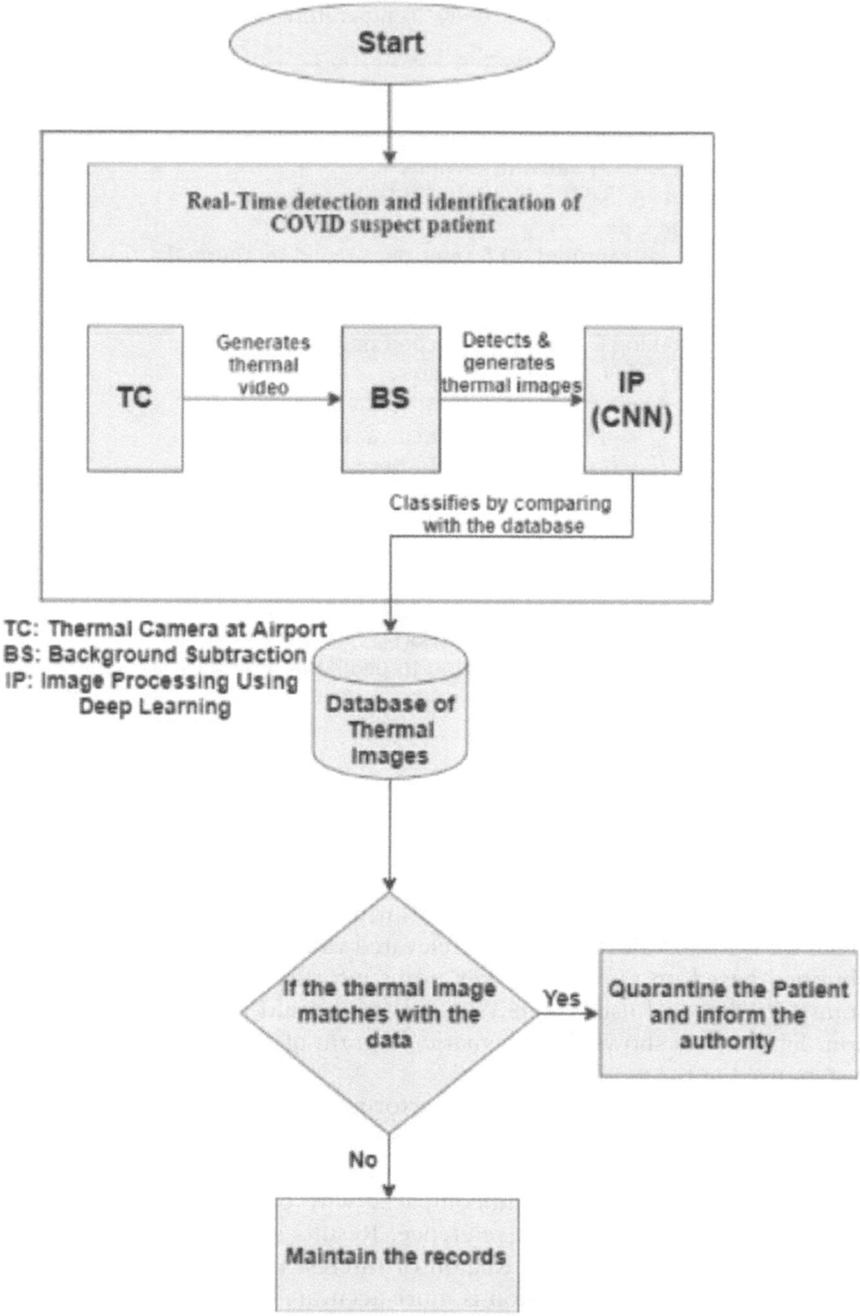

Figure 14.5 Algorithm for detection of people infected by COVID-19 as proposed in work [38].

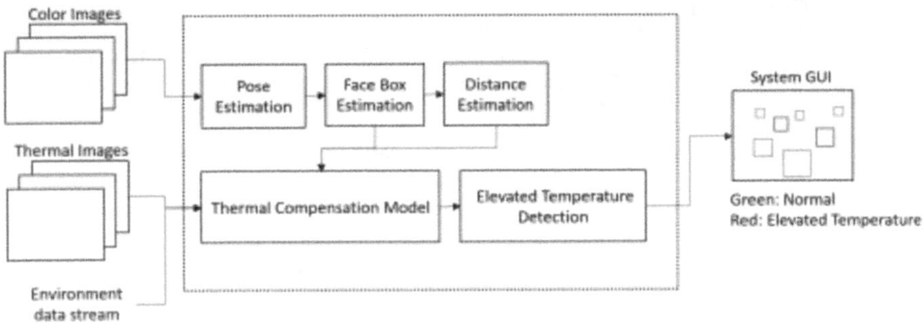

Figure 14.6 Schematic diagram of temperature screening system used in work [39].

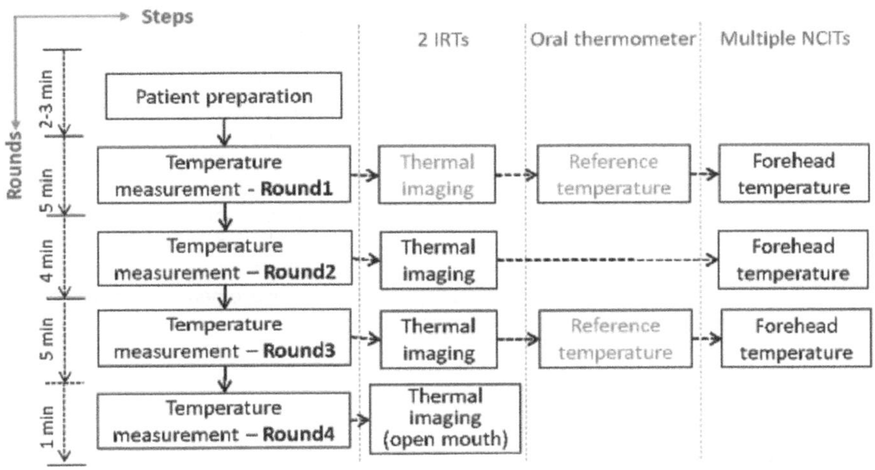

Figure 14.7 Flowchart of temperature measurement procedure used in work [40].

Table 14.1 Systems based thermal cameras for body temperature measurement

Reference	*Cameras used*	*FPS/resolution of cameras*	*ROI*	*Software*	*Accuracy assessment*
[31]	IR thermal camera (Mi-Fraht-800)	–	Facial region	–	–
[32]	5MP CMOS Camera + FLIR Thermal Camera	30fps/2592 × 1944 (CMOS Camera)	Facial region + forehead	Pre-trained TensorFlow and OpenCV models for ROI detection	90% accuracy

(Continued)

Table 14.1 (Continued)

Reference	*Cameras used*	*FPS/resolution of cameras*	*ROI*	*Software*	*Accuracy assessment*
[34]	RGB + thermal camera	8fps/1280 × 960(RGB) and 336 × 252(thermal)	Eyes and forehead, face and head	Feed-forward neural network	100% sensitivity and 96.9% specificity
[36]	Thermal camera + optical camera	–	Facial region	EmguCV cross-platform, OpenCV, C#. Net	–
[37]	Thermal camera	–	Human body	Background subtraction model + (CNN)	–
[38]	Logitech BRIO color camera + FLIR E8-XT thermal camera	3840 × 2160(RGB) and 320 × 240(thermal)	Subject's forehead	The OpenPose real-time multi-person 2D pose estimation algorithm + proposed thermal compensation model for temperature measurement	The Pearson correlation coefficient (PCC) is 0.9295, average mean absolute error (MAE) is 0.588, and average root-mean-square error (RMSE) is 0.677

14.5 DISCUSSION

This review aims to present a state of the art of body temperature screening using telemonitoring during the COVID-19 pandemic. The accuracy of body temperature measurement using thermal imaging can be impacted by multiple factors that should be taken into consideration in any future work dedicated to body temperature screening using thermal imaging. One of the most important factors is ambient temperature. As mentioned in Section 2.1, human body temperature is affected by ambient temperature. When it is too high or too low, the thermoregulatory system tries to calibrate it. The accuracy of measurement of body temperature using thermal imaging also depends on ambient temperature. In [41] authors have performed an experience in two different conditions. In the first one, a constant temperature air-conditioned environment is used and in the other, an environment without air-conditioning is used. Results show that with the increase of outdoor environment temperature, the human body temperature measured by infrared imaging is gradually rising, which makes it difficult to distinguish between the persons who have fever from the normal ones. To mitigate the influence of ambient temperature, Rayanasukha et al. have developed a self-compensation technique built into a thermal camera. The system is based on a 3D depth sensor, an electronic temperature sensor, and a reference temperature. Results of the experience show a low standard deviation of 0.10°C under 21.0–40.0°C ambient temperature [42]. The second impact factor is the distance between the person and the camera which directly impacts the accuracy of measurement. In work [35], authors have measured body temperature using thermal cameras at 100 cm and 150 cm from the subject, the results show that the farther the IR thermal camera from the object is, the more inaccurate. For this reason, multiple works have proposed methods for reducing the impact of measuring distance on temperature measurement accuracy. Zhang et al. [43] have developed a method that reduces the effect of the distance on the accuracy of measurement. The method acts on three aspects influencing principles including the angular field of the infrared camera, the contrast between tested object temperature and environmental temperature, and the atmospheric transmittances. Those aspects change along with measuring distance.

14.6 CONCLUSION

Since the onset of the virus, scientific research has tried to develop mechanisms to curb the rapid outbreak of the pandemic. Researchers therefore took a great interest in the subject of remote monitoring of vital signs. In this review, we have shown the important role that remote body temperature detection tools play (infrared thermometers and thermal cameras) to prevent the spread of COVID-19. But first, it was necessary to understand the physiological aspect of body temperature, both core and peripheral body temperatures are impacted by several factors (age, sex, ambient temperature, physical activities, etc.), and this is where the thermoregulatory system comes in. This chapter also presents briefly the historical development of medical thermometers.

ACKNOWLEDGMENT

We owe a debt of gratitude to the Ministry of National Education, Vocational Training, Higher Education and Scientific Research (MENFPESRS) and National Centre for Scientific and Technical Research of Morocco (CNRST) for its financial support for the project Cov/2020/109.

REFERENCES

1. A. Bella, R. Latif, A. Saddik, and L. Jamad, "Review and Evaluation of Heart Rate Monitoring Based Vital Signs, A Case Study: Covid-19 Pandemic," in *2020 6th IEEE Congress on Information Science and Technology (CiSt)*, Jun. 2020, pp. 79–83, doi: 10.1109/CiSt49399.2021.9357302.

2. A. Bella, R. Latif, A. Saddik, and F. Z. Guerrouj, "Monitoring of physiological signs and their impact on the Covid-19 pandemic: Review," *E3S Web Conf.*, vol. 229, p. 01030, 2021, doi: 10.1051/e3sconf/202122901030.
3. A. Al-Naji, G. A. Khalid, J. F. Mahdi, and J. Chahl, "Non-contact SpO2 prediction system based on a digital camera," *Appl. Sci.*, vol. 11, no. 9, Art. no. 9, Jan. 2021, doi: 10.3390/app11094255.
4. A. Somboonkaew *et al.*, "Mobile-Platform for Automatic Fever Screening System Based on Infrared Forehead Temperature," in *2017 Opto-Electronics and Communications Conference (OECC) and Photonics Global Conference (PGC)*, Jul. 2017, pp. 1–4, doi: 10.1109/OECC.2017.8114910.
5. J.-W. Lin, M.-H. Lu, and Y.-H. Lin, "A Thermal Camera Based Continuous Body Temperature Measurement System," in *2019 IEEE/CVF International Conference on Computer Vision Workshop (ICCVW)*, Oct. 2019, pp. 1681–1687, doi: 10.1109/ICCVW.2019.00208.
6. A. H. Ornek, M. Ceylan, and S. Ervural, "Health status detection of neonates using infrared thermography and deep convolutional neural networks," *Infrared Phys. Technol.*, vol. 103, p. 103044, Dec. 2019, doi: 10.1016/j.infrared.2019.103044.
7. P. A. Mackowiak and G. Worden, "Carl Reinhold August Wunderlich and the evolution of clinical thermometry," *Clinical thermometry. Clinical infectious diseases*, vol. 18, no. 3, pp. 458–467, Mar. 1994, doi: 10.1093/clinids/18.3.458.
8. K. Kräuchi, "How is the circadian rhythm of core body temperature regulated?," *Clinical Autonomic Research*, vol. 12, no. 3, pp. 147–149, Jun. 2002, doi: 10.1007/s10286-002-0043-9.
9. S.-H. Lu and Y.-T. Dai, "Normal body temperature and the effects of age, sex, ambient temperature and body mass index on normal oral temperature: A prospective, comparative study," *Int. J. Nurs. Stud.*, vol. 46, no. 5, pp. 661–668, May 2009, doi: 10.1016/j.ijnurstu.2008.11.006.
10. L. Nummenmaa, E. Glerean, R. Hari, and J. K. Hietanen, "Bodily maps of emotions," *Proc. Natl. Acad. Sci.*, vol. 111, no. 2, pp. 646–651, Jan. 2014, doi: 10.1073/pnas.1321664111.
11. A. Nixdorf-Miller, D. M. Hunsaker, and J. C. Hunsaker III, "Hypothermia and hyperthermia medicolegal investigation of morbidity and mortality from exposure to environmental temperature extremes," *Arch. Pathol. Lab. Med.*, vol. 130, no. 9, pp. 1297–1304, Sep. 2006, doi: 10.5858/2006-130-1297-HAHMIO.
12. D. M. Nierman, "Core temperature measurement in the intensive care unit," *Crit. Care Med.*, vol. 19, no. 6, pp. 818–823, Jun. 1991, doi: 10.1097/00003246-199106000-00015.
13. D. S. Moran and L. Mendal, "Core temperature measurement: Methods and current insights," *Sports medicine*, vol. 32, no. 14, pp. 879–885, 2002, doi: 10.2165/00007256-200232140-00001.
14. C. M. Lee, S.-P. Jin, E. J. Doh, D. H. Lee, and J. H. Chung, "Regional variation of human skin surface temperature," *Ann. Dermatol.*, vol. 31, no. 3, pp. 349–352, Jun. 2019, doi: 10.5021/ad.2019.31.3.349.
15. G. Pezzagno, "[Heat exchange between human body and environment (theoretical bases of physiological measurement and evaluation)]," *G. Ital. Med. Lav. Ergon.*, vol. 21, no. 3, pp. 159–205, 1999.
16. J. A. Stolwijk, "Heat exchangers between body and environment," *Bibl. Radiol.*, no. 6, pp. 144–150, 1975.
17. R. F. Hellon, "Central thermoreceptors and thermoregulation," in *Enteroceptors*, B. Andersson, M. Fillenz, R. F. Hellon, A. Howe, B. F. Leek, E. Neil, A. S. Paintal, J. G. Widdicombe, and E. Neil, Eds., in Handbook of Sensory Physiology. Berlin, Heidelberg: Springer, 1972, pp. 161–186, doi: 10.1007/978-3-642-65252-3_5.

18. B. K. Alba, J. W. Castellani, and N. Charkoudian, "Cold-induced cutaneous vasoconstriction in humans: Function, dysfunction and the distinctly counter-productive," *Exp. Physiol.*, vol. 104, no. 8, pp. 1202–1214, 2019, doi: 10.1113/EP087718.
19. E. A. Tansey and C. D. Johnson, "Recent advances in thermoregulation," *Adv. Physiol. Educ.*, vol. 39, no. 3, pp. 139–148, Sep. 2015, doi: 10.1152/advan.00126.2014.
20. K. Nakamura and S. F. Morrison, "Central efferent pathways for cold-defensive and febrile shivering," *J. Physiol.*, vol. 589, no. Pt 14, pp. 3641–3658, Jul. 2011, doi: 10.1113/jphysiol.2011.210047.
21. M. A. Francisco and C. T. Minson, "Chapter 12 – Cutaneous active vasodilation as a heat loss thermoeffector," in *Handbook of Clinical Neurology*, A. A. Romanovsky, Ed., in Thermoregulation: From Basic Neuroscience to Clinical Neurology Part I, vol. 156. Elsevier, 2018, pp. 193–209, doi: 10.1016/B978-0-444-63912-7.00012-6.
22. M. Shibasaki, T. E. Wilson, and C. G. Crandall, "Neural control and mechanisms of eccrine sweating during heat stress and exercise," *J. Appl. Physiol.*, vol. 100, no. 5, pp. 1692–1701, May 2006, doi: 10.1152/japplphysiol.01124.2005.
23. D. Sherry, "Thermoscopes, thermometers, and the foundations of measurement," *Stud. Hist. Philos. Sci. Part A*, vol. 42, no. 4, pp. 509–524, Dec. 2011, doi: 10.1016/j.shpsa.2011.07.001.
24. E. Grodzinsky and M. S. Levander, "History of the thermometer," *Underst. Fever Body Temp.*, pp. 23–35, Aug. 2019, doi: 10.1007/978-3-030-21886-7_3.
25. J. M. S. Pearce, "A brief history of the clinical thermometer," *QJM Int. J. Med.*, vol. 95, no. 4, pp. 251–252, Apr. 2002, doi: 10.1093/qjmed/95.4.251.
26. I. Blumenthal, "Should we ban the mercury thermometer? Discussion paper," J. R. Soc. Med., vol. 85, no. 9, pp. 553–555, Sep. 1992, doi: 10.1177/014107689208500915.
27. D. C. Crawford, B. Hicks, and M. J. Thompson, "Which thermometer? Factors influencing best choice for intermittent clinical temperature assessment," *J. Med. Eng. Technol.*, vol. 30, no. 4, pp. 199–211, Jan. 2006, doi: 10.1080/03091900600711464.
28. G. Wang, W. Wang, K. Li, and H. Liu, "A Digital Thermometer with Fast Response and High Precision," in *2014 7th International Conference on Biomedical Engineering and Informatics*, Oct. 2014, pp. 504–510, doi: 10.1109/BMEI.2014.7002827.
29. F. Piccinini, G. Martinelli, and A. Carbonaro, "Reliability of body temperature measurements obtained with contactless infrared point thermometers commonly used during the COVID-19 pandemic," *Sensors*, vol. 21, no. 11, p. 3794, May 2021, doi: 10.3390/s21113794.
30. A. Ramelan, G. S. Ajie, M. H. Ibrahim, S. Pramono, and M. A. Rizqulloh, "Design low cost and contactless temperature measurement gate based on the Internet of Things (IoT)," *IOP Conf. Ser. Mater. Sci. Eng.*, vol. 1096, no. 1, p. 012060, Mar. 2021, doi: 10.1088/1757-899X/1096/1/012060.
31. S. Paramasivam, C. H. Shen, A. Zourmand, A. K. Ibrahim, A. M. Alhassan, and A. F. Eltirifl, "Design and Modeling of IoT IR Thermal Temperature Screening and UV Disinfection Sterilization System for Commercial Application Using Blockchain Technology," in *2020 IEEE 10th International Conference on System Engineering and Technology (ICSET)*, Nov. 2020, pp. 250–255, doi: 10.1109/ICSET51301.2020.9265363.
32. P. Ulleri, M. S, S. K, K. Zenith, and Sai Shibu, N.B., "Development of Contactless Employee Management System with Mask Detection and Body Temperature Measurement Using TensorFlow," in *2021 Sixth International Conference on Wireless Communications, Signal Processing and Networking (WiSPNET)*, Mar. 2021, pp. 235–240, doi: 10.1109/WiSPNET51692.2021.9419418.

33. Z. Zeng, G. Mei, T. Liao, and Y. huang, "The temperature difference method for screening patients with COVID-19 fever symptoms," *J. Phys. Conf. Ser.*, vol. 2226, no. 1, p. 012010, Mar. 2022, doi: 10.1088/1742-6596/2226/1/012010.
34. K. Rao *et al.*, "F3S: Free Flow Fever Screening," in *2021 IEEE International Conference on Smart Computing (SMARTCOMP)*, Aug. 2021, pp. 276–285, doi: 10.1109/SMARTCOMP52413.2021.00060.
35. P. Netinant, P. Vasprasert, and M. Rukhiran, "Evaluations of Effective on LWIR Micro Thermal Camera IoT and Digital Thermometer for Human Body Temperatures," in *Proceedings of the 5th International Conference on E-Commerce, E-Business and E-Government*, in ICEEG '21. New York, NY, USA: Association for Computing Machinery, juillet 2021, pp. 20–24, doi: 10.1145/3466029.3466043.
36. Assoc. Prof. Dr. M. Abdulrazaq, H. Zuhriyah, S. Al-Zubaidi, S. Karim, R. Ramli, and E. Yusuf, "Novel COVID-19 detection and diagnosis system using IOT based smart helmet," *Int. J. Psychosoc. Rehabil.*, vol. 24, pp. 2296–2303, Mar. 2020.
37. A. Barnawi, P. Chhikara, R. Tekchandani, N. Kumar, and B. Alzahrani, "Artificial intelligence-enabled Internet of Things-based system for COVID-19 screening using aerial thermal imaging," *Future Gener. Comput. Syst.*, vol. 124, pp. 119–132, Nov. 2021, doi: 10.1016/j.future.2021.05.019.
38. B. Peddinti, A. Shaikh, Bhavya. K. R., and Nithin Kumar. K. C., "Framework for real-time detection and identification of possible patients of COVID-19 at public places," *Biomed. Signal Process. Control*, vol. 68, p. 102605, Jul. 2021, doi: 10.1016/j.bspc.2021.102605.
39. J. W. Chin, K. Long Wong, T. T. Chan, K. Suhartono, and R. H. Y. So, "An Infrared Thermography Model Enabling Remote Body Temperature Screening Up to 10 Meters," in *2021 IEEE/CVF Conference on Computer Vision and Pattern Recognition Workshops (CVPRW)*, Nashville, TN, USA: IEEE, Jun. 2021, pp. 3870–3876, doi: 10.1109/CVPRW53098.2021.00429.
40. Y. Zhou *et al.*, "Clinical evaluation of fever-screening thermography: Impact of consensus guidelines and facial measurement location," *J. Biomed. Opt.*, vol. 25, no. 9, p. 097002, Sep. 2020, doi: 10.1117/1.JBO.25.9.097002.
41. G. Mei, S. Peng, Z. Zeng, T. Liao, and Y. Huang, "The influence of high temperature weather on human body temperature measurement by infrared thermal imaging thermometer," *J. Phys. Conf. Ser.*, vol. 2112, no. 1, p. 012024, Nov. 2021, doi: 10.1088/1742-6596/2112/1/012024.
42. S. Rayanasukha *et al.*, "Self-compensation for the influence of working distance and ambient temperature on thermal imaging-based temperature measurement," *IEEE Trans. Instrum. Meas.*, vol. 70, pp. 1–6, Aug. 2021, doi: 10.1109/TIM.2021.3103242.
43. Y. Zhang, Y. Chen, X. Fu, and C. Luo, "A method for reducing the influence of measuring distance on infrared thermal imager temperature measurement accuracy," *Appl. Therm. Eng.*, vol. 100, pp. 1095–1101, May 2016, doi: 10.1016/j.applthermaleng.2016.02.119.

Chapter 15

Signal processing system for Heart Rate Extraction via LabVIEW

Zakaria El khadiri, Rachid Latif, and Amine Saddik

15.1 INTRODUCTION

Remote vital measurement is particularly helpful for biomedical diagnostics and monitoring, and it has a significant role in maintaining cardiovascular health by anticipating the illness associated with cardiovascular disease and immediately monitoring the circumstances of chronic disorders [1, 2]. However, some physiological parameters, such as heart rate, breathing rate, arterial oxygen saturation, blood pressure, and others, might be precisely calculated through physiological waves such as photoplethysmography (PPG) and ballistocardiography concepts to offer adequate details about the cardiovascular system and to predict the physiological and pathological conditions of a human being [3–14]. Typically, the heart is one of the major physical organs that permits the body to expel blood by beating between 60 and 100 times per minute [15]. Contrarily, it can be regarded as an abnormality whether it is below 60 bpm (known as bradycardia) or beyond 100 bpm (known as tachycardia) [16]. The latter change depends on several factors that might affect physiological parameters, such as age, air temperature, body position, and emotions.

Because it can provide helpful hints to medical diseases, including cardiovascular disease, sleep difficulties, or anomalies, vital sign monitoring has become more crucial. Technologies that enable contactless, simple deployment, and long-term vital sign monitoring are urgently needed for the medical industry. However, the first health monitoring system, developed by Costa et al. [17] to extract physiological information from an RGB camera, was released in 1995. As time goes on, the discipline is renowned for its fast evolution, and several reliable methods have been developed to detect the majority of these parameters from PPG. Moreover, the blood volume pulse (BVP) can be measured using a variety of methods. Verkruysse et al. [18] used the G channel of movies recorded by a consumer camera to show the measurement of BVP under ambient light. Additionally, Poh et al. created a remote BVP measurement method based on blind source separation utilizing a low-cost camera [19]. Furthermore, in the non-invasive monitoring of the automatic nervous system, the heart rate variability spectrograms (HRVSs)

DOI: 10.1201/9781032664293-17

are very helpful; the latter control the heartbeat, blood pressure, and other involuntary bodily functions. Given that it rises during cognitive stress, the low-frequency (LF) power in HRVS is regarded as one of the most accurate measures of sympathetic activity [20].

The current paper introduces several signal processing techniques and approaches to estimate the most adequate physiological parameters, continuously track a patient's status, provide relevant details regarding the cardiovascular system, and anticipate the physiological and pathological information of human beings. Our system is mainly based first on the normalization signal to center the raw input signal and not assess it with its first level that has been gathered from the image processing; the latter centralization is notably based on its mean and the standard deviation. Afterward, filtering the noisy PPG signal is the most prominent step to remove the undesired parts of the signal and to prevent the unexpected noises generated by the user's environment. In our case, we will proceed with the median filter (MF) as a technique to reduce unwanted parts. Then, the singular value decomposition (SVD) method is applied to the signals to reduce the signal's dimensionality and isolate the wanted part of the PPG signal. The SVD method can effectively reduce the number of input variables or columns in modeling data [21]. Finally, the last step is to extract the frequency of the heartbeat. To achieve this purpose, we will employ a Fourier transform (FT) method to convert the signal from the time domain to the frequency domain, and then, by identifying the index with the greatest spectral power, the HR values can be calculated [22].

The remainder of this paper is organized as follows: Section 15.2 describes some related works. Section 15.3 outlines the common techniques employed in the present work and the proposed methodology for physiological parameter estimation. Section 15.4 shows the experiments conducted. Then, a brief conclusion and a look at the future directions in Section 15.5.

15.2 RELATED WORKS

In order to assess a person's health status, measurements of physiological markers are essential. Contactless monitoring of vital signs may be advantageous in a variety of contexts, including healthcare, clinical settings, occupational settings, and sporting events. These recent studies have concentrated on the possibilities of camera-based systems that operate in the visible spectrum for measuring vital signs through the detection of minute color changes or motions brought on by physiological activity but imperceptible to the human eye. In 2022, N. Molinaro et al. [23] proposed an overview of the contactless vital signs monitoring algorithm from videos recorded through digital cameras. The authors provided the rationale for the measurement and classification of the different techniques implemented for post-processing the original videos, as well as the main outcomes achieved during several applications or validation studies.

In 2022, S. Liu et al. [24] proposed a novel vital signs detection and extraction method with high efficiency, high precision, high sensitivity, and a high signal-to-noise ratio based on the NVA6100 pulse radar system. They also adopted the SVD to perform signal denoising and decomposition after preprocessing, and the temporal and spatial eigenvectors of each principal component have been obtained. Typically, the algorithm proposed includes signal preprocessing, signal denoising and decomposition, vital signs extraction and restoration, and vital signs reconstruction. In 2023, Z. Liang et al. [25] described the variety of techniques employed for vital signals extraction and their accuracy and efficiency; they also stated the utility of unifying several approaches and algorithms, such as wavelet analysis and mode decomposition, which can offer great opportunities to measure vital signals. Also, they provided the basic structure of the radar system and the mathematical model of human cardiopulmonary activity monitoring; moreover, they explained the extraction of the cardiopulmonary signals and the evaluation standards for the algorithms, as well as the data used for processing.

Most recently, in 2023, S. Liu et al. [24] proposed singular spectrum analysis (SSA) to reconstruct the eigenvalues of noisy vital signs to eliminate noise peaks around the heartbeat rate, reduce the noise of vital signs, and eliminate noise interference. When combined with variational modal decomposition (VMD), the target vital signs can be extracted with high accuracy. However, the main contribution is that SVD is proposed to adaptively implement Ultra-wideband (UWB) radar echo to eliminate static clutter and reduce echo noise. Then, SSA is proposed to eliminate noise peaks around the heartbeat rate. Their experiment results confirmed that the target vital signs information can be extracted with high accuracy from ten subjects at different distances, which can play an important role in short-distance human detection and vital sign monitoring.

15.3 PROPOSED METHODOLOGIES

15.3.1 Methods and tools

15.3.1.1 Median filter (MF)

A non-linear digital filtering method called the MF is frequently used to eliminate noise from an image or signal. Indeed, it is a typical preprocessing stage to enhance the output of later processing. The MF's primary principle is to iteratively replace each element in the signal with the median of its nearby entries. The window is the neighborhood pattern, which moves entry by entry over the global signal. The simplest window for one-dimensional signals is the first few preceding and following entries, while the window for higher-dimensional data must contain all entries within a certain radius or elliptical region [26–28]. Figure 15.1 shows an example of the filtered signal as per the original noisy input signal.

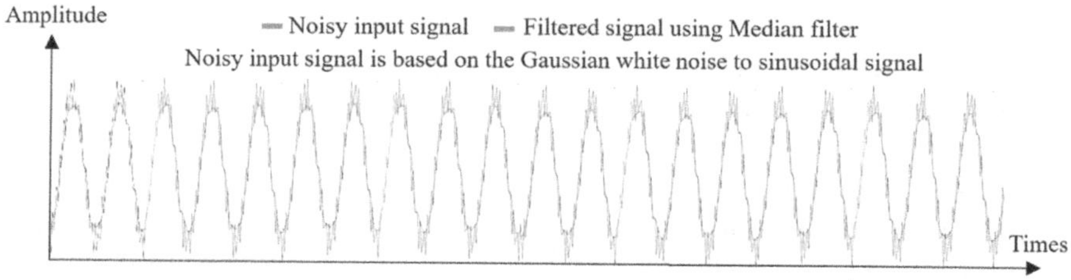

Figure 15.1 Filtered signal vs noisy signal using median.

15.3.1.2 Singular value decomposition (SVD)

One of many methods for reducing a data set's dimensionality, or the number of columns, is SVD. More columns typically signify that building models and scoring data requires more effort in predictive analytics. If certain columns lack predictive value, time will be wasted and the model's quality or predicted accuracy will suffer because of the noise those columns introduce. Therefore, reducing the number of input features is frequently desirable, but this decreases the feature space's dimensions. Additionally, it helps to eliminate redundant characteristics and noisy features from a huge data set. This can be achieved by decreasing the overall matrix's dimension while retaining the majority of the features that have a significant impact on the model. In the end, the SVD technique can make it easier to plot and visualize the features and to store, analyze, and process data. While SVD can be used for dimensionality reduction, it is often used in digital signal processing for noise reduction, image compression, and other fields [29, 30].

15.3.1.3 Fourier transform (FT)

A mathematical technique known as the FT converts time-domain data into frequency domains in order to determine the signal strength information at particular frequencies. The discrete FT (DFT) formulation applies the FT to discrete signals as outlined in the following equation:

$$X[m] = \sum_{n=0}^{N-1} X[n] e^{-\frac{2j\pi mn}{N}} \tag{16.1}$$

where:

- n = index in the time domain. n = 0, 1, …, N-1.
- m = index in the frequency domain. m = 0, 1, …, N-1.

DFT is a Fourier representation for periodic discrete signals. DFT computing involves several intricate procedures. The fast FT (FFT), an algorithm, was created to lessen this complexity. Ordinarily, the FFT algorithm breaks N points into N/2 points and then splits them into N/4 points, and so on up to 1 point. Additionally, the functionality of the FFT technique is to confirm the outcomes of frequency-domain signal decomposition.

15.3.1.4 LabVIEW (NI)

The Virtual Instrument Engineering Workbench (LabVIEW) is a system-design platform and development environment for a visual programming language from National Instruments. It's a graphical programming language that enables users to control devices, gather, manipulate, and display data when used with a data acquisition device and personal computer. Written code is not employed in LabVIEW; instead, graphical representations of the circuits have been created, which are termed virtual instruments (VIs). The latter have been manipulated to perform the desired tasks at hand, and they are run from their front panels, which contain all of the controls and displays. Each one has an associated block diagram; the graphical programming language G was used to create this block diagram. Block diagram wiring illustrates the data flow among these components, whereas block diagram components represent various structures, loops, and functions.

15.3.2 Proposed system

As depicted below in the following flow chart (Figure 15.2), the first step in our signal processing system is signal normalization (outlined in step (a) in Figure 15.2) to center the raw input signal.

This centralization is based on the following equation:

$$X_i(t) = \frac{Y_i(t) - \mu_i(t)}{\delta_i} \tag{16.2}$$

where:

- Y_i(t) is the brute PPG signal.
- X_i(t) is the normalized PPG signal.
- μ_i(t) is the mean of *Yi*.
- δi is the standard deviation of *Yi*.

Then, the MF technique is yet to be applied to filter the noisy PPG signal, remove the undesired parts of the signal, and prevent the unexpected noises generated by the user's environment. From our side, the MF is depicted in step (b) in Figure 15.2. Afterward, the SVD method is employed on the signals to isolate the desired portion of the PPG signal and decrease the signal's dimensionality. Indeed, the SVD has been established using significant mathematical tools and background; this method is depicted in step (c) in

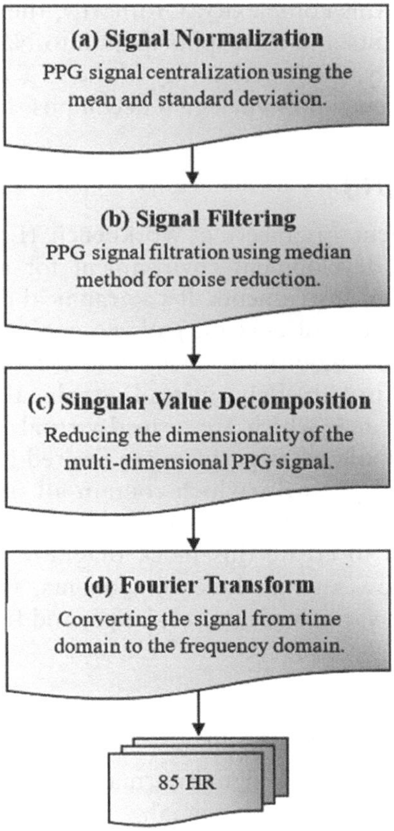

Figure 15.2 Proposed signal processing systems.

Figure 15.2. The last step is to determine the heartbeat's frequency; for this purpose, the signal will be transformed using the FT method from the time domain to the frequency domain. The HR values can then be determined by locating the index with the highest spectral power; this step is outlined in step (d) in Figure 15.2.

15.4 RESULT

Our physiological signs estimated values using signal processing methods are primarily tested using the LabVIEW platform using several experimental PPG multi-dimensional signals. In our case, we will be based on a PPG signal with n dimensions, which are accordingly, for example, to different image sequence channels (RGB band, Luv band, or others), Figure 15.3 shows the PPG signals loading.

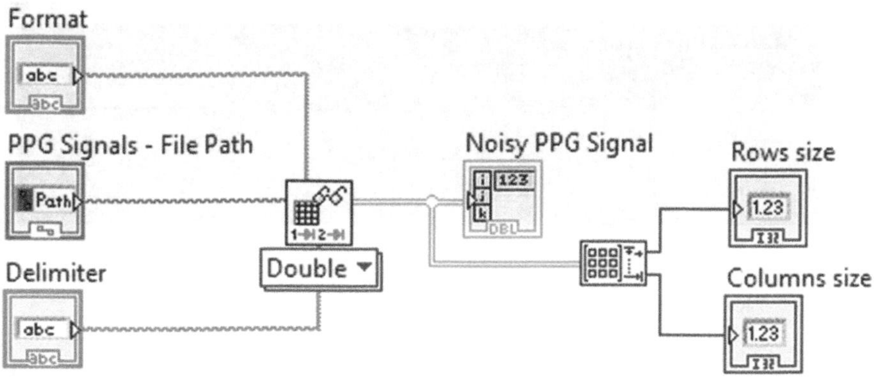

Figure 15.3 PPG signals loading.

After loading the needed PPG signal, the next step is to normalize the input signal following Equation (15.2) in the time domain and the filtered signal by applying the MF method, then reduce the signal dimension using the SVD method. These steps can be briefly shown in the following block diagram VI in Figure 15.4.

A typical first step is to centralize or normalize the raw signals because the level of raw signals has no significance when evaluating periodicity. Figures 15.5 and 15.6 show typical PPG normalized signals and noisy PPG signals, respectively.

As depicted in Figure 15.2, after extracting the first noisy PPG signal from the sequence of human face frames, the next step is to filter the signal and remove unwanted high, illumination noises, and LF noise using the MF, the latter technique represents an equivalent low-pass using a rolling window that averages a specific number of values, Figures 15.7–15.9 illustrate the noisy pulse and median filtered pulse with different values of filter rank (i.e. window size) used to compute the MF.

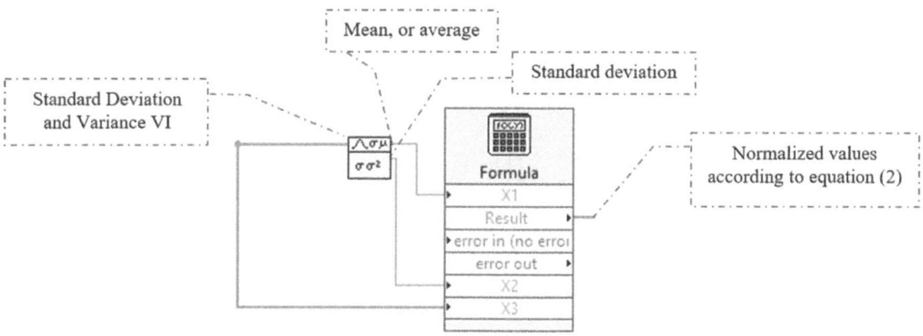

Figure 15.4 Example of signal normalization.

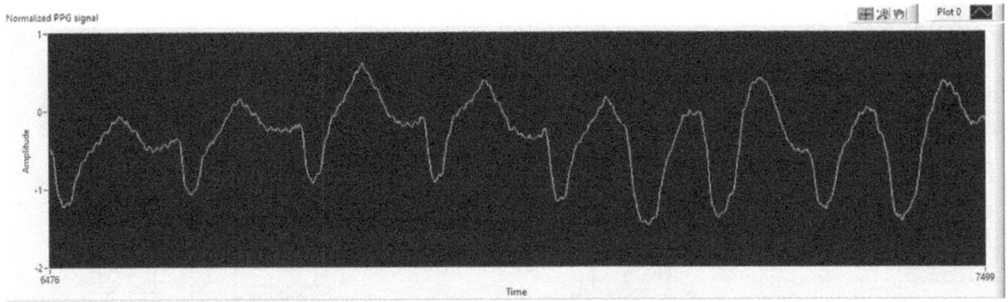

Figure 15.5 Normalized PPG signal.

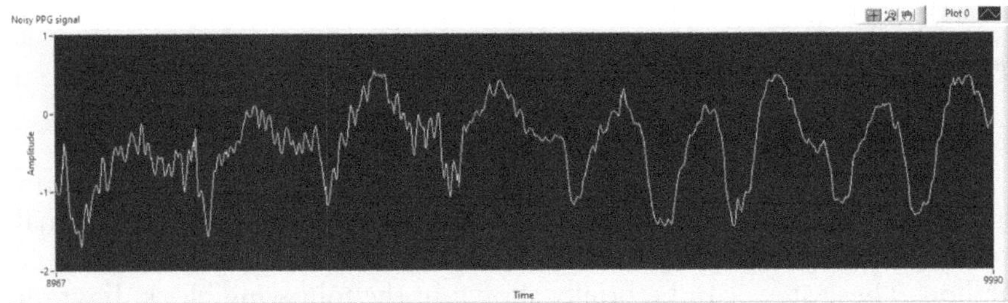

Figure 15.6 Noisy PPG signal.

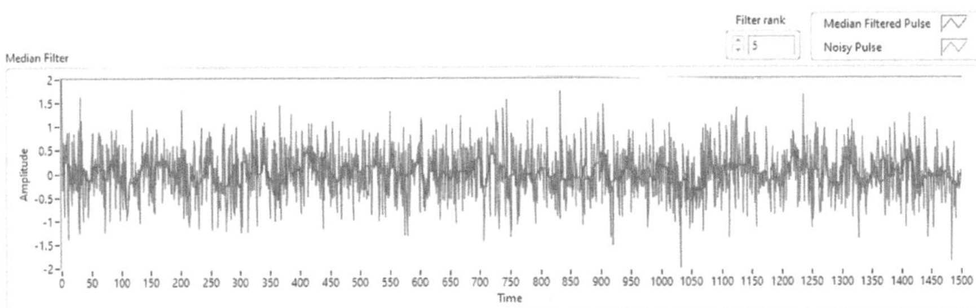

Figure 15.7 Median-filtered pulse and noisy pulse with filter rank equal 5.

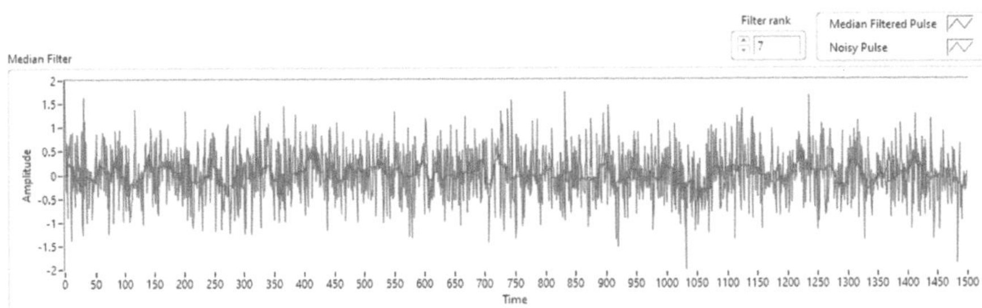

Figure 15.8 Median-filtered pulse and noisy pulse with filter rank equal 7.

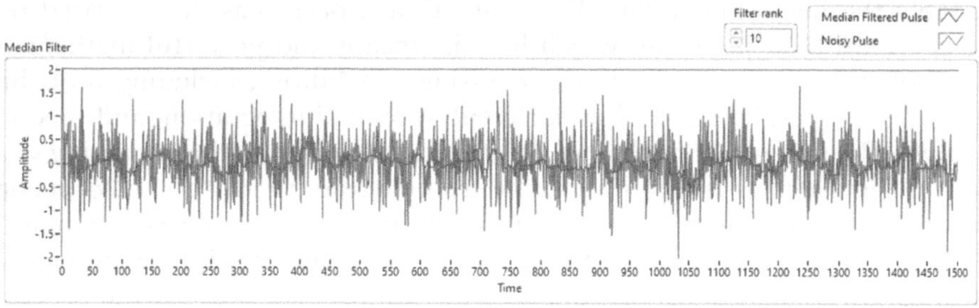

Figure 15.9 Median-filtered pulse and noisy pulse with filter rank equal 10.

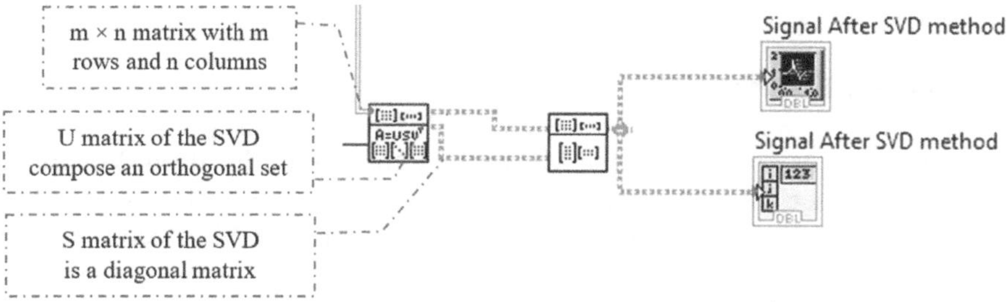

Figure 15.10 Singular value decomposition.

Afterward, an SVD is yet applied for lowering the dimensionality of a data set, Figure 15.10 outlines SVD decomposition VI.

The above block diagrams illustrate briefly the prominently used functions to gather the heart rate values. The first VI takes as the input argument the global PPG signal and provides the mean, standard deviation, and variance values in the output parameter. The outcomes of the latter function construct the normalization formula in accordance with Equation (15.2); after that, the next function is the MF, which takes the noisy signal in the input parameter and provides the filtered signal on the output. However, the next function is the SVD, which takes a matrix type as an input parameter and provides three matrices on the output parameter (i.e. Matrix U, S, and V). Furthermore, several advanced biomedical toolkits have been offered for bio-signals and biomedical image analysis and processing purposes. Based on these VIs, they can provide useful data for recognizing, visualizing, and understanding biomedical characteristics in human bodies. They also include further tools that can be used to acquire, preprocess, extract, and analyze bio-signals and biomedical images.

15.5 CONCLUSION

This research describes the stacking layers of the most widely employed signal-processing techniques by the biomedical and signal-processing communities. Our system is mainly dedicated to determining physiological parameters, notably the heart rate value. The simulation process has been carried out on the LabVIEW platform, which has significant and powerful mathematical tools for analysis and signal processing capabilities, including probability and statistics, linear algebra, differentiation and integration, and spectral analysis. The proposed method is also helpful in collecting the remaining vital parameters, such as breathing rate, arterial blood oxygen, and HRV. In conclusion, this research may help illuminate a different method that can handle PPG signal noises. In our upcoming study, we intend to select an additional strategy involving further techniques while taking into account the processing time versus other implementation architectures and the real-time requirement, including our validation experiment results, displaying performance metrics and offering comparisons with current contact-based techniques.

ACKNOWLEDGMENT

We owe a debt of gratitude to the Ministry of National Education, Vocational Training, Higher Education and Scientific Research (MENFPESRS) and the National Centre for Scientific and Technical Research of Morocco (CNRST) for their financial support for the project Cov/2020/109.

REFERENCES

1. Allen, J. (2007). Photoplethysmography and its application in clinical physiological measurement. Physiological Measurement, 28(3), R1.
2. Cook, S., Togni, M., Schaub, M. C., Wenaweser, P., & Hess, O. M. (2006). High heart rate: A cardiovascular risk factor?. European Heart Journal, 27(20), 2387–2393.
3. Yan, Y., Ma, X., Yao, L., & Ouyang, J. (2015). Non-contact measurement of heart rate using facial video illuminated under natural light and signal weighted analysis. Bio-Medical Materials and Engineering, 26(s1), S903–S909.
4. El Khadiri, Z., Latif, R., & Saddik, A. (2023, March). Breathing Pattern Assessment through the Empirical Mode Decomposition and the Empirical Wavelet Transform Algorithms. In The 3rd International Conference on Artificial Intelligence and Computer Vision (AICV2023), March 5–7, 2023 (pp. 262–271). Cham: Springer Nature Switzerland.
5. El Boussaki, H., Latif, R., & Saddik, A. (2022, November). A Review on Video-Based Heart Rate, Respiratory Rate and Blood Pressure Estimation. In International Conference of Machine Learning and Computer Science Applications (pp. 129–140). Cham: Springer Nature Switzerland.
6. Hassan, M. A., Malik, A. S., Fofi, D., Saad, N., Karasfi, B., Ali, Y. S., & Meriaudeau, F. (2017). Heart rate estimation using facial video: A review. Biomedical Signal Processing and Control, 38, 346–360.
7. El Boussaki, H., Latif, R., Saddik, A., El Khadiri, Z., & El Boujaoui, H. (2023). Non-contact respiratory rate monitoring based on the principal component analysis. International Journal of Advanced Computer Science and Applications, 14(9).
8. Latif, R., Addaali, B., & Saddik, A. (2023, January). Real-Time SPO2 Monitoring Based on Facial Images Sequences. In International Conference on Digital Technologies and Applications (pp. 474–483). Cham: Springer Nature Switzerland.

9. El khadiri, Z., Latif, R., & Saddik, A. (2023, April). Remote Heart Rate Measurement Using Plethysmographic Wave Analysis. In Advances in Machine Intelligence and Computer Science Applications: Proceedings of the International Conference ICMICSA'2022 (pp. 254–267). Cham: Springer Nature Switzerland.
10. El Boussaki, H., Latif, R., & Saddik, A. (2023). Video-based heart rate estimation using embedded architectures. International Journal of Advanced Computer Science and Applications, 14(5).
11. Hoda El Boussaki, Rachid Latif and Amine Saddik, "Video-based Heart Rate Estimation using Embedded Architectures" International Journal of Advanced Computer Science and Applications(IJACSA), 14(5), 2023. http://dx.doi.org/10.14569/IJACSA.2023.01405119
12. Rachid, L., Khadija, J., & Amine, S. (2023, January). OpenCL Kernel Optimization Metrics for CPU-GPU Architecture. In International Conference on Digital Technologies and Applications (pp. 773–781). Cham: Springer Nature Switzerland.
13. Jenkal, W., Mejhoudi, S., Saddik, A., & Latif, R. (2023). Embedded systems in biomedical engineering: Case of ECG signal processing using multicores CPU and FPGA aqrchitectures. Smart Embedded Systems and Applications, 103.
14. Bella, A., Latif, R., Saddik, A., & Guerrouj, F. Z. (2021). Monitoring of Physiological Signs and Their Impact on the Covid-19 Pandemic. In E3S Web of Conferences (Vol. 229, p. 01030). EDP Sciences.
15. Bauer, A., Malik, M., Schmidt, G., Barthel, P., Bonnemeier, H., Cygankiewicz, I., ..., & Zareba, W. (2008). Heart rate turbulence: Standards of measurement, physiological interpretation, and clinical use: International Society for Holter and Noninvasive Electrophysiology Consensus. Journal of the American College of Cardiology, 52(17), 1353–1365.
16. Kazemi, S., Ghorbani, A., Amindavar, H., & Li, C. (2014). Cyclostationary approach to Doppler radar heart and respiration rates monitoring with body motion cancelation using Radar Doppler System. Biomedical Signal Processing and Control, 13, 79–88.
17. Da Costa, G. 1995. Optical remote sensing of heartbeats. Optics Communications, 117(5–6), 395–398.
18. Verkruysse, W., Svaasand, L. O., & Nelson, J. S. (2008) Remote plethysmographic imaging using ambient light. Optics Express 16(26), 21434–21445.
19. Poh, M.-Z., McDuff, D. J., & Picard, R. W. (2010) Non-contact, automated cardiac pulse measurements using video imaging and blind source separation. Optics Express 18(10), 10762–10774.
20. Pagani, M., Furlan, R., Pizzinelli, P., Crivellaro, W., Cerutti, S., & Malliani, A. (1989) Spectral analysis of R–R and arterial pressure variabilities to assess sympatho-vagal interaction during mental stress in humans. Journal of Hypertension 7(Suppl), S14–S15.
21. Leskovec, J., Rajaraman, A., Ullman, J. D., Leskovec, J., Rajaraman, A., & Ullman, J. D. (2014). Dimensionality reduction. Mining of Massive Datasets, 415–447.
22. Rouast, P. V., Adam, M. T., Cornforth, D. J., Lux, E., & Weinhardt, C. (2017). Using Contactless Heart Rate Measurements for Real-Time Assessment of Affective States. In Information Systems and Neuroscience: Gmunden Retreat on NeuroIS 2016 (pp. 157–163). Springer International Publishing.
23. Molinaro, N., Schena, E., Silvestri, S., Bonotti, F., Aguzzi, D., Viola, E., ..., & Massaroni, C. (2022). Contactless vital signs monitoring from videos recorded with digital cameras: An overview. Frontiers in Physiology, 13, 160.
24. Liu, S., Qi, Q., Cheng, H., Sun, L., Zhao, Y., & Chai, J. (2022). A vital signs fast detection and extraction method of UWB impulse radar based on SVD. Sensors, 22(3), 1177.

25. Liang, Z., Xiong, M., Jin, Y., Chen, J., Zhao, D., Yang, D., ..., & Mo, J. (2023). Non-contact human vital signs extraction algorithms using IR-UWB radar: A review. Electronics, 12(6), 1301.
26. Justusson, B. I. (2006). Median filtering: Statistical properties. Two-dimensional digital signal prcessing II: transforms and median filters, 161–196.
27. Stone, D. C. (1995). Application of median filtering to noisy data. Canadian Journal of chemistry, 73(10), 1573–1581.
28. Huang, T., Yang, G. J. T. G. Y., & Tang, G. (1979). A fast two-dimensional median filtering algorithm. IEEE Transactions on Acoustics, Speech, and Signal Processing, 27(1), 13–18.
29. Anowar, F., Sadaoui, S., & Selim, B. (2021). Conceptual and empirical comparison of dimensionality reduction algorithms (pca, kpca, lda, mds, svd, lle, isomap, le, ica, t-sne). Computer Science Review, 40, 100378.
30. Kanth, K. R., Agrawal, D., El Abbadi, A., & Singh, A. (1999). Dimensionality reduction for similarity searching in dynamic databases. Computer Vision and Image Understanding, 75(1–2), 59.

Chapter 16

Investigating the impacts of COVID-19 over time using sentiment analysis and topic modeling

Mustapha Hankar, Toufik Mzili, Mohammed Kasri, and Abderrahim Beni-Hssane

16.1 INTRODUCTION

The new coronavirus, also called SARS-CoV-2, has created a calamitous situation around the globe by causing millions of deaths, and infections and putting other lives at stake. The pandemic has severely affected and challenged public healthcare systems, governments, and societies in an unimaginable way. Moreover, its impacts still go on and will remain for many years to come [1, 2]. The COVID-19 outbreak had a negative impact on every aspect of life including public health, economy, education, travel, social life, and the labor market [3–6]. As the pandemic continues to spread exponentially and new variants emerge, it is clear that the world will be facing a long-term crisis and disruption mostly on the economic level.

In the periods of quarantine and stay-in-home measures, people have suffered from mental health disorders such as stress, anxiety, depression, feelings of isolation, and worry. Many conducted studies [7–9] showed that protective measures which have been imposed by governments such as travel restrictions, lockdowns, and economic shutdowns have exposed people to psychological disorders while being isolated in their places, socially distanced from one another, and living at constant risk of losing their work. These effects include symptoms of post-traumatic stress disorder, long quarantine suppress symptoms, and many others. Social media outlets like Facebook and Twitter, news platforms, blogs, and forums, during the periods of the pandemic have been experiencing reactions, feelings, emotions, and thoughts of people toward the COVID-19 pandemic and its impacts [10].

Natural language processing (NLP) techniques are broadly applied to retrieve and analyze relevant information from these huge amounts of generated content. For instance, topic models are used in this context to extract the main topics related to the outbreak. Sentiment analysis (SA) is another NLP technique that is often applied to identify sentiments and emotions in a given text as a way to investigate the effects of COVID-19 on people's normal lives. SA is widely applied across various domains like business intelligence to analyze customers' feedback regarding a service or a product [11], and social media monitoring to analyze people's opinions about public policies

DOI: 10.1201/9781032664293-18

or certain figures [12, 13]. SA approaches are mostly categorized into three main levels: document level, sentence level, and aspect level [14]. At the document level, the whole document is classified as negative, neutral, or positive. At the sentence level, subjective sentences are filtered out before classifying their polarity. The aspect-level approach focuses on performing SA regarding a specific aspect [15].

In this chapter, we performed a topic-based SA on a corpus of collected comments related to the COVID-19 outbreak. The dataset that is made available for this study consists of comments scraped from the online newspaper Hespress[1] written in both Modern Standard Arabic (MSA) and Moroccan dialect (MD). Our approach follows a process of two steps: topic extraction and SA. For topic modeling, we applied an embedding-based topic model, namely BERTopic, to retrieve topics from the corpus and generated document clusters based on the extracted themes. Second, a topic-based SA is carried out to identify polarities within the texts and classify them into positive, neutral, or negative. Finally, a time series analysis is performed on a monthly basis to track changes in sentiments in the period of three years of the pandemic. The visualization of these changes shows how people's reactions and sentiments evolve over time regarding the COVID-19 situation. Time series SA can relate events like quarantine, COVID-19 spikes of infections and deaths, and travel restrictions; to the changes in sentiments and emotions of people. For example, our study showed that negative sentiments of fear, anxiety, and depression are increased during quarantine periods and when a new wave of COVID-19 variant emerges and spreads. Negative sentiments also increased regarding the vaccine topic indicating that people hold suspicious sentiments toward vaccination campaigns.

The key contributions to this chapter are: (i) collecting a real dataset containing thousands of comments written in Arabic, (ii) designing a semi-supervised model that integrates both topic modeling and sentiments analysis techniques to examine the effects of COVID-19 in Morocco, and (iii) tracking the changes of SA results from March 2020 until the end of December 2022 in correlation to the epidemiological situation. The rest of the chapter is organized as follows: Section 16.2 is dedicated to reviewing and discussing related works to the study. In Section 16.3, we introduce the methods and the materials used to perform this research. Results are presented and discussed in Section 16.4. Finally, we summarize our chapter and discuss some feature works in Section 16.5.

16.2 RELATED WORKS

In recent years, academic researchers have shown growing interest in the field of NLP in order to design advanced techniques for processing human languages. However, Arabic NLP is still facing many challenges as a result of some specific and intrinsic features of the language. This section will present and examine some relevant research that relates to our study.

Chandrasekaran et al. [16] analyzed trends, topics, and sentiments in a dataset that contains tweets about the COVID-19 outbreak. They first applied Latent Dirichlet Allocation (LDA) to extract topics within the dataset and then implemented a dictionary-based method, namely valence aware dictionary and sentiment reasoner (VADER), to compute sentiment scores. In another work, Xue et al. [17] analyzed COVID-19 tweets using LDA and machine-learning-based SA to examine public discourse and the reactions of tweeters to the pandemic. Their results showed that people expressed feelings of fear and threat regarding the unknown nature of the coronavirus virus. Boon-Itt and Skunkan [18] conducted a study by utilizing LDA and SA techniques to explore and identify topic discussions in a dataset of collected Tweets over time. In their experiment results, they claimed that people expressed negative feelings toward the COVID-19 outbreak.

Yin et al. performed [19] an in-depth analysis of social media posts and discussions related to COVID-19 vaccines on Twitter. In the first step, the authors used LDA to extract the main topics about various vaccines and then applied the VADER model to compute sentiment polarities. Reported results showed that people feel positively confident to take vaccines. Qorib et al [20] conducted a study to examine COVID-19 vaccine hesitancy among people through their social media posts. They applied several machine-learning algorithms combined with different vectorization methods (term frequency-inverted document frequency (TF-IDF), Doc2vec, and bag of words (BoW)) to predict hesitancy regarding vaccines. In their experiment results, they suggest that the combination of TextBlob, TF-IDF, and LinearSVC outperformed other models in sentiment classification. They also concluded that people are feeling optimistic about taking vaccines and hesitancy decreases over time.

Abdul-Mageed et al. [21] developed a hybrid model for subjectivity detection and SA. The model first detects subjectivity within a text and then classifies its polarity into neutral, positive, or negative. They tested the model on a dataset collected from social media outlets like Twitter and other web platforms such as Wikipedia Talk Pages, mini blogs, chat apps, and web forums. The model utilizes a machine-learning algorithm (SVMlight) to classify sentiments. However, they faced challenging and complex issues in their experiments due to the specific characteristics of the Arabic language. In [22], Zarra et al. introduced a semi-supervised model that integrates topic modeling with SA to conduct aspect-oriented opinion mining. They used an unsupervised method to retrieve the main themes within Facebook comments written in colloquial Maghreb and then applied a supervised technique to compute sentiment polarity. Shelke et al. [23] designed an aspect-oriented model to conduct SA on a dataset of product reviews. They used the SentiWordNet lexicon and some specific features of reviews to identify their polarity. Madani et al. [24] designed a SA recommender to classify tweets as positive, negative, or neutral. The dataset they used for the study contains COVID-19-related tweets from Morocco, collected between March 2020 and October 2020. Their experimental results showed that their proposed model reached 86% accuracy outperforming baseline machine algorithms. They found out that changes in sentiments over time are affected by the COVID-19 epidemic situation. In our previous works [25, 26], we used LDA to extract topics from a collected dataset that contains more than 20,000 comments from Hespress, and then we applied a pre-trained transformer model to classify sentiments into negative, positive, and neutral. The findings showed that negative sentiments were high regarding all extracted topics in the first year of the pandemic.

16.3 PROPOSED APPROACH

16.3.1 Datasets

16.3.1.1 COVID-19 time series

Many sources such as the Moroccan Ministry Of Health, WHO, and the University of Johns Hopkins contributed to the collection of this dataset. Since January 22, 2020, the team at the Johns Hopkins Center for Systems Science and Engineering (CSSE) has been responsible for recording and updating COVID-19 data from across the globe. They have been cleansing, transforming, and normalizing data to make it easier for analysis and further processing. They arranged dates and consolidated several files to transform data into normalized time series. The dataset is made available in a GitHub data repository as a comma separated values (CSV) file and can be accessed via this link.[2] The dataset contains six columns: the dates that confirmed cases or fatalities were recorded, cumulative confirmed cases, cumulative deaths, recovered cases, Region/Country, and finally Province/state. We filtered out the time series based on the country column to retrieve cases and fatalities based in Morocco from March 2, 2020, the date in which the first case appeared, to December 31, 2022. We transformed the dataset into a time series of daily confirmed cases and daily fatalities indexed by DateTime.

16.3.1.2 Hespress comments

The dataset contains more than 20,000 comments sourced from the prominent news outlet, Hespress. These comments are written in both MSA and MD. For data collection, we developed a Python crawler using the Selenium library to extract comments from news articles related to the COVID-19 pandemic across various domains, including health, economy, politics, society, and vaccines. The scraping process was conducted during different time frames between March 2020 and December 2022 to ensure comprehensive coverage of the COVID-19 timeline. Comments in languages other than Arabic, such as English, French, or Spanish, were filtered out to retain only Arabic comments. The resulting dataset is stored as a CSV file, containing columns for the publication date, the user's name, and the comment text. The dataset can be accessed on GitHub via this link.[3]

Arabic language processing is still facing numerous challenges due to some inherited characteristics of the language itself like metaphor, diglossia, ambiguity, and various other difficulties related to Arabic morphology [27]. Moreover, researchers frequently encounter a mixture of three versions of the Arabic script on the web: MSA, Classical Arabic, and Dialectical Arabic. The differentiations in these scripts make processing and understanding tasks more difficult. The Arabic language has a different structure and morphology compared to Latin-based languages. For instance, the Arabic script is written from right to left and has 28 different letters. The Arabic letters can have diacritics such as hamza (همزة:ء), sukun (سكون: ْ), fatha (فتحة: َ), kasra (كسرة: ِ), and tanwin (تنوين: ً). These diacritics can directly affect the semantics of words in the text and increase ambiguity to capture its meaning.

The morphology of Arabic letters can change according to their location (initial, medial, final, or isolated) in the text which makes the processing task even more challenging. Cleaning texts in Arabic includes text normalization and removing punctuations, diacritics, numbers, links, and elongations [28]. The preprocessing task starts with tokenization to split the text into tokens, stop words removal to eliminate non-informative words such as fi (في), min (من), and ala (على), and stemming to transform words into their base form by removing suffixes and prefixes [27]. Table 16.1 shows a sample of texts in the dataset before and after preprocessing and Figure 16.1 illustrates the main steps to preprocess Arabic text.

Arabic text may include Arabic numerals, Arabic-specific characters, or mixed Arabic words with words in languages. Handling these challenges involves making decisions based on the specific requirements of the SA task.

Table 16.1 Examples of comments before and after preprocessing

Raw text	Preprocessed text
كلنا متضررين و لكن أحسن حاجة هي تمديد الحجر الصحي مع مراعاة الظروف الإجتماعية لينا كاملين. و الله يحد الباس.	متضررين احسن حاجة تمديد حجر صحي مراعاة ظروف اجتماعية كاملين الله يحد باس
مغرب بمثابة كثافة، هي الدار البيضاء هي اكثر المدن يجب انه اظن. الوطني الاقتصاد قلب و هي مصغر تسجل تعد لم التي الجهات باقي على الحجر تخفيف خاصة و البيضاء على و التركيز عديدة اصابات فهو للرفع اصابة 0 تسجيل انتظار اما الشركات المستحيل ضروب .من ضرب	اقتصاد قلب مصغر مغرب بمثابة كثافة مدن البيضاء الدار تسجل تعد لم جهات باقي حجر تخفيف يجب اظن وطني رفع اصابة تسجيل انتظار شركات تركيز عديدة اصابات مستحيل ضروب ضرب
بالنسبة للناس اللي ما خداوش جرعة اللقاح راه ما عندهمش اي نية للسفر سواء داخل المغرب او خارجه فاغلب الناس المهتمين بالسفر تلقحو هادشي راه ماشي معقول لا حول ولا قوة إلا بالله	ناس ماخداوش جرعة لقاح ما عندهمش نية سفر سواء داخل المغرب خارج اغلب ناس مهتمين سفر تلقحو ماشي معقول

Figure 16.1 Arabic text preprocessing pipeline.

For example, we might choose to replace Arabic numerals with their written forms or decide how to handle mixed language text and code-switching. Arabic text may contain spelling errors or typos. Applying spell-checking and correction techniques can improve the accuracy of subsequent analysis steps. Arabic-specific spell checkers or general-purpose spell checkers with Arabic language support can be used. Additional preprocessing steps may be required for the Arabic language. For example, for aspect-based SA, we may try identifying and extracting aspect terms or entities from the text using named entity recognition or part-of-speech tagging.

16.3.1.3 TBCOV tweets

TBCOV, also known as 2 billion COVID-19, is a large-size Twitter collection [29] that contains more than 2B multilingual COVID-19-related tweets. Specifically, TBCOV contains over 2 billion tweets using more than 800 keywords from different languages. The collection was gathered within a period of 14 months beginning from February 1, 2020, until March 31, 2021. The tweets are written in 67 different languages (including Arabic) and posted by more than 87 million unique users on Twitter across 218 countries all over the world. Due to the largeness of the original multilingual dataset, the collectors have offered different filters based on location, language, and time window on their website.[4] We used language and location filters to retrieve more than 40,000 Arabic-written tweets that were published in Morocco during the same period of collection. The Arabic tweets are cleaned, preprocessed, and prepared for topic modeling and SA using NLP tools.

16.3.2 Topic extraction using BERTopic

Topic modeling is a text-mining method that is commonly used to discover hidden themes within a corpus of textual documents. Topic models represent documents in the collection as a mixture of different topics, and in turn, each topic is considered as a distribution of various words weighted by their respective scores. Topic models are widely applied in several domains like information retrieval, text classification, SA, and recommendation systems. Classic topic models like LDA [30] and non-negative matrix factorization (NMF) [31] are limited to representing documents as a BoW which leads to ignoring the order of words, semantic relationships, and contextuality between words [32, 33]. In response to this issue, word embedding models have emerged in the NLP field to address the problems of BoW [34, 35]. For instance, Bidirectional Encoder Representations from Transformers (BERT) [36] have shown promising results in generating contextual text representations that capture the semantic structures within sentences and word vectors. Modern topic models take great advantage of word embeddings to build coherent models powered with centroid-based techniques. For example, BERTopic generates representations of topics in documents in a way that each topic is assigned to a cluster of documents. The TF-IDF is computed by multiplying the term frequency $tf_{t,d}$ of a word t in document d by the inverse document frequency (IDF). The value of IDF is derived by calculating the logarithm of the corpus size N divided by the total number of documents containing the term t. The formula is shown in Equation (16.1):

$$\mathrm{tfidf}_{t,d} = \mathrm{tf}_{t,d} \times \log\left(\frac{\mathrm{N}}{\mathrm{df}_t}\right) \tag{16.1}$$

This TF-IDF formula is adjusted to measure the importance of a term to a topic instead of a document. Equation (16.2) measures the class-based TF-IDF of a term in a given class:

$$W_{t,c} = tf_{t,c} \cdot \log\left(1 + \frac{A}{tf_t}\right) \tag{16.2}$$

where $tf_{t,c}$ measures the frequency of a term t in class c representing a set of documents grouped into one document known as a cluster. This term frequency is then multiplied by the inverse class frequency (ICF) to assess the importance of a term within a class. The value of ICF is obtained by computing the logarithm of the average number of terms in each class A divided by the frequency of the term t across classes. The one inside the logarithm is added to the division to ensure the output score values remain positive. The class-based TF-IDF assesses the relevance of terms in document clusters allowing to extract topic-term distributions for each cluster.

In this chapter, we applied a word embedding-based technique, namely BERTopic, to extract topics from Hespress comments and tweets to generate document clusters. Afterward, each document cluster is assigned to each extracted topic based on the aspect representations. As for that, we used BERTopic due to its good results in generating coherent and diverse topics when compared to the performance of conventional topic models such as LDA [30], NMF [31], Doc2vec [37], and Top2vec [38] in various Arabic datasets [39] including Hespress comments dataset. Table 16.2 shows the experimental results of four topic models in terms of topic coherence (TC) and topic diversity (TD). TC measures the rate of semantic similarity between the most important words in a topic while TD measures the degree of diversity among topics.

From Table 16.2, we can see that BERTopic outperforms other topic models with high scores of TC and TD. The results shown in the table were recorded

Table 16.2 Performance of topic models leveraged with topic coherence and topic diversity across different topic numbers (K)

	K = 6		*K = 10*		*K = 15*	
Topic model	*TC*	*TD*	*TC*	*TD*	*TC*	*TD*
LDA	–0.021	0.42	–0.029	0.28	–0.024	0.33
NMF	0.158	0.95	0.18	0.93	0.196	0.92
Topic2Vec	0.093	0.84	0.099	0.72	0.154	0.672
BERTopic	0.22	0.96	0.211	0.992	0.20	0.99

after five iterations of training across three different topic numbers (K = 6, K = 10, K = 15) to avoid overfitting. The word embeddings of texts were generated using an Arabic pre-trained word embedding model, namely AraBERT [40]. Afterward, we trained the BERTtopic model on top of the constructed word embeddings to extract the themes that are related to the COVID-19 pandemic. The extraction of topics is performed in the Algorithm 1.

Algorithm 1: Topic extraction

```
Input Data: corpus of documentsc
Result: document_clusters
clened_text ← clean_diacritics()& nomalize_text();
preprocessed_text = remove_stopwords()&prprocess();
word_embeddings ← AraBERT(preprocessed_text);
extracted_topics ← BERTopic(word_embeddings);
enhanced_topics ← AraBERT_similarity(extracted_topics);
for topic in enhanced_topics do:
       scan text;
       if topic_words in text then
                    document_cluster ← add(text);
       else
             continue;
       endif
endfor
```

16.3.3 Sentiment classification using CAMeL

SA is an NLP technique that is often used to identify sentiments within the subjective text and classify its polarity into negative, neutral, or positive. SA techniques are typically categorized into three primary types: lexicon-based techniques, machine-learning techniques, and hybrid approaches [15]. Machine-learning techniques combine statistical and probabilistic algorithms with specific linguistic characteristics to identify sentiments in a given text. However, Arabic sentiment analysis (ASA) is particularly facing numerous challenges because of various language-specific characteristics. To address

Table 16.3 CAMeL sentiment analyzer accuracy using AraBERT and mBERT compared to Mazajak over three benchmark datasets [41]

	CAMeL(AraBERT)	*CAMeL(mBERT)*	*Mazajak*
ArSAS	0.92	0.89	0.90
ASTD	0.73	0.66	0.72
SemEval	0.69	0.60	0.63

these challenges, researchers have contributed to developing various tools and resources, such as CAMeL [41]. This open-source toolkit is used to perform various Arabic NLP tasks including SA, part-of-speech tagging, and named entity recognition. CAMeL sentiment analyzer is a pre-trained model that is fine-tuned on AraBERT and multilingual BERT word embeddings to detect sentiments in Arabic texts. The classifier was trained and evaluated on different datasets including the Arabic Speech-Act dialectical dataset and the ArSAS tweet sentiment corpus. Table 16.3 shows the evaluation results of the CAMeL sentiment analyzer over three datasets.

16.3.4 Topic-based sentiment analysis

The conventional models employed for ASA often focus on computing text polarity on the document level while ignoring the aspect/topic level that holds an opinion toward a specific entity within the text. In our approach, we consider that subjective texts express multiple topics, and each one of them has its own sentiment. Given this, we chose to perform SA on a topic-based level to capture people's opinions and examine their reactions regarding many different COVID-19-related aspects or topics. Our approach follows a process of two steps: we first extract topics from the corpus of documents using BERTopic and then generate word distributions of extracted topics based on word scores representing each specific topic. Afterward, document clusters are constructed and assigned to map each extracted topic with a corpus of documents using the c-TF-IDF formula. Table 16.4 presents the final topics and their word distributions while Figure 16.2 shows the distribution of document clusters for each topic.

In the second step of our approach, we transformed constructed document clusters and saved them as data frames and each data frame represents a document cluster of a specific topic. We mention that a text in the corpus can belong to multiple clusters at once, and thus, it may hold many opinions about different topics. After that, we used CAMeL to classify texts of document clusters into positive, neutral, or negative. The results of the whole process from text representation to topic clustering to SA are a refined aspect-based infodemic SA. Algorithm 2 represents the performed sentiment classification task.

Table 16.4 High-scored words for each topic

<table>
<tr><th>Topic</th><th>High-scored topic words</th></tr>
<tr><td>Economy</td><td>أرباب | شركات | اغلاق | عمل | اقتصاد | معامل | صناعة | شغل | متاجر | محلات</td></tr>
<tr><td>Education</td><td>| منصة | مدرسة | امتحان | استاذ | صحي حجر | بعد عن | عمومي | تعليم | دراسة
تلميذ | وزارة | تربية</td></tr>
<tr><td>Fear</td><td>|انتحار | سقم | الله | شفاء | عدوى | خوف | قلق | بلاء | موت | خطر | جوع | ملل
صحة | تشاؤم | مصيبة</td></tr>
<tr><td>Support</td><td>حجر | مساعدات | ضمان | بطالة | ازمة | دعم | صندوق | راميد | تعويضات | تمديد
حظر | مواطن | حكومة | دولة | التزام | طوارئ | داخلية | صحي</td></tr>
<tr><td>Health</td><td>| صحة | حالات | وقاية | اختناق | نقص | تنفس | ذوق | شم | فقدان | اعراض | تعقيم
كمامة | اصابات | تحاليل</td></tr>
<tr><td>Vaccine</td><td>| فيروس | فايزر | استرازينيكا | سينوفارم | اوميكرون | كوفيد | دلتا|لقاح | جواز
| ماسونية | مؤامرة | اوهام | كذب | مسرحية | خطة جرعة | الغاء | متحور | تلقيح
تأثير | ابادة</td></tr>
</table>

Figure 16.2 Topics distribution of document clusters for each year.

Algorithm 2: Topic-based sentiment analysis

```
Data: document_clusters
Result: sentiment classification
for document_cluster in document_clusters do:
        for text in document_cluster do:
              polarity_score ← CAMeLSentimentAnalyzer(text);
              if polarity_score > 0 then
                    sentiment ← positive;
              elseif    polarity_score < 0 then
                    sentiment ← negative;
              else
                    sentiment ← neutral;
              endif
        endfor
endfor
```

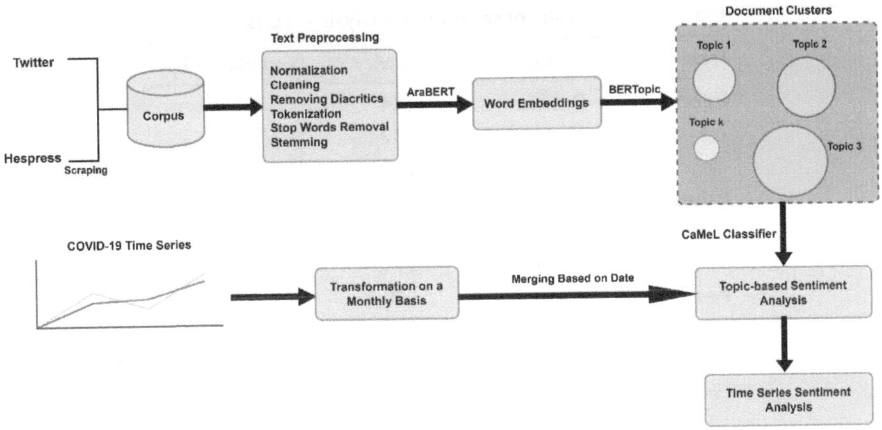

Figure 16.3 The proposed approach workflow.

Finally, we merged the data frames with the COVID-19 time series to track sentiment changes over three periods of time and to examine any correlations between the epidemiological situation in Morocco and the variation in sentiment results. Figure 16.3 shows the overall workflow of our proposed approach.

16.4 RESULTS AND DISCUSSION

This section outlines the experiment results of applying our proposed approach. While most approaches in the literature review have examined the question of COVID-19 pandemic effects on people by simply applying baseline SA methods to identify sentiments in the user-generated content from Twitter, Facebook, and other social media outlets; we combined many techniques including topic modeling, word embeddings, SA, and time series to realize this work. Table 16.5 shows the frequency of sentiment polarities and the polarity rate regarding each topic.

The overall findings showed that the COVID-19 outbreak has dramatically affected most aspects of normal life in Morocco, meaning that people had reacted with high rates of negative sentiments regarding all COVID-19-related topics. For example, more than 80% of respondents were feeling very deficient on how the outbreak has calamitously affected the economy and job market, 70% expressed feelings of anger, fear, and anxiety toward the novel coronavirus spread, and 80% of them expressed negative feelings about education, especially during lockdowns where students had to stay at home and learn online. People have worried about health systems collapse because of the huge number of new cases received every day at hospitals. The majority of Moroccans reacted to vaccine campaigns (80%) and expressed concerns of worry and hesitancy about the ones that the health authorities suggested for people to take, especially Sinopharm and AstraZeneca vaccines.

Table 16.5 Topic-level sentiment analysis results

Topic	*Polarity*	*Frequency*	*Percentage (%)*
Economy	Positive	254	6.3
	Neutral	537	13.3
	Negative	3285	80.5
Education	Positive	480	8.8
	Neutral	849	15.6
	Negative	4114	75.6
Fear	Positive	1550	17.09
	Neutral	1118	12.4
	Negative	6400	70.6
Health	Positive	631	8.85
	Neutral	1198	16.8
	Negative	5300	74.5
Support	Positive	344	7.7
	Neutral	524	11.7
	Negative	3604	80.6
Vaccine	Positive	283	3.8
	Neutral	1252	17
	Negative	5815	80.2

In the first year of the pandemic in Morocco (2020), people have been mostly concerned about health and education. This led them to express emotions of fear and disruption as illustrated in Figure 16.4. However, high rates of negative concerns and reactions have emerged about the vaccine in the second year of the pandemic (Figure 16.5). The widespread dissemination of misleading information and fake news about the COVID-19 outbreak and

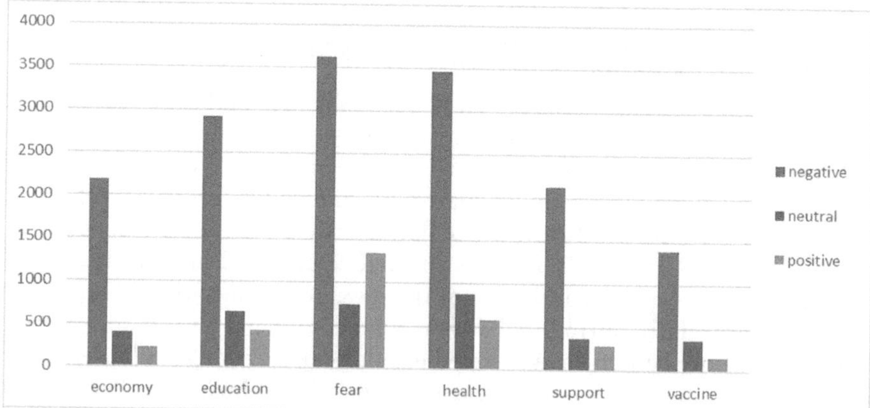

Figure 16.4 Topic-based sentiments in 2020.

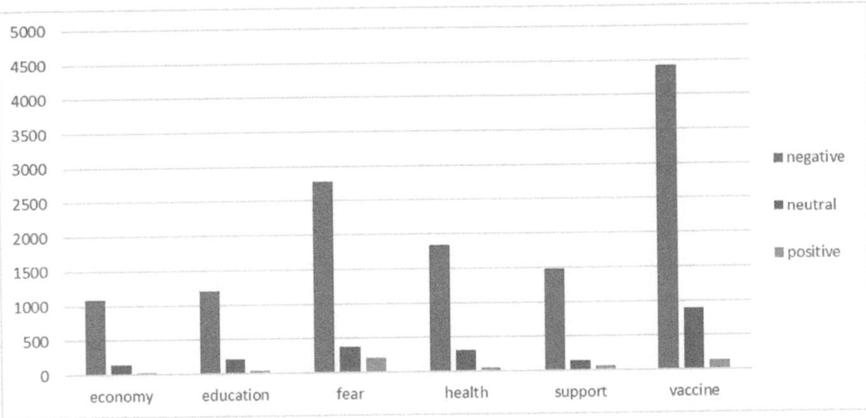

Figure 16.5 Topic-based sentiments in 2021.

the vaccines through social media and the internet has led to a growing sense of fear and distrust among people, making them hesitant to take the vaccine. This can be seen clearly in Figure 16.6 which shows the impacts of COVID-19 regarding all other topics except vaccine. Sentiments of fear were highly negative for these topics. Consequently, some people have formed groups of anti-vaccination to manifest against vaccines, and imposing mandatory vaccine passes for accessing government buildings, and other public places. As a result, vaccine hesitancy has become a pressing issue that drives people to delay in accepting or refusing vaccines despite their availability. It is often fueled by concerns, doubts, and misinformation shared on social media and the internet. This hesitancy can lead to lower vaccination rates, hindering efforts to achieve herd immunity and control the spread of the virus.

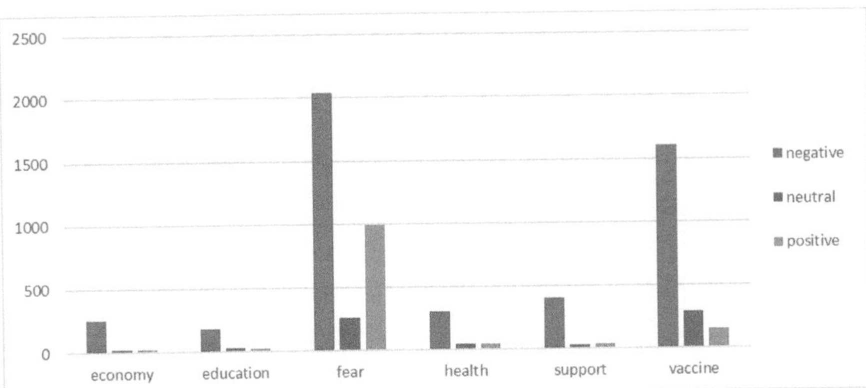

Figure 16.6 Topic-based sentiments in 2022.

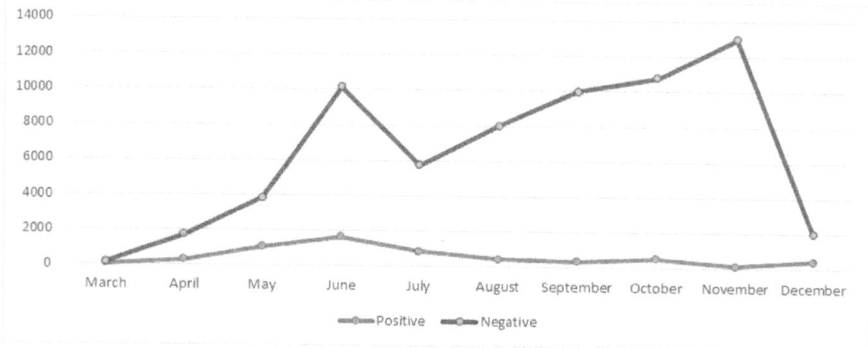

Figure 16.7 Sentiment variation in 2020.

We intended to separate our study into three different periods of the pandemic to track sentiment changes and compare these variations from March 2020 until the end of 2022 in the history of the pandemic. Figure 16.7 shows that negative sentiment rates started to increase exponentially from March 2020, the very beginning of the outbreak, to peak in November 2020. After this period, people's general concerns about the COVID-19 situation started to fade. In the second period, negative sentiment feedback was not as high as in 2020. However, it is obvious from Figure 16.8 that negative sentiments in the time window of March 2021 to May 2021 have spiked and after that started to decrease. After June 2021, negative polarities started to increase slightly until November and decreased at the end of the year. In the beginning of 2022 as showed in Figure 16.9, negative sentiments continued to decrease until March and then started to increase and peaked in June and started to decrease until the end of the year while rates of positive sentiments slightly increased in the last months of the year.

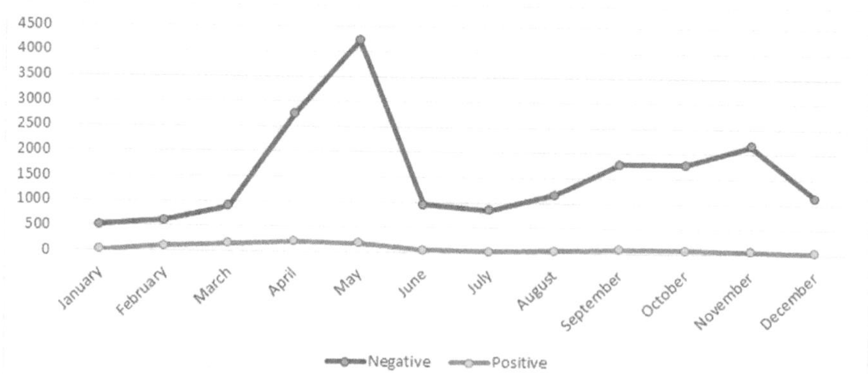

Figure 16.8 Sentiment variation in 2021.

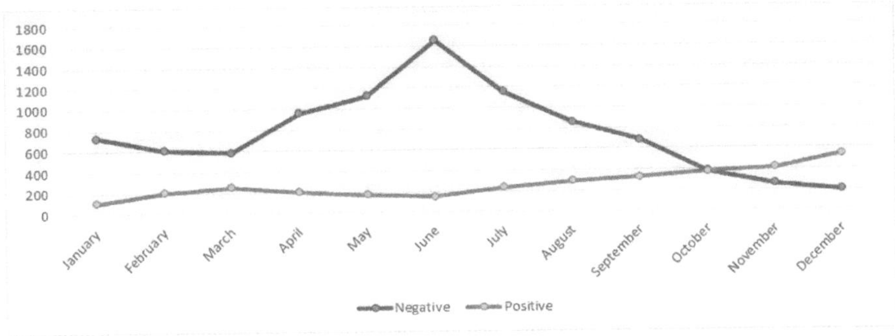

Figure 16.9 Sentiment variation in 2022.

The variation of negative sentiments over time is well explained in Figures 16.10–16.12. The epidemiological situation, represented by monthly records of COVID-19 cases and fatalities, directly influences how people react to the pandemic. The line plots show that changes in COVID-19 situations directly affect feedback sentiments during a given period of time. High rates of negative sentiments are explained by the spikes of coronavirus spread (cases and fatalities) among people and the subsequent actions taken, such as lockdowns, wearing masks, social distancing, and other protective measures to slow down the outbreak.

As the infection rate decreases and the government slightly eases protective measures, negative sentiments also decrease, and people gradually return to their normal lives. When the situation improves, and there is a lower risk of COVID-19 transmission, the negative sentiments may start to subside. People might feel more relieved and less anxious as they perceive a lower threat level. As a result, they may gradually return to their pre-pandemic routines and lifestyles. The interplay between the epidemiological situation, government

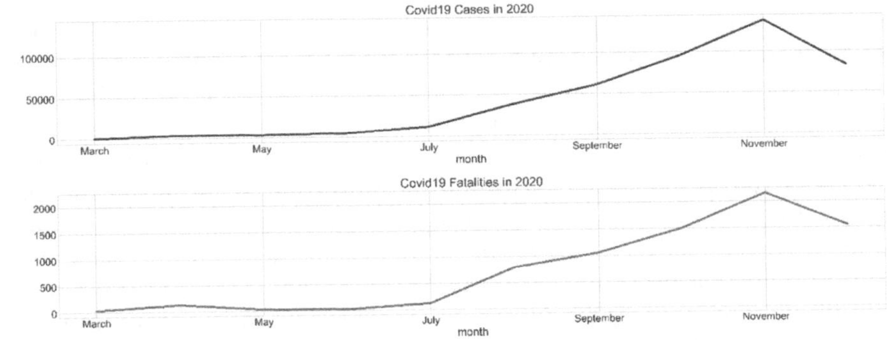

Figure 16.10 Monthly COVID-19 cases and fatalities in Morocco (2020).

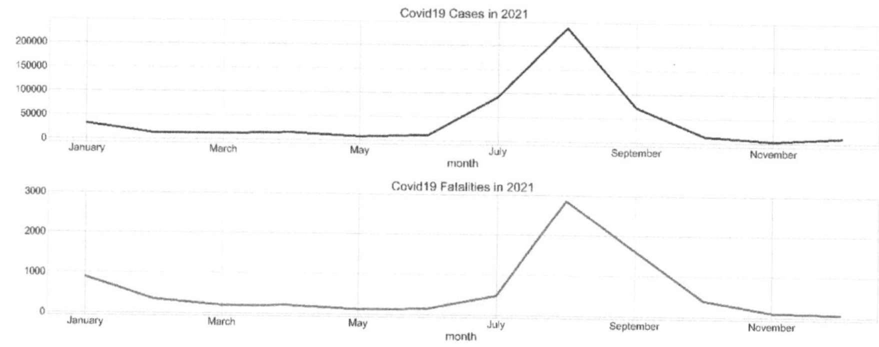

Figure 16.11 Monthly COVID-19 cases and fatalities in Morocco (2021).

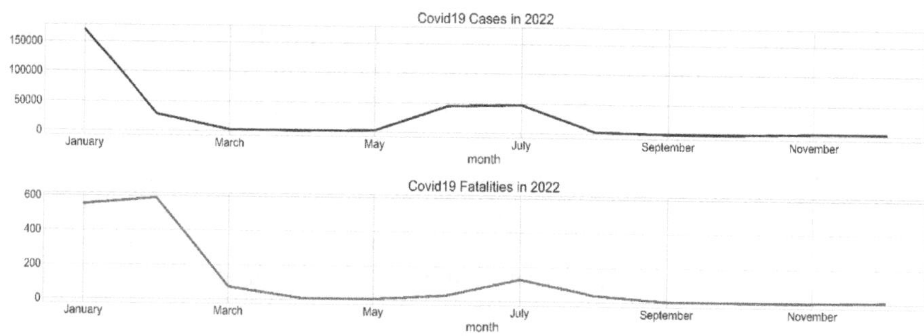

Figure 16.12 Monthly COVID-19 cases and fatalities in Morocco (2022).

actions, and people's reactions can significantly impact the changes in sentiments over time during the course of the COVID-19 pandemic.

16.5 CONCLUSION

In this chapter, we proposed an approach that combines many different NLP techniques to examine COVID-19's effects on Moroccan citizens through their generated content on social media and other web platforms. The system takes textual data from different sources (Twitter and the online newspaper Hespress) and performs necessary text cleaning and preprocessing, then extracts main topics within the corpus using the BERTopic model, and finally identifies sentiments on a topic-based level using the pre-trained Arabic CAMeL analyzer. The results showed that the pandemic has drastically affected all aspects of normal life. We also used COVID-19 times series data of monthly cases and fatalities to examine sentiment variation within three different periods in the outbreak's history, and we found that the epidemiological situation highly impacts SA results.

NOTES

1 **Hesspress** news website (accessed 26/5/2024): https://www.hespress.com/
2 C0VID-19 time series data repository (accessed 25/5/2024): https://github.com/CSSEGISandData/COVID-19
3 **Hespress** dataset repository (accessed 26/5/2024): https://github.com/HankarM88/Hespress_COVID-19_Dataset
4 TBCOV dataset (accessed 26/5/2024): https://crisisnlp.qcri.org/tbcov

REFERENCES

1. A. Haleem, M. Javaid, R. Vaishya, Effects of COVID-19 pandemic in daily life, Curr Med Res Pract. 10 (2020). https://doi.org/10.1016/j.cmrp.2020.03.011.
2. H. Onyeaka, C.K. Anumudu, Z.T. Al-Sharify, E. Egele-Godswill, P. Mbaegbu, COVID-19 pandemic: A review of the global lockdown and its far-reaching effects, Sci Prog. 104 (2021). https://doi.org/10.1177/00368504211019854.
3. M. Škare, D.R. Soriano, M. Porada-Rochoń, Impact of COVID-19 on the travel and tourism industry, Technol Forecast Soc Change. 163 (2021). https://doi.org/10.1016/j.techfore.2020.120469.
4. S.S. Priya, E. Cuce, K. Sudhakar, A perspective of COVID 19 impact on global economy, energy and environment, Int J Sustain Eng. 14 (2021). https://doi.org/10.1080/19397038.2021.1964634.
5. A.Y. Chang, M.R. Cullen, R.A. Harrington, M. Barry, The impact of novel coronavirus COVID-19 on noncommunicable disease patients and health systems: A review, J Intern Med. 289 (2021). https://doi.org/10.1111/joim.13184.
6. S. Pokhrel, R. Chhetri, A literature review on impact of COVID-19 pandemic on teaching and learning, High Educ Future. 8 (2021). https://doi.org/10.1177/2347631120983481.
7. M. Passavanti, A. Argentieri, D.M. Barbieri, B. Lou, K. Wijayaratna, A.S. Foroutan Mirhosseini, F. Wang, S. Naseri, I. Qamhia, M. Tangerås, M. Pelliciari, C.H. Ho, The psychological impact of COVID-19 and restrictive measures in the world, J Affect Disord. (2021). https://doi.org/10.1016/j.jad.2021.01.020.
8. G. Serafini, B. Parmigiani, A. Amerio, A. Aguglia, L. Sher, M. Amore, The psychological impact of COVID-19 on the mental health in the general population, QJM: Int J Meds. 113 (2020). https://doi.org/10.1093/qjmed/hcaa201.
9. S.K. Brooks, R.K. Webster, L.E. Smith, L. Woodland, S. Wessely, N. Greenberg, G.J. Rubin, The psychological impact of quarantine and how to reduce it: Rapid review of the evidence, Lancet. (2020). https://doi.org/10.1016/S0140-6736(20)30460-8.
10. L. Nemes, A. Kiss, Social media sentiment analysis based on COVID-19, J Inform Telecommun. 5 (2021). https://doi.org/10.1080/24751839.2020.1790793.
11. T. Quyyam, H. Ghous, Sentiment analysis of amazon customer product reviews: A review, Int J Sci Res Eng Dev. 4 (2021).
12. K. Chakraborty, S. Bhattacharyya, R. Bag, A survey of sentiment analysis from social media data, IEEE Trans Comput Soc Syst. 7 (2020). https://doi.org/10.1109/TCSS.2019.2956957.
13. M. Kasri, A. El-Ansari, M. El Fissaoui, B. Cherkaoui, M. Birjali, A. Beni-Hssane, Public sentiment toward renewable energy in Morocco: Opinion mining using a rule-based approach, Soc Netw Anal Min. 13 (2023) 124. https://doi.org/10.1007/s13278-023-01119-3.
14. M. Birjali, M. Kasri, A. Beni-Hssane, A comprehensive survey on sentiment analysis: Approaches, challenges and trends, Knowl Based Syst. 226 (2021) 107134. https://doi.org/10.1016/j.knosys.2021.107134.

15. M. Birjali, M. Kasri, A. Beni-Hssane, A comprehensive survey on sentiment analysis: Approaches, challenges and trends, Knowl Based Syst. 226 (2021). https://doi.org/10.1016/j.knosys.2021.107134.
16. R. Chandrasekaran, V. Mehta, T. Valkunde, E. Moustakas, Topics, trends, and sentiments of tweets about the COVID-19 pandemic: Temporal infoveillance study, J Med Internet Res. 22 (2020). https://doi.org/10.2196/22624.
17. J. Xue, J. Chen, C. Chen, C. Zheng, S. Li, T. Zhu, Public discourse and sentiment during the COVID 19 pandemic: Using Latent Dirichlet Allocation for topic modeling on twitter, PLoS One. 15 (2020). https://doi.org/10.1371/journal.pone.0239441.
18. S. Boon-Itt, Y. Skunkan, Public perception of the COVID-19 pandemic on Twitter: Sentiment analysis and topic modeling study, JMIR Public Health Surveill. 6 (2020). https://doi.org/10.2196/21978.
19. H. Yin, X. Song, S. Yang, J. Li, Sentiment analysis and topic modeling for COVID-19 vaccine discussions, World Wide Web. 25 (2022). https://doi.org/10.1007/s11280-022-01029-y.
20. M. Qorib, T. Oladunni, M. Denis, E. Ososanya, P. Cotae, Covid-19 vaccine hesitancy: Text mining, sentiment analysis and machine learning on COVID-19 vaccination Twitter dataset, Expert Syst Appl. 212 (2023). https://doi.org/10.1016/j.eswa.2022.118715.
21. M. Abdul-Mageed, S. Kübler, M. Diab, SAMAR: A system for subjectivity and sentiment analysis of Arabic social media, in: 12 Proceedings of the 3rd Workshop in Computational Approaches to Subjectivity and Sentiment Analysis, 2012.
22. T. Zarra, R. Chiheb, R. Moumen, R. Faizi, A. El Afia, Topic and sentiment model applied to the colloquial Arabic: A case study of Maghrebi Arabic, in: ACM International Conference Proceeding Series, 2017. https://doi.org/10.1145/3128128.3128155.
23. N. Shelke, S. Deshpande, V. Thakare, Domain independent approach for aspect oriented sentiment analysis for product reviews, in: Advances in Intelligent Systems and Computing, 2017. https://doi.org/10.1007/978-981-10-3156-4_69.
24. Y. Madani, M. Erritali, B. Bouikhalene, A new sentiment analysis method to detect and analyse sentiments of Covid-19 Moroccan tweets using a recommender approach, Multimed Tools Appl. (2023). https://doi.org/10.1007/s11042-023-14514-x.
25. M. Hankar, M. Birjali, A. El-Ansari, A. Beni-Hssane, COVID-19 impact sentiment analysis on a topic-based level, J ICT Stand. 10 (2022). https://doi.org/10.13052/jicts2245-800X.1027.
26. M. Hankar, M. Birjali, A. El-Ansari, A. Beni-Hssane, Arabic topic modeling-based sentiment analysis on COVID-19 feedback comments, in: Lecture Notes in Networks and Systems, 2022. https://doi.org/10.1007/978-3-030-91738-8_9.
27. M.O. Hegazi, Y. Al-Dossari, A. Al-Yahy, A. Al-Sumari, A. Hilal, Preprocessing Arabic text on social media, Heliyon. 7 (2021). https://doi.org/10.1016/j.heliyon.2021.e06191.
28. I. Abu Farha, W. Magdy, A comparative study of effective approaches for Arabic sentiment analysis, Inf Process Manag. 58 (2021). https://doi.org/10.1016/j.ipm.2020.102438.
29. M. Imran, U. Qazi, F. Ofli, TBCOV: Two billion multilingual COVID-19 tweets with sentiment, entity, geo, and gender labels, Data (Basel). 7 (2022). https://doi.org/10.3390/data7010008.
30. D.M. Blei, A.Y. Ng, M.I. Jordan, Latent Dirichlet Allocation, J Mach Learn Res. 3 (2003). https://doi.org/10.1016/b978-0-12-411519-4.00006-9.
31. P.O. Hoyer, Non-negative matrix factorization with sparseness constraints, J Mach Learn Res. 5 (2004).

32. E. Rudkowsky, M. Haselmayer, M. Wastian, M. Jenny, Š. Emrich, M. Sedlmair, More than bags of words: Sentiment analysis with word embeddings, Commun Methods Meas. 12 (2018). https://doi.org/10.1080/19312458.2018.1455817.
33. M. Kasri, M. Birjali, A. Beni-Hssane, A comparison of features extraction methods for Arabic sentiment analysis, in: Proceedings of the 4th International Conference on Big Data and Internet of Things, ACM, New York, NY, USA, 2019: pp. 1–6. https://doi.org/10.1145/3372938.3372998.
34. M. Kasri, M. Birjali, A. El Ansari, A. Beni-Hssane, Enhanced Word Embeddings with Sentiment Contextualized Vectors for Sentiment Analysis, in: 2022: pp. 77–86. https://doi.org/10.1007/978-3-030-91738-8_8.
35. M. Kasri, M. Birjali, M. Nabil, A. Beni-Hssane, A. El-Ansari, M. El Fissaoui, Refining word embeddings with sentiment information for sentiment analysis, J ICT Stand. (2022). https://doi.org/10.13052/jicts2245-800X.1031.
36. J. Devlin, M.W. Chang, K. Lee, K. Toutanova, BERT: Pre-training of deep bidirectional transformers for language understanding, in: NAACL HLT 2019 – 2019 Conference of the North American Chapter of the Association for Computational Linguistics: Human Language Technologies – Proceedings of the Conference, 2019.
37. Q. Le, T. Mikolov, Distributed representations of sentences and documents, in: 31st International Conference on Machine Learning, ICML 2014, 2014.
38. R. Egger, J. Yu, A topic modeling comparison between LDA, NMF, Top2Vec, and BERTopic to Demystify Twitter posts, Front Sociol. 7 (2022). https://doi.org/10.3389/fsoc.2022.886498.
39. A. Abuzayed, H. Al-Khalifa, BERT for Arabic topic modeling: An experimental study on BERTopic technique, Procedia Comput Sci. 189 (2021) 191–194. https://doi.org/10.1016/J.PROCS.2021.05.096.
40. W. Antoun, F. Baly, H. Hajj, AraBERT: Transformer-Based Model for Arabic Language Understanding, Proceedings of the 4th Workshop on Open-Source Arabic Corpora and Processing Tools, with a Shared Task on Ossffensive Language Detection, 2021.
41. O. Obeid, N. Zalmout, S. Khalifa, D. Taji, M. Oudah, B. Alhafni, G. Inoue, F. Eryani, A. Erdmann, N. Habash, CAMeL tools: An open-source python toolkit for Arabic natural language processing, in: LREC 2020 – 12th International Conference on Language Resources and Evaluation, Conference Proceedings, 2020.

Part III

Internet of Things (IoT) and big data

Chapter 17

Smart EV routing to charging stations for traffic optimization in smart cities

A case study in Agadir

Nour-Eddine Moumni, Rachid Alaoui, Driss Kiouach, and Ibrahim El-Fedany

17.1 INTRODUCTION

The beginning of the 21st century has undergone a notable change, with a clear emphasis on sustainability and sustainable development, with electric vehicles (EVs) playing a leading role in this axis, increasing resilience to sustainable environmental impacts, and supporting society's shift to green mobility solutions [1].

However, the integration of EVs is a complex challenge in infrastructure and urban planning [2]. Efficient and extensive charging infrastructure is essential for the perfect integration of EVs into urban environments. But establishing charging stations (CSs) is not a complete solution. The strategic location and the fluid access of vehicles to these stations are more important, specifically during peak hours. Because congestion in metropolitan areas may reduce the advantages of EVs.

Agadir, a bustling city in Morocco, embodies the challenges expected of the urban center. It is known for its beautiful beaches, modern architecture, and tourist sites, as well as forests with rare trees such as Argan. It embodies a blend of tradition and modernity. The integration of EVs is an important aspect of its environmental commitment and needs to adapt its urban infrastructure [2, 3]. As a result, the road network, traffic patterns, placement, and access to CSs become key points for study and optimization.

This chapter examines the complex interaction of EVs on the streets of Agadir to optimize the effectiveness and harmony of this process. We investigate how real-time simulations of urban mobility and intelligent systems can revolutionize intelligent routing systems [4, 5]. These systems not only guide EVs along the most efficient path but also facilitate simple access to CSs, supporting Agadir's goal of promoting sustainable transportation.

17.2 RELATED WORK

A comprehensive summary of the developments in the integration of EVs into smart city ecosystems, which is a critical element in improving urban

DOI: 10.1201/9781032664293-20

sustainability, can be found in the "Related Work" part of our chapter. This exploration synthesizes a wide range of research in related fields, highlighting the multifaceted nature of EV adoption, their interaction with urban infrastructure [6], and corresponding optimization challenges [7, 8]. It delves into the technology's evolution of EVs, their role in transforming urban mobility, and the development of supporting infrastructures such as charging networks and smart electrical grids. This section examines how the integration of EVs intersects with the broader goals of smart cities such as reducing carbon emissions, improving energy efficiency, and promoting sustainable urban development [9, 10]. It also considers the socio-economic factors influencing the adoption of EVs, including political initiatives, market trends, and public perception, which collectively shape the effectiveness of EV integration strategies. This review looks at how technology, urban planning, policies, and user behavior affect EVs in smart cities [11].

- Improving EV Driving Range: Pioneering research in this area has focused on addressing EV range anxiety. Notable work [11–13], and [14] has explored dynamic wireless charging solutions, offering innovative methods to extend EV driving range without extensive infrastructure changes.
- Optimizing EV Charging Networks: [4, 15], and [16] have made a significant contribution to the optimization of EV charging in residential grids through the utilization of fuzzy logic control systems. This research streamlines the charging process, thereby increasing grid efficiency.
- The integration of the Internet of Vehicles (IoV) has been widely studied to improve the mobility of EVs. Studies [17, 18], and [19] present token-based incentive systems developed to reduce congestion at CSs, demonstrating the potential of IoV in smart mobility.
- Leveraging artificial intelligence (AI) and traffic forecasting: Recent studies have focused on refining EV CS recommendations by incorporating traffic forecasts and AI algorithms. These approaches aim to optimize EV battery charging, reduce transmission delays, and improve quality of service (QoS). Key contributions in this area include the work of [20], which explored the safe and efficient prediction of EV charging demand using machine learning algorithms, and [21], which performed short-term prediction of EV load using AI techniques.
- Advanced CS selection systems: the study by [22] introduces an intelligent system that combines real and predicted data to recommend optimal CSs [23]. This system is particularly effective in reducing total travel time during peak hours.

To list the most important things added by these studies, we suggest the contents of Table 17.1. It shows the research conducted on their optimization strategies, energy management methods, IoV integration solutions, and their impact on urban mobility. This in-depth look at the present and future of EV

Table 17.1 Comparative analysis of EV charging optimization strategies in smart cities

Reference	*Optimization strategy*	*Energy management method*	*IoV integration solution*	*Impact on urban mobility*
[11–14]	Dynamic wireless charging	Mobile Energy Disseminators (MED)	Not applicable	Enhanced EV range
[4, 15, 16]	EV charging prioritization	Fuzzy logic control system	Not applicable	Increased grid efficiency
[17–19]	Token-based incentive system	Not applicable	Tokenization	Optimized traffic and reduced station wait times
[22, 23]	Real and predicted data system	Smart algorithms for CS selection	Not applicable	Minimized total trip duration during peak times
[20, 21]	Optimized station recommendations	Traffic forecasts and AI	Not applicable	Reduced transmission delays and improved QoS

integration in smart cities shows how technology can help solve problems and create opportunities in this rapidly evolving field.

17.3 METHOD

17.3.1 Model system

In the conceptual framework of this chapter, detailed state diagrams are provided which delineate the various stages and operational sequences for the Mobile Charging Station (MCS).

These diagrams, as illustrated in Figure 17.1, furnish a profound insight into the states that the MCS navigates throughout its charging and discharging cycle.

The specific states of the diagram are explained as follows:

Driving State: In this state, EVs are in motion within the city, heading toward their respective destinations.

- Low Battery State: Indicates the battery requires an imminent recharge.
- Driver Decision States: These states reflect the decisions of drivers regarding the recharging of their EVs.
- Reservation State: This occurs when an EV initiates a charging reservation request.
- Moving to CS State: The driver confirms the reservation and heads toward the chosen CS.

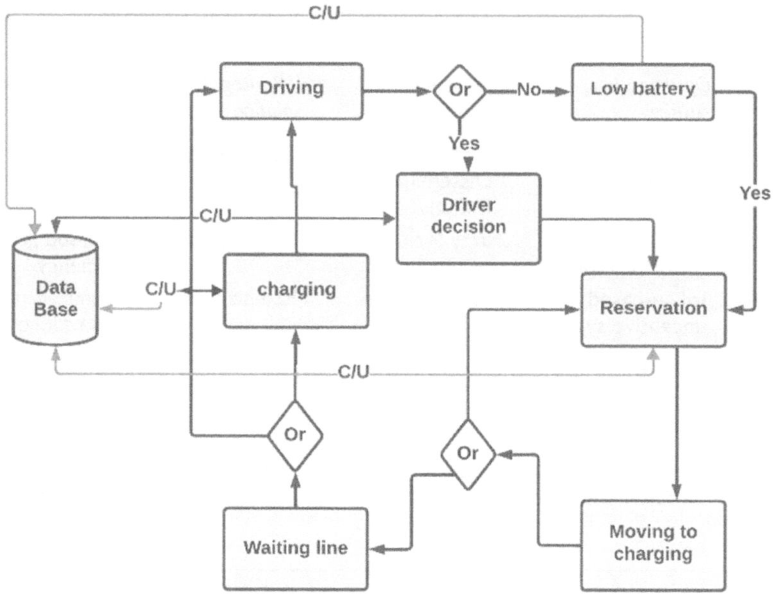

Figure 17.1 Flowchart of EV charging transition state.

- Waiting Line State: Upon arrival at the charging station, the EV joins a queue.
- Charging State: The EV is actively being charged.
- Database: Stores outcomes and measurements derived from the simulation.

17.3.2 Communication system architecture

The most valuable aspect of this concept lies in the seamless connection between the EVs and the CSs. The Internet of Vehicles (IoV) facilitates uninterrupted bidirectional communication, allowing EVs to receive up-to-the-minute data regarding the accessibility, whereabouts, and condition of the CSs.

Concurrently, the CSs collect information from the EVs, including their current level of charge and estimated time of arrival (ETA). This data enables the CSs to strategize and enhance the charging procedure.

Figure 17.2 provides a summary of EV charging options inside a smart city infrastructure for the Internet of Electric Vehicles (IoV).

17.3.3 Assignment algorithm

Our system is based on two different but closely related algorithms. The primary algorithm is the EV charging planning algorithm, which is used at CSs.

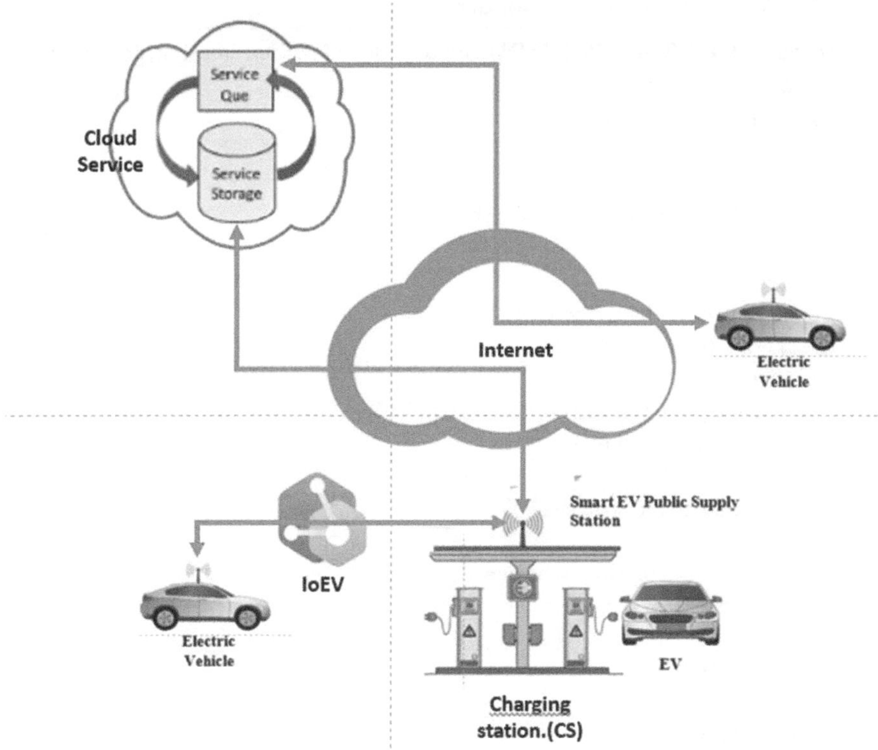

Figure 17.2 Smart city with IoV interconnections.

It generates a comprehensive list of estimated charging and queuing times for each CS along the route. The **FCFS** (first-come, first-served) approach is employed to consider the energy requirements of EVs in line and allocate charging slots accordingly.

When a new reservation request for an EV arrives, the decision algorithm for selecting the charging station, represented in Table 17.2, is activated. After determining the most suitable CS for the electric vehicle, this algorithm uses the first algorithm to pinpoint the exact location of the chosen station where the vehicle will be recharged.

Before planning a charge, it is crucial to select the most appropriate CS for each electric vehicle. This is precisely what this algorithm does.

Depending on the current battery level of each electric vehicle, the vehicle is directed either to the nearest CS (in the case of a critical battery level) or to an optimal station based on various criteria. The process begins by defining a critical battery level, known as Ecritical. Each EV is then examined individually. If its battery level is at or below Ecritical, the **findClosestCS** function (which could be part of the first algorithm) is called to locate the nearest charging

Table 17.2 EV charging scheduling algorithm

```
procedure CHARGINGSCHEDULING (Nslot, Nev, Tcur, EmaxList, EcurList, β)
   PICs ← empty list
   # Initialize the schedule list
   for i = 1 to Nslot do
      PICs.add(Tcur)
      # Add the current time to the schedule
   end for
   if Nev > 0 then
      Sort vehicles by arrival time
      # Sort EVs by arrival time
      for j = 1 to Nev do
         Tendj ← PICs.first() + (EmaxList[j] - EcurList[j]) / β
         # Estimate end time
         Replace PICs.first() with Tendj
         # Update the schedule
         Sort PICs
         # Sort the updated schedule
      end for
   end if
   return PICs
   # Return the final schedule
end procedure
```

station. Otherwise, the **findOptimalCS** function (which could also be part of the first algorithm) is used to determine the optimal CS based on other criteria.

The lists of maximum and current vehicle capacities are called **EmaxList** and **EcurList**, respectively. Vehicles are ranked according to their arrival time before evaluating the charging end time. The distances between the electric vehicle's location and all CSs are calculated using the **FINDCLOSESTCS** procedure (Table 17.3), which returns the nearest one.

A logic is implemented in the **FINDOPTIMALCS** procedure (Table 17.4) to determine the optimal CS for EVs that do not have a critical battery condition. This approach can be based on various criteria such as waiting times and charge levels. Although presented separately, the algorithms are intrinsically linked in a logical sequence. The decision algorithm must be run first to select the charging station.

After directing a vehicle to a suitable CS and recording this decision in the TCR register, the EV charging planning algorithm is responsible for determining the specific charging slot for this vehicle at the designated station.

In other words, Table 17.5 of the second algorithm calls Table 17.2 of the first algorithm once it has identified and registered the appropriate CS for a specific electric vehicle. Table 17.6 shows the variables used in these algorithms.

Table 17.3 FINDCLOSESTCS procedure

```
procedure FINDCLOSESTCS (location, charging_stations)
    min_distance ← ∞
    closest_station ← null
    for each station in charging_stations do
        distance ← calculateDistance(location, station.location)
        if distance < min_distance then
            min_distance ← distance
            closest_station ← station
        end if
    end for
    return closest_station
end procedure
```

Table 17.4 FINDOPTIMALCS procedure

```
procedure FINDOPTIMALCS(EV, charging_stations)
    best_score ← -∞
    optimal_station ← null
    for each station in charging_stations do
        score ← station.charging_rate / (station.waiting_time + 1)
        if score > best_score then
            best_score ← score
            optimal_station ← station
        end if
    end for
    return optimal_station
end procedure
```

Table 17.5 Decision algorithm for charging station selection

```
procedure CHARGINGDECISION (TR, TCR, Ecritical, CS)
    Define Ecritical = 17
    # Critical battery level in kWh
    Initialize TR with all EVs and their current battery levels
    Initialize TCR (To Charge Registry) as empty
    while TR is not empty do
        EV = TR.getNextEV()
        # Fetch the next EV
        if EV.batteryLevel ≤ Ecritical then
            CS_closest = FINDCLOSESTCS(EV.location, CS)
            # Find closest charging station
            TCR.add(EV, CS_closest)
```

(Continued)

Table 17.5 (Continued)

```
            #Add to charging registry
        Else
            CS_optimal = FINDOPTIMALCS(EV, CS)
            # Find optimal charging station based on other criteria
            TCR.add (EV, CS_optimal)
        end if
        TR.remove(EV)
        # Remove EV from TR
    end while
return TCR
end procedure
```

Table 17.6 List of symbols and descriptions

Notation	*Meaning*
Nslot	Total number of time slots available for scheduling.
PICs	Planned charging slot
Tcur	list. Current time.
Nev	Total number of electric vehicles (EVs) needing a charge.
Eevjmax	Maximum battery capacity of the jth EV.
Eevjcur	Current battery level of the jth EV.
β	Charging rate (kWh/min).
Tevjfin	Finish time for charging the jth EV.
Ecritical	Critical battery level necessitating urgent recharging (in our case, 17 kWh).
TR	List containing all EVs and their current battery levels.
TCR	Registry tracking which EV is to be charged at which charging station.
CSj	The jth charging station.
NCSi	Total number of charging stations available to the ith EV.
Eicur	Current battery level of the ith EV.
Eijtran	Energy required for the ith EV to reach the jth charging station.
Tijchar	Time to charge the ith EV at the jth charging station.
Tijarr	Arrival time of the ith EV at the jth charging station.

17.4 SIMULATION OF ROAD TRAFFIC

17.4.1 Preparation of the simulation

Simulating road traffic accurately is a critical preliminary step in assessing the efficacy of any proposed solution, especially in the realm of intelligent vehicle routing. Given the dynamic nature of urban environments, especially in a city as lively as Agadir, constructing a virtual model that accurately mimics real-world conditions is no trifling feat. Below we outline the key components and methodologies involved in our simulation of road traffic in Agadir.

Utilizing the Simulation of Urban Mobility (SUMO) platform, we can mimic the real-world behavior of vehicles\cite{Behrisch2001}, including EVs, on Agadir city roads. This simulation considers:

- Traffic densities at different times of the day
- Availability and distribution of CSs
- Road conditions and events (like accidents or maintenance)

17.4.2 Simulation area selection

In the first step, we delineated a specific area of Agadir for our study. This area was chosen because of its relevance to EV routes and the presence of CSs. Once the area was defined, we extracted the relevant data from OpenStreetMap (OSM). This includes not only roads but also points of interest such as CSs for EVs. The graphical representation of the extracted data is shown in Figure 17.3.

With the OSM data in hand, we undertook the task of converting this information into route files suitable for SUMO. This step ensures that roads, intersections, and other components are formatted correctly for optimal simulation. The outcome of this conversion is showcased in Figure 17.4.

Table 17.4 enumerates some of the primary EV CSs located in the city of Agadir [11]. For each charging station, the type of charging method available

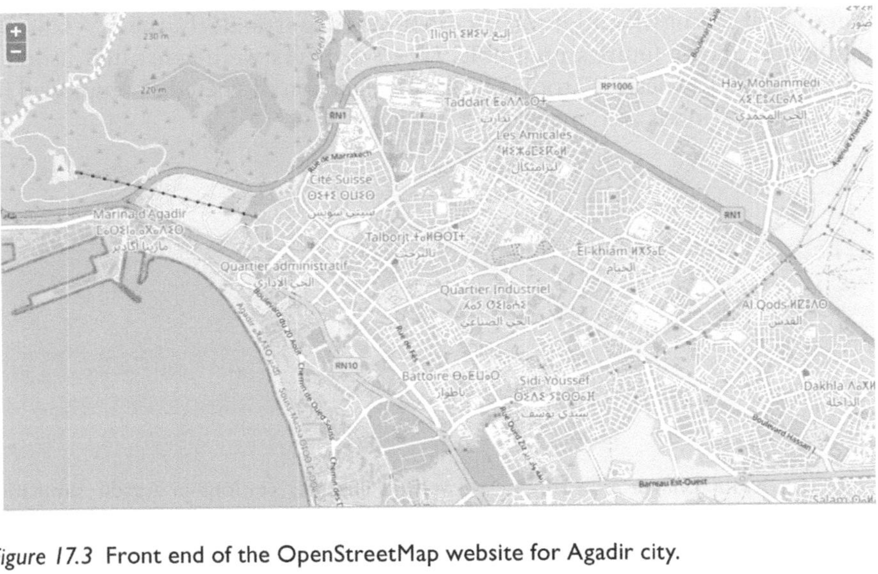

Figure 17.3 Front end of the OpenStreetMap website for Agadir city.

```xml
<?xml version="1.0" encoding="UTF-8"?>
<!-- generated on 2023-08-18 19:00:52 by Eclipse SUMO sumo Version 1.18.0
-->
<configuration xmlns:xsi="http://www.w3.org/2001/XMLSchema-instance" xsi:noNamespaceSchemaLocation=
"http://sumo.dlr.de/xsd/sumoConfiguration.xsd">

    <input>
        <net-file value="agadir.net.xml"/>
        <route-files value="agadir.rou.xml"/>
        <additional-files value="agadir.add.xml"/>
    </input>

    <processing>
        <ignore-route-errors value="true"/>
    </processing>

    <routing>
        <device.rerouting.adaptation-steps value="18"/>
        <device.rerouting.adaptation-interval value="10"/>
    </routing>

    <report>
        <verbose value="true"/>
        <duration-log.statistics value="true"/>
        <no-step-log value="true"/>
    </report>

    <gui_only>
        <gui-settings-file value="osm.view.xml"/>
    </gui_only>

</configuration>
```


Figure 17.4 Configuration file and running the network.

is specified. These methods represent the power rating (in kW) and the nature of the charging (Alternating Current - AC or Direct Current - DC).

The charging methods are categorized into TYPE 2, COMBO CCS EU, and CHADEMO, with their respective power ratings. The list serves as an informative tool for EV users to decide on which station to opt for, based on their vehicle's compatibility and proximity to these stations.

Figure 17.5 illustrates the XML configuration for a traffic simulation scenario, detailing the additional components, particularly the EV CSs. As highlighted in the figure's preamble, these elements were generated utilizing Eclipse SUMO's "netedit" tool. The specific version of "netedit" employed for this configuration is also mentioned in Figure 17.5.


```xml
<?xml version="1.0" encoding="UTF-8"?>

<!-- generated on 2023-08-26 09:10:41 by Eclipse SUMO netedit Version 1.18.0
-->
<additional xmlns:xsi="http://www.w3.org/2001/XMLSchema-instance" xsi:noNamespaceSchemaLocation=
"http://sumo.dlr.de/xsd/additional_file.xsd">
    <!-- StoppingPlaces -->
    <chargingStation id="cs_0" name="Station-service Total Relais" lane="745150552#1_0" startPos="10.83" endPos="55.83"/>
    <chargingStation id="cs_1" name="Fairmont Taghazout" lane="38087057#0_0" startPos="10.00" endPos="20.00"/>
    <chargingStation id="cs_2" name="Afriquia Agadir-Centre-ville" lane="100113564#1_0" startPos="13.94" endPos="23.94"/>

</additional>
```


Figure 17.5 XML configuration of electric vehicle charging stations in Agadir simulation scenario.

17.5 RESULT AND DISCUSSION

The optimization of EV routes has significantly reduced waiting times at CSs while increasing the energy efficiency of the trip. These results confirm the positive impact of our approach in promoting sustainable electric mobility in urban environments.

Figure 17.6, which graphically illustrates the execution of this simulation using the OSM-generated Agadir Road network, summarizes these results.

The simulation results of this study are very promising and confirm the effectiveness of our approach. The optimized routes proposed by our algorithms have shown a significant reduction in travel time for EVs, while at the same time minimizing waiting time at CSs. This dual improvement not only accelerates the transition toward electric transportation but also optimizes the utilization of the charging infrastructure powered by renewable energy, so encouraging a more ecologically friendly and efficient ecosystem.

In evaluating our approach, we based it on two different simulation scenarios to assess the effectiveness of the proposed algorithms. The first scenario, "**Scenario1**" replicated a traditional operation by randomly assigning cars to CSs without optimization. In contrast, "**Scenario2**" included the use of our optimized algorithms to intelligently direct EVs to the most appropriate CSs. We designed these scenarios to present various charging strategies, enabling us to measure the tangible effects of our algorithms on charging efficiency and overall system congestion. The analysis of data collected during these simulations revealed a significant reduction in CO_2 emissions associated with EV trips. This observation highlights the positive environmental impact of

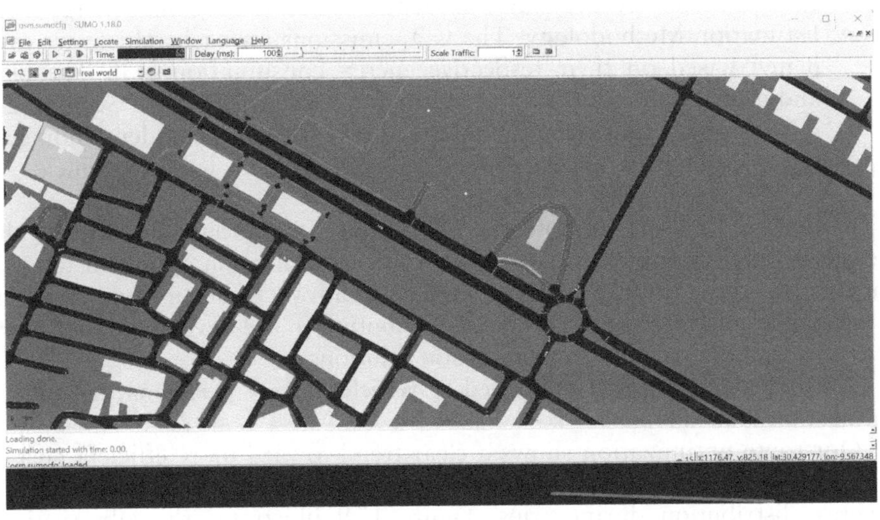

Figure 17.6 Vehicular traffic in SUMO.

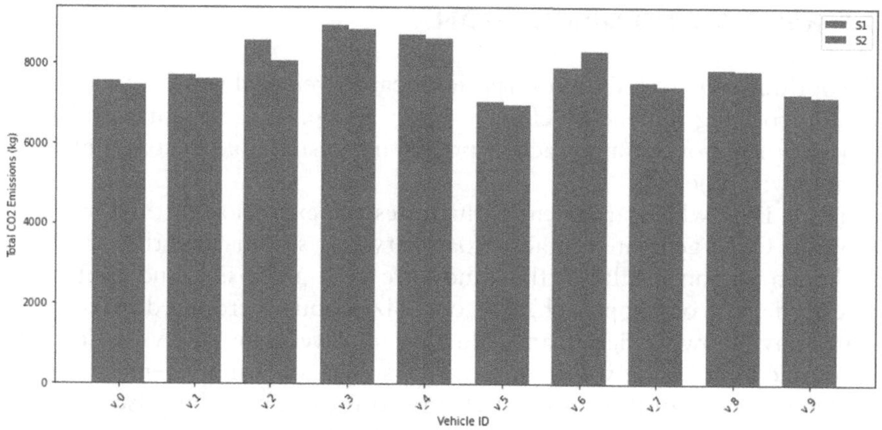

Figure 17.7 CO_2 emissions for each vehicle.

our approach and further supports the increased adoption of EVs in urban environments.

Figure 17.7 illustrates the estimated CO_2 emissions per vehicle based on its energy consumption in two distinct scenarios:

- Underlying Assumption: Every unit of energy consumed results in a specific amount of CO_2 emissions, primarily determined by the electricity source and production method. In this analysis, we consider a representative emission of 0.5 kg of CO_2 per kWh of energy consumed [12]. This value is indicative and may vary depending on the regional energy mix.
- Estimation Methodology: The CO_2 emissions for each vehicle are estimated based on their respective energy consumption [13]. The blue and green bars represent the results for "Scenario1" and "Scenario2", respectively. Comparing these bars allows for a clear understanding of the emission differences between the two scenarios for each vehicle.

In summary, the results obtained from our simulations demonstrate the feasibility and tangible benefits of our route optimization approach for EVs in Agadir. These findings mark a significant advancement in the pursuit of sustainable electric urban mobility, contributing to the development of more environmentally friendly transportation solutions.

We further analyzed data related to vehicle battery performance and CS utilization Figure 17.8.

Our route optimization shows a clear trend toward more efficient energy utilization. EVs exhibited smarter energy management, leading to improved energy distribution during trips. Figure 17.9 illustrates this observation, revealing a significant reduction in battery charge state variations during optimized trips [13].

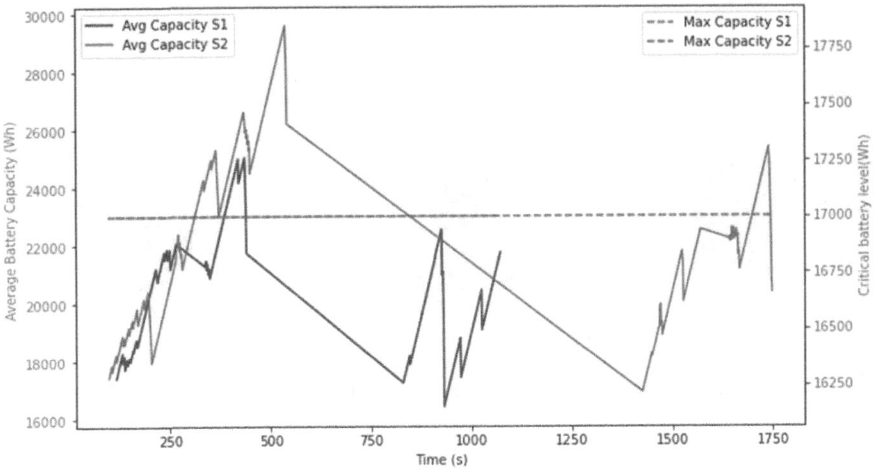

Figure 17.8 Average battery capacity versus critical battery level over time.

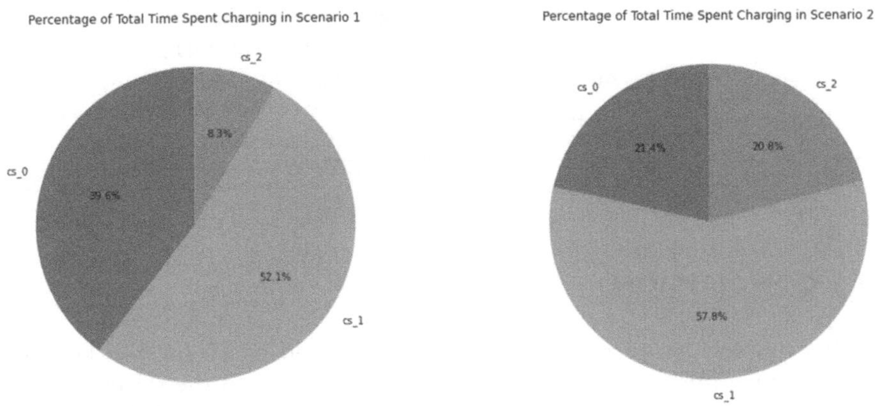

Figure 17.9 Percentage of total time spent charging.

Regarding CS usage, Figure 17.10 presents a comparison of waiting times at CSs before and after the application of our optimization algorithms. A distinct decrease in waiting times effectively demonstrates the efficiency of our approach in minimizing charging delays.

These comprehensive analyses of battery output data and the use of CSs confirm the effectiveness of our approach in all crucial aspects of electric mobility. They highlight the importance of our optimization algorithms and strengthen our argument based on two scenarios comparing battery capacities, facilitating a smoother and more environmentally responsible transition to EVs in Agadir (Table 17.7). These results perfectly align with our objectives of promoting a more sustainable and accountable urban mobility landscape.

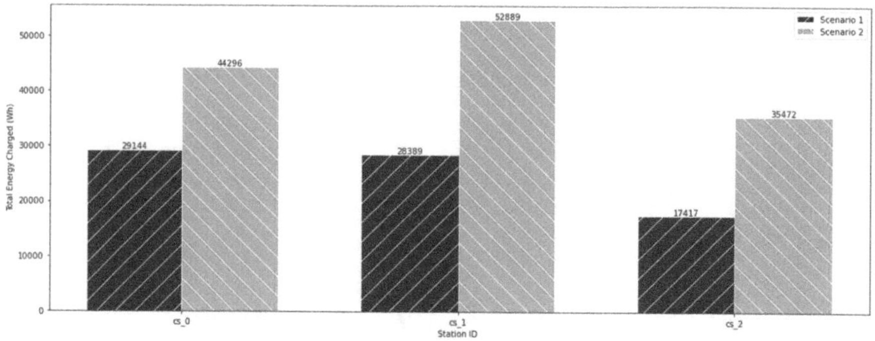

Figure 17.10 Total energy charged per station for both scenarios.

Table 17.7 List of electric vehicle charging stations in Agadir city

Charging station	*Charging method*
Afriquia Agadir-Centre-ville	TYPE 2: 22kW/AC – three phases COMBO CCS EU: 50kW/ DC CHADEMO: 50kW/DC
Station-service Total Relais Agadir	TYPE 2: 22kW/AC – three phases
FastVolt Afriquia Imouran Taghazout	COMBO CCS EU 50kW/AC – three phases CHADEMO:- 50kW/DC TYPE 2: 22kW/AC – three phases

17.6 CONCLUSION

In this chapter, we present our intelligent solution based on two key algorithms. One aims to optimize EV charging sessions and the selection of the most optimal route, while the other effectively guides these EVs directly to the CS blindly. Thanks to real-time simulations on Agadir's road networks, these strategies have demonstrated their ability to reduce traffic congestion, reduce waiting times at CSs, and, ultimately, improve the experience of using EVs in urban areas. The innovative aspect of our study lies precisely in this real-time simulation, integrating our algorithmic solution into the complexity of Agadir's traffic. This approach has allowed us to achieve tangible results and identify important potential areas for future improvement. Indeed, the data collected from these simulations offers great potential in the fields of AI [24, 25] and machine learning [21]. This could lead to more accurate predictions of driver behavior, better route planning, and better integration with smart city components through vehicle-to-everything (V2X) technology [22]. In short, as EVs become more widespread, our work on real-time optimization and simulation will encourage governments and urban planners to adopt smarter and more sustainable strategies to meet future needs.

REFERENCES

1. M. S. Hossain, L. Kumar, M. M. Islam, and J. Selvaraj, "A comprehensive review on the integration of electric vehicles for sustainable development," Journal of Advanced Transportation, vol. 2022, pp. 1–26, Oct. 2022.
2. N.-E. Moumni, R. Alaoui, D. Kiouach, and I. El-Fedany, "The Smart City of Tomorrow: A Simulation Platform for Optimizing EV Routes to Charging Points–A Moroccan Perspective," 2023 14th International Conference on Intelligent Systems: Theories and Applications (SITA), Casablanca, Morocco, 2023, pp. 1–7, doi: 10.1109/SITA60746.2023.10373710.
3. I. El-Fedany, D. Kiouach, and R. Alaoui, "A smart management system of electric vehicles charging plans on the highway charging stations," Indonesian Journal of Electrical Engineering and Computer Science, vol. 23, no. 2, p. 752, Aug. 2021.
4. N. Odkhuu, M. Ahmed, and Y.-C. Kim, "Priority Determination Based On Fuzzy Logic For Charging Electric Vehicles," In : 2019 Eleventh International Conference on Ubiquitous and Future Networks (ICUFN), IEEE, pp. 295–299.
5. M. Bilal, I. Alsaidan, M. Alaraj, F. M. Almasoudi, and M. Rizwan, "Techno-economic and environmental analysis of grid-connected electric vehicle charging station using AI-based algorithm," Mathematics, vol. 10, no. 6, p. 924, Mar. 2022.
6. Wikipedia contributors, "Smart city—wikipedia, the free encyclopedia," 2023, [Online; accessed 10-November-2023]. [Online]. Available: https://en.wikipedia.org/wiki/Smartcity
7. B. Anthony Jr., "Integrating electric vehicles to achieve sustainable energy as a service business model in smart cities," Frontiers in Sustainable Cities, vol. 3, 2021.
8. Calearo, L., Marinelli, M., & Ziras, C. (2021). A review of data sources for electric vehicle integration studies. Renewable and Sustainable Energy Reviews, 151, 111518.
9. F. Alanazi, "Electric vehicles: Benefits, challenges, and potential solutions for widespread adaptation," Applied Sciences, vol. 13, no. 10, p. 6016, May 2023.
10. C. Z. El-Bayeh, K. Alzaareer, A.-M. I. Aldaoudeyeh, B. Brahmi, and M. Zellagui, "Charging and discharging strategies of electric vehicles: A survey," World Electric Vehicle Journal, vol. 12, no. 1, p. 11, Jan. 2021.
11. D. Kosmanos, L. A. Maglaras, M. Mavrovouniotis, S. Moschoyiannis, A. Argyriou, A. Maglaras, and H. Janicke, "Route optimization of electric vehicles based on dynamic wireless charging," IEEE Access, vol. 6, pp. 42 551–42 565, 2018.
12. ElGhanam, E. A., Hassan, M. S., & Osman, A. H. (2020, September). Deployment optimization of dynamic wireless electric vehicle charging systems: A review. In 2020 IEEE International IOT, Electronics and Mechatronics Conference (IEMTRONICS) (pp. 1–7). IEEE.
13. L. Maglaras, J. Jiang, A. Maglaras, and F. Topalis, "Mobile energy disseminators increase electrical vehicles range in a smart city," 2014.
14. L. A. Maglaras, J. Jiang, A. Maglaras, F. V. Topalis, and S. Moschoyiannis, "Dynamic wireless charging of electric vehicles on the move with mobile energy disseminators," International Journal of Advanced Computer Science and Applications, vol. 6, no. 6, pp. 239–251, 2015.
15. V. Boglou, C.-S. Karavas, K. Arvanitis, and A. Karlis, "A fuzzy energy management strategy for the coordination of electric vehicle charging in low voltage distribution grids," Energies, vol. 13, no. 14, p. 3709, Jul. 2020.
16. S. Nag and K. Y. Lee, "Optimized fuzzy logic controller for responsive charging of electric vehicles," IFAC-PapersOnLine, vol. 52, no. 4, pp. 147–152, 2019.

17. N. Aung, W. Zhang, K. Sultan, S. Dhelim, and Y. Ai, "Dynamic traffic congestion pricing and electric vehicle charging management system for the internet of vehicles in smart cities," Digital Communications and Networks, vol. 7, no. 4, pp. 492–504, Nov. 2021.
18. J. P. Martins, J. C. Ferreira, V. Monteiro, J. A. Afonso, and J. L. Afonso, "IoT and blockchain paradigms for EV charging system," Energies, vol. 12, no. 15, p. 2987, Aug. 2019.
19. V. Chamola, A. Sancheti, S. Chakravarty, N. Kumar, and M. Guizani, "An IoT and edge computing based framework for charge scheduling and EV selection in V2G systems," IEEE Transactions on Vehicular Technology, vol. 69, no. 10, pp. 10 569–10 580, Oct. 2020.
20. M. Bharat, R. Dash, K. J. Reddy, A. S. R. Murty, C. Dhanamjayulu, and S. M. Muyeen, "Secure and efficient prediction of electric vehicle charging demand using α2-LSTM and AES-128 cryptography," Energy and AI, vol. 16, p. 100307, 2024.
21. P. B. Pati, G. Vishnu, D. Kaliyaperumal, A. Karthik, N. Subbanna, and A. Ghosh, "Short-term forecasting of electric vehicle load using time series, machine learning, and deep learning techniques," World Electric Vehicle Journal, vol. 14, p. 266, Sep. 2023.
22. I. EL-Fedany, D. Kiouach, and R. Alaoui, "A smart system combining real and predicted data to recommend an optimal electric vehicle charging station," Indonesian Journal of Electrical Engineering and Computer Science, vol. 30, no. 1, p. 394, Apr. 2023.
23. B. Alinia, M. H. Hajiesmaili, Z. J. Lee, N. Crespi, and E. Mallada, "Online EV scheduling algorithms for adaptive charging networks with global peak constraints," IEEE Transactions on Sustainable Computing, vol. 7, no. 3, pp. 537–548, 2020.
24. A. K. Kale, M. S. Shahriar, A. Islam, and K. Chang. "Deep learning based multi-zone AVP system utilizing V2I communications," IEEE Access, vol. 11, 90510–90525, 2023, doi: 10.1109/ACCESS.2023.3307571.
25. A. Mekrache, A. Bradai, E. Moulay, and S. Dawaliby. "Deep reinforcement learning techniques for vehicular networks: Recent advances and future trends towards 6G," Vehicular Communications, vol. 33, 100398, Jan. 2022, doi: 10.1016/j.vehcom.2021.100398.

Chapter 18

Dual-scored dimensionality reduction and spectral unmixing for hyperspectral data analysis

Vijaya Sindhoori Kaza and Rithika Badam

18.1 INTRODUCTION

Hyperspectral imaging has emerged as a revolutionary technique that facilitates the capture of intricate spectral information from a scene, unveiling hidden details and offering insights beyond the capabilities of traditional imaging systems. This unique capability has found applications in diverse fields such as agriculture, environmental monitoring, remote sensing, and medical imaging. However, the sheer complexity and dimensionality of hyperspectral data pose significant challenges in extracting meaningful information and facilitating accurate classification tasks. As a consequence, the development of advanced analytical methodologies is imperative to unlock the full potential of hyperspectral imaging and translate its richness into actionable insights.

In this chapter, we present a pioneering approach that amalgamates cutting-edge techniques from the realms of dimensionality reduction, scoring mechanisms, and classification algorithms to enhance the efficacy of hyperspectral data analysis. We recognize that efficient dimensionality reduction is pivotal to alleviate the curse of dimensionality, enabling improved classification performance and interpretation. Our method encompasses an innovative dual-scoring strategy that evaluates dimensionality reduction methods based on both variance and neighborhood characteristics. This multifaceted scoring approach enhances our understanding of the impact of various reduction techniques on the underlying data structure. Furthermore, our approach extends its reach beyond mere dimensionality reduction by incorporating Linear Spectral Unmixing (NMF) to decipher the underlying endmember proportions and compositions within the data. The inherent non-negativity constraint of NMF aligns with the physical constraints present in hyperspectral imagery, making it an apt tool for spectral unmixing. This chapter presents a literature review of this research work followed by the hyperspectral and multispectral data characteristics, dimensionality reduction techniques, spectral unmixing, feature selection and extraction as well as performance evaluation of our proposed model, concluding with future scope in this area.

DOI: 10.1201/9781032664293-21

18.2 PROPOSED WORK

The dataset employed in this research comprises hyperspectral imagery of the Indian Pines region, acquired through a hyperspectral sensor. This dataset encompasses spatial dimensions of 145 × 145 pixels and spectral information spanning 199 bands. Each pixel in the dataset represents a high-dimensional vector, with each band capturing spectral reflectance values. Additionally, ground truth labels are provided for each pixel, allowing supervised classification evaluations. The Indian Pines dataset presents a challenging scenario due to its high dimensionality and complex spectral characteristics, making it an ideal testbed for our proposed methodology. The application of these dimensionality reduction techniques on the dataset is represented in Figure 18.1.

On this dataset, our proposed methodology integrates various stages, including dimensionality reduction, scoring mechanisms, spectral unmixing, feature selection, and classificationas the algorithm in Table 18.1 proposes. The workflow commences with the application of dimensionality reduction techniques such as Principal Component Analysis (PCA) and Truncated Singular Value Decomposition (TruncatedSVD). Subsequently, we introduce a dual-scoring strategy, evaluating the effectiveness of dimensionality reduction methods based on both variance and neighborhood characteristics. This innovative scoring mechanism enriches our understanding of the impact of reduction techniques on data preservation. Expanding upon the dimensionality reduction phase, we integrate Linear Spectral Unmixing (NMF) to uncover endmember proportions and compositions. The NMF algorithm decomposes the hyperspectral data into non-negative endmember proportions and

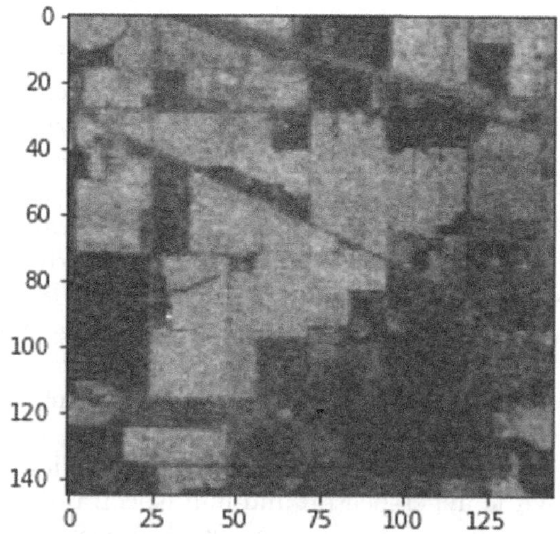

Figure 18.1 Application of dimensionality reduction techniques on the Indian Pines hyperspectral image.

Table 18.1 Algorithm for hyperspectral data analysis and classification

1. Data Loading and Visualization:
(a) Load Indian pines and ground truth data.
(b) Display the hyperspectral image and the ground truth labels.

2. Preprocessing and Dimensionality Reduction:
(a) Flatten the hyperspectral data:
imagex = flatten(indiana pines).
(b) Flatten the ground truth labels:
flattened ground = flatten(ground truth).
(c) Apply Standard Scaling to
imagex: norm flattened = standard(imagex)
(d) Perform PCA on norm flattened:
pca image = PCA fit(norm flattened)
(e) Perform Truncated SVD:
Truncated SVD fit(norm flattened).

3. Neighborhood Scoring & NMF:
(a) Compute neighborhood scores for the image:
neighborhood scores = neighbor score(imagex).
(b) Compare these scores with variance-based scores.
(c) Apply NMF to imagex:
endmember proportions, endmembers = NMF fit(imagex).

4. Feature Selection & Classification:
(a) Perform Mutual Information-based Selection:
selected bands mi = mutual info select(imagex, ground).
(b) Split data into training and testing sets:
(c) Train SVM classifier:
svm classifier = train SVM(X train, y train).
Similarly Train Random Forest classifier:
(d) Predict using SVM and RF
(e) Calculate accuracy:
calculate accuracy(y test, y predictions)

5. Visualization:
(a) Visualize PCA results & classification results
(b) Calculate ROC curve and visualize
(c) Visualize ground truth spatial distribution
(d) Create density plot and 3D PCA plot

endmembers, yielding a more interpretable representation of the data. This information-rich step lays the groundwork for informed feature selection.

Feature selection is executed through Recursive Feature Elimination (RFE) and Mutual Information-based Feature Selection. These methodologies, driven by mathematical metrics, strategically identify spectral bands that contribute significantly to classification accuracy. The resultant reduced-dimensional feature space enhances classification efficiency while minimizing the risk of overfitting. For classification, we deploy support vector machines (SVM) and Random Forest algorithms. SVM harnesses the power of hyperplanes to optimally separate classes, while Random Forest leverages an ensemble of decision trees to enhance robustness and accuracy. Classification accuracy serves as an evaluative metric, alongside visualizations that provide deeper insights into the classification outcomes. SVM aims to find the hyperplane that maximizes the margin between classes and Random Forest combines multiple decision trees to enhance accuracy and robustness.

18.3 RESULTS AND DISCUSSION

Figure 18.1 shows the results of applying PCA and Truncated SVD dimensionality reduction techniques on the Indian Pines hyperspectral image. The original image has 220 spectral bands, but the PCA and Truncated SVD reduced the dimensionality to 50 bands each. The PCA results (top row) show that the first two principal components capture most of the variance in the image. The Truncated SVD results (bottom row) show that the first two singular values capture most of the energy in the image.

Figure 18.2 shows a binary segmentation of an image. The foreground pixels are labeled 1, and the background pixels are labeled 0. The unique values in the ground truth array are 0 and 1. This image could be used to train a machine-learning model to segment images.

Figure 18.3 shows the reduced-dimensional data of a hyperspectral image after applying PCA. The data is well-separated into different clusters, which indicates that PCA was successful in reducing the dimensionality of the data while preserving the important information. The different clusters correspond to different ground truth labels, which means that the PCA transformation was able to capture the underlying structure of the data. This could be used to classify the different objects in the image.

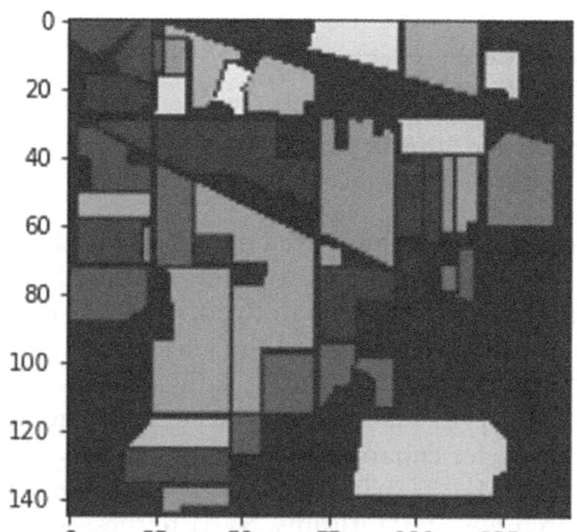

Figure 18.2 Application of binary segmentation using ground truthing on the Indian Pines hyperspectral image.

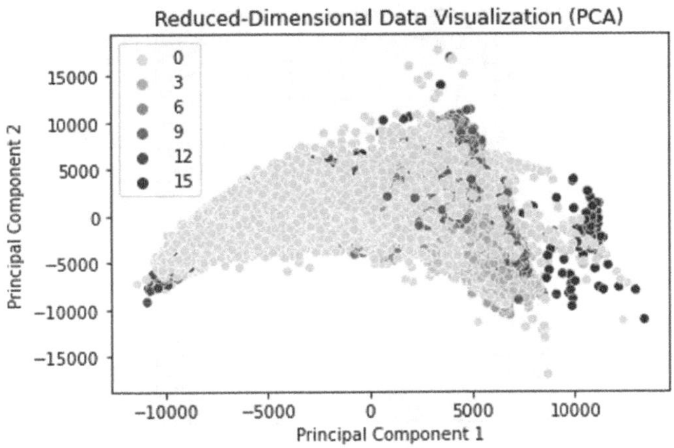

Figure 18.3 Reduced-Dimensional data visualization (PCA).

Figure 18.4 shows the Receiver-operating characteristic (ROC) curves for the three classes in the multi-class SVM classifier. The ROC curve is a measure of the classifier's performance. The area under the curve. (AUC) for the three classes are all above 0.75, which indicates that the classifier is performing well.

Figure 18.5 shows a spatial distribution map of a city. The different spectral bands of hyperspectral imagery can be used to distinguish between different

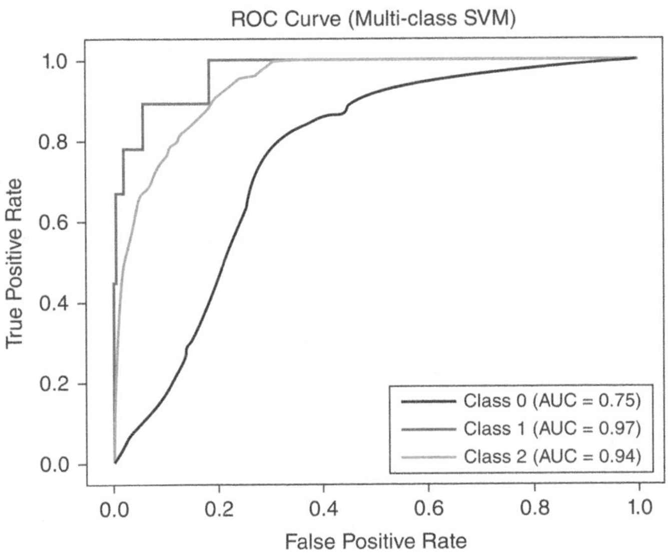

Figure 18.4 ROC curve.

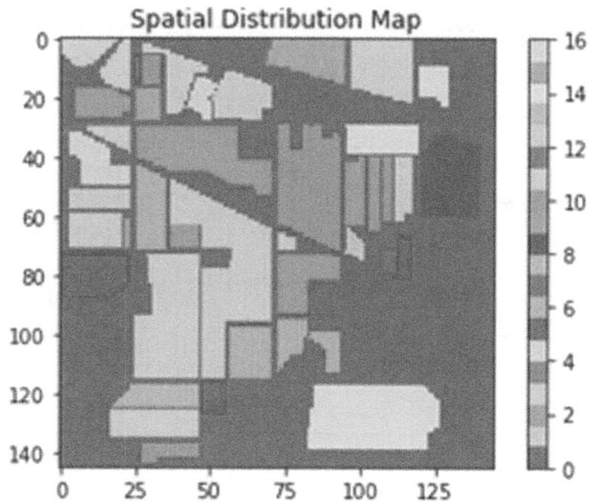

Figure 18.5 Spatial distribution map.

land covers, such as agriculture, forest, water, and urban areas. In the image, each square represents a pixel of the image. The color of the square represents the land cover in that pixel. The different colors in the image correspond to the following land covers as Red, Green, Blue, Yellow, Purple, Orange, and White for Agriculture, Forest, Water, Urban areas, Bare soil, Roads, and Unclassified, respectively. The image shows that the city is mostly covered by urban areas, with some agriculture and forestland. There is also a significant amount of water in the city, which is likely due to the presence of rivers and lakes. The image also shows some roads and bare soil.

Figure 18.6 shows a PCA plot of a hyperspectral image of a city. The PCA plot is a two-dimensional representation of the data in the image, where the axes are the first two principal components. The principal components are the directions in which the data varies the most. In the image, each point represents a pixel of the image. The PCA plot shows that the data is well-separated into different clusters. The first cluster is mostly composed of agriculture pixels, the second cluster is mostly composed of forest pixels, the third cluster is mostly composed of water pixels, and the fourth cluster is mostly composed of urban areas. The other clusters are smaller and less well-defined. The PCA plot shows that PCA was successful in reducing the dimensionality of the data while preserving the important information. The different clusters correspond to different land covers, which means that the PCA transformation was able to capture the underlying structure of the data.

Figure 18.7 shows a 3D PCA plot which shows that the data is well-separated into different clusters. The first cluster is mostly composed of agriculture pixels, the second cluster is mostly composed of forest pixels, the third cluster is mostly composed of water pixels, and the fourth cluster is mostly composed of urban areas. The other clusters are smaller and less well-defined. The 3D

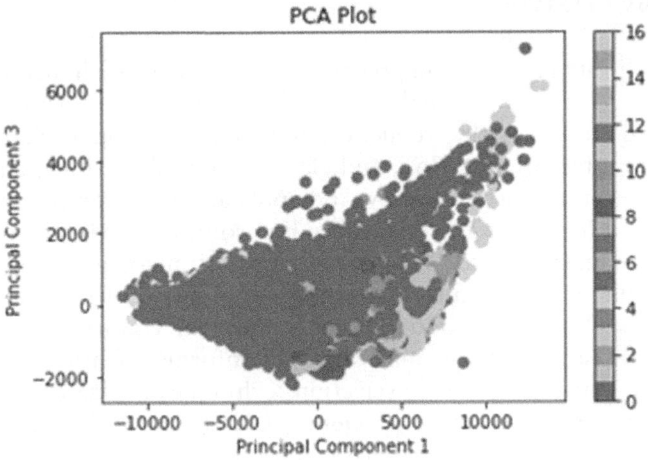

Figure 18.6 PCA plot.

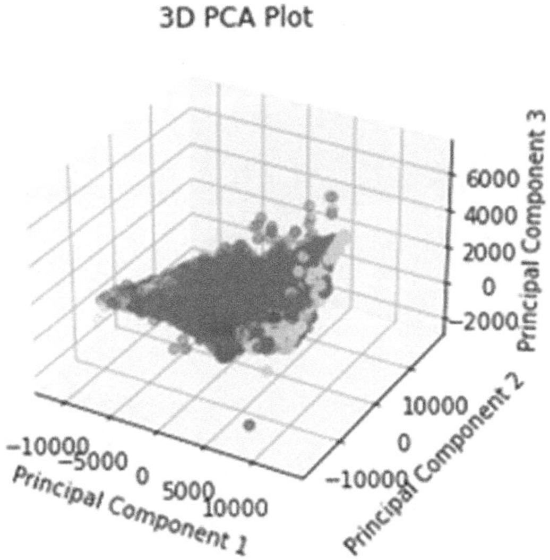

Figure 18.7 3D PCA plot.

PCA plot shows that PCA was successful in reducing the dimensionality of the data while preserving the important information. The different clusters correspond to different land covers, which means that the PCA transformation was able to capture the underlying structure of the data. This information could be used to classify the different objects in the image. For example, the image could be classified into agriculture, forest, water, and urban areas.

18.4 CONCLUSION

Our research introduces a comprehensive framework for hyperspectral data analysis that addresses the challenges of high-dimensional data. By combining advanced dimensionality reduction, innovative scoring mechanisms, spectral unmixing, feature selection, and classification algorithms, our approach enhances interpretability, accuracy, and applicability in complex scenarios. It can be applied in Precision Agriculture, Environmental Monitoring, Natural Disaster management, etc. The dual-scoring strategy provides a nuanced understanding of reduction technique impact, aiding in the selection of optimal methods. The integration of Linear Spectral Unmixing enriches data interpretation, laying the groundwork for informed feature selection and classification. Informed feature selection techniques mitigate dimensionality issues and boost classification efficiency. Our framework's efficacy is substantiated through extensive experimentation, featuring SVM and Random Forest algorithms, and comprehensive visualizations. ROC curve analysis and spatial visualizations provide a holistic evaluation. In conclusion, this chapter presents advanced hyperspectral data analysis, offering a solution that enhances accuracy, interpretability, and practical usability across various domains.

Chapter 19

Counterfeit medicine detection system

Waqar Hussain, Mubashar Ali, Abdullah Akbar, and Muhammad Saleem

19.1 INTRODUCTION

The fight against counterfeit products began in the 1980s, primarily targeting renowned brands. However, over time, it extended into the realm of medicine. Counterfeiters found the medicine industry lucrative due to lower chances of detection and the difficulty in discerning between counterfeit and authentic medicines [1]. Counterfeit medicines, while appearing genuine, pose significant health risks. According to a report by the World Health Organization (WHO), counterfeit drugs amounting to nearly $83 billion are sold worldwide annually, with a high likelihood of one in ten medicines being counterfeit. These medicines not only fail to provide therapeutic benefits but also exacerbate illnesses and can lead to fatalities. Africa, for instance, witnesses approximately 120,000 deaths annually due to counterfeit drugs. Despite stringent measures in countries like the United States to curb counterfeit medicine trade, dishonest individuals continue to smuggle these products from illicit sources, perpetuating the problem [2].

In today's era, counterfeit medicines present a growing challenge, particularly in underdeveloped countries with limited technological advancements compared to developed nations. While developed countries like the United States possess sophisticated technology, underdeveloped Asian countries struggle to distinguish between authentic and counterfeit medicines. Although organizations such as drug authorities oversee drug-related matters, the situation remains largely unchecked, especially concerning smuggled counterfeit medicines that evade official records. Consequently, the onus falls on consumers to verify the authenticity of medicines they purchase, as counterfeit medicines often contain dangerous substances that can lead to fatalities [2].

Given this scenario, there is a pressing need to develop a system to assist individuals in making informed choices. Such a system would enable people to scan the barcode on medicine packets, which contains essential information about the product. Designing this system involves implementing the 3T's of the Drug Supply Chain Security Act (DSCSA): transaction history (TH), transaction information (TI), and transaction statement (TS). TH ensures medicine authentication, TI provides detailed information about the medicine's

DOI: 10.1201/9781032664293-22

production and movement, and TS offers an overview of the medicine's journey [3]. Additionally, GS1 standards, incorporating the Global Trade Item Number (GTIN) in barcodes, will be integral to the Authenticare system. This process will play a crucial role in authenticating and detecting counterfeit medicines [4].

19.2 LITERATURE STUDY

This section is about the literature study performed to extract limitations related to counterfeit medicine detection using different techniques, which are as follows:

Kalliroi S. Ziavrou describes the concept of the WHO, which defines counterfeit medicine as deliberately and fraudulently mislabeled. Counterfeit medicine could be branded or generic drugs [5]. The WHO uses the following three terms to categorize them: (a) "falsified" is the medicine that has misprinted its identity, e.g., misspelling, mislabeling of the content or origin of the country authorizing. (b) "substandard" is also called "out of the specification"; these medicines fail to fulfill their standard requirement. They are manufactured or stored that have expired. (c) "Unregistered/Unlicensed" are medicines that have not undergone the government official's or drug authority's official license [6]. WHO set these standards for counterfeit medicines. Unfortunately, the problem of counterfeit medicine detection has not been resolved. However, "Authenticare" has developed a solution that uses GS1, which will be implemented within the "Authenticare" system. GS1 utilizes the GTIN to set standards for storing information in barcodes. This unique identifier helps identify counterfeit medicines and will play a crucial role in detecting counterfeit medicines during authentication [7]. The counterfeit medicine and pharmaceutical products market is booming in the international and local markets. These products will cause serious health issues like cancer. Pharmaceutical companies represent the most crucial role in the supply chain, pharmaceuticals, purchase, and distribution of counterfeit medicine, and they also play critical roles in detecting and reducing the circulation of counterfeit medicines. [8]. Counterfeit medicines are on the boom, according to data (Figure 19.1). From pharmaceutical companies to manufacturing. The nonprofit recorded almost 6,000 counterfeit crime incidents in 2021, which shows a 38% increase from last year; in 2020, the highest number of recorded incidents of making counterfeit products. In terms of the international market and global distribution of counterfeit medicine outbreaks, the greatest number was recorded in North America (1,579), Asia Pacific (1,151), Latin America (620), Eurasia (591), Near East (501), Europe (364), and Africa (138) [9] (Figure 19.2). This order is mainly because the countries in these regions effectively identify counterfeit medicines crimes through law enforcement activities and drug regulatory agency inspections. The distribution of different medicines and identification of counterfeits is possible using "Authenticare." Authenticare will use the 3Ts of the DSCSA for identifying counterfeit medicines. 3T will use pharmaceutical products

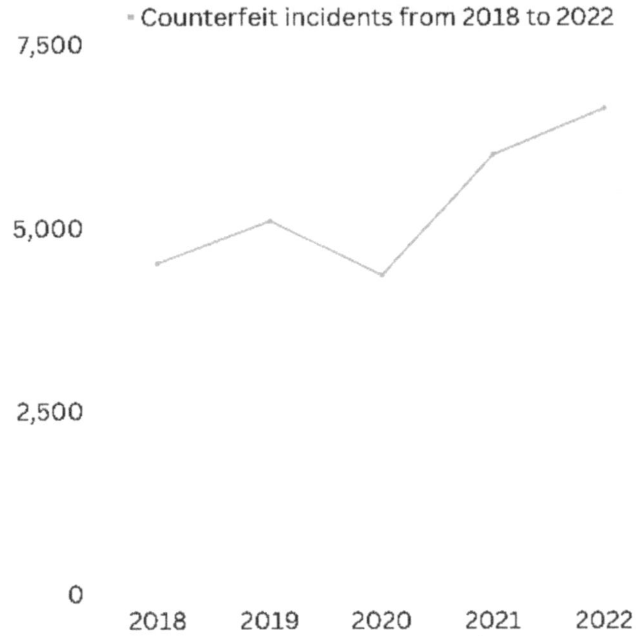

Figure 19.1 Counterfeit incidents from 2018 to 2022.

data like 1. TH, 2. TI and 3. TS [10]. The TH ensures medicine authentication; TI will contain detailed details on production and transportation, and the TS summarizes the entire product journey.

19.3 COUNTERFEIT MEDICINE DETECTION SYSTEM

Counterfeit medicine, also known as falsified or substandard drugs, poses a significant challenge. Detecting counterfeit medicine is a critical task, and leveraging technology for detection can be immensely beneficial in this regard. Currently, counterfeit medicines have a global impact, particularly affecting developing countries in Asia, Africa, and Latin America [11]. These regions are often associated with the sale of counterfeit drugs, and distributors and manufacturers can report such instances if identified. According to the WHO, approximately 40% of drugs sold worldwide are counterfeit, leading to substantial economic losses for developing nations, amounting to billions of dollars [12]. Consumers and pharmaceutical companies play pivotal roles in the drug supply chain [13].

Our application, “Authenticare,” depicted in Figure 19.3, integrates with all organizations involved in the pharmaceutical supply chain, including manufacturers, exporters, warehouses, hospitals, and pharmacies. “Authenticare”

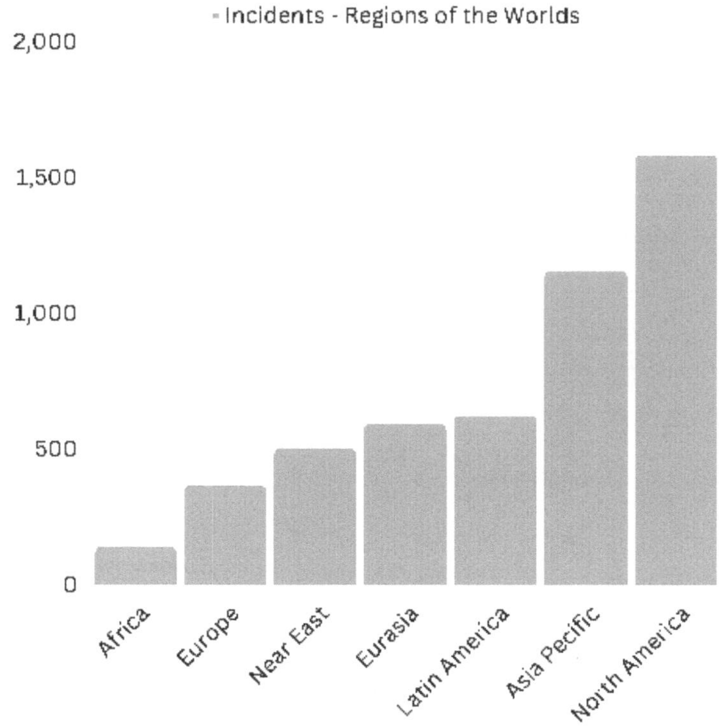

Figure 19.2 Incidents—Regions of the world.

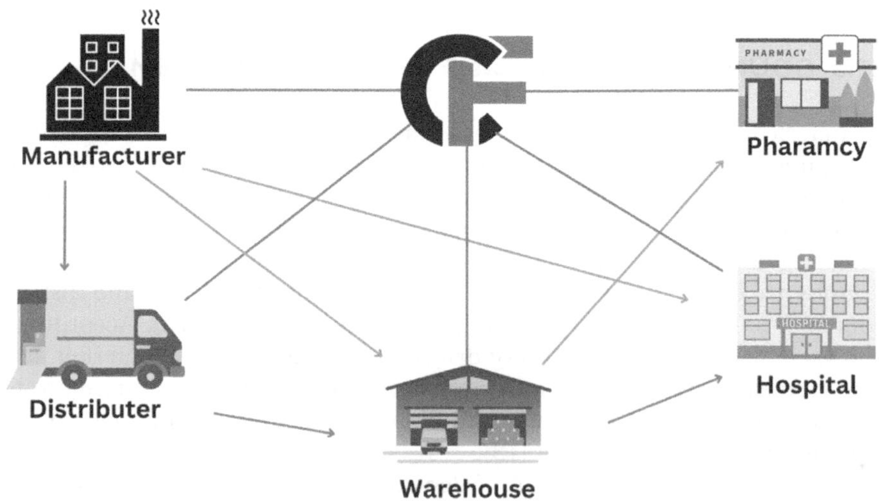

Figure 19.3 Authenticare.

analyzes data from various pharmaceutical companies to distinguish between counterfeit and genuine medicines. The application utilizes GS1 technology to identify the unique item number or barcode on medicine packets, which aids in combating illegal drug importation and distribution. Serving as a central hub connected to all pharmaceutical companies, "Authenticare" ensures transparency in the drug supply chain.

19.3.1 Authentication

In 2019, a law was implemented in the American European that requires each medicine to have a special code on top of it. This code enables identification of whether the medicine is genuine or counterfeit [14]. Following the implementation of this law, a study was conducted in a hospital in the United Kingdom to assess the system's functionality. Over 4,000 medicines were included in this study, all of which were pre-validated. However, upon re-validation, it was found that 4% of the 4,000 medicines had expired.

Incorporating this special code along with the supply chain into the system will ensure the authenticity of the medicines being administered. Some of the key elements of the supply chain include TH, TI, and TS. TH tracks details, such as the manufacturer of the drug, the ingredients used, and the involved manufacturers. It also records the machinery used to mix the ingredients and form the medicine.

TI ensures that pertinent details, such as the medicine's name, quantity produced, transfer date from the manufacturer for sale, and recipient information, are recorded and stored.

A TS verifies whether the person or business manufacturing or selling a particular drug possesses a DSCSA license. Consequently, the buyer is provided with a letter of authority approved by DSCSA, ensuring the authenticity of all purchased medicines for further sale at pharmacies or medical stores. Additionally, an application will be available for pharmacy customers to scan the special barcode on medicines using their mobile camera, thereby confirming the authenticity of the medicine [15] (Figure 19.4).

19.3.2 User management

The user management dashboard serves as a gateway for users to personalize their profiles. It allows users to access and modify information, such as their username, email address, and password. Additionally, users can change their password directly from the user dashboard. Moreover, users have the option to update their profile picture and create multiple medication lists, as depicted in Figure 19.5.

The user has the ability to modify their username directly from the dashboard. Each user possesses a unique username, and if the system generates a random one, the user has the option to change it to an available one. Additionally, users can update their email address from the dashboard,

Figure 19.4 Verification process for authenticity of medicines via mobile barcode scanning.

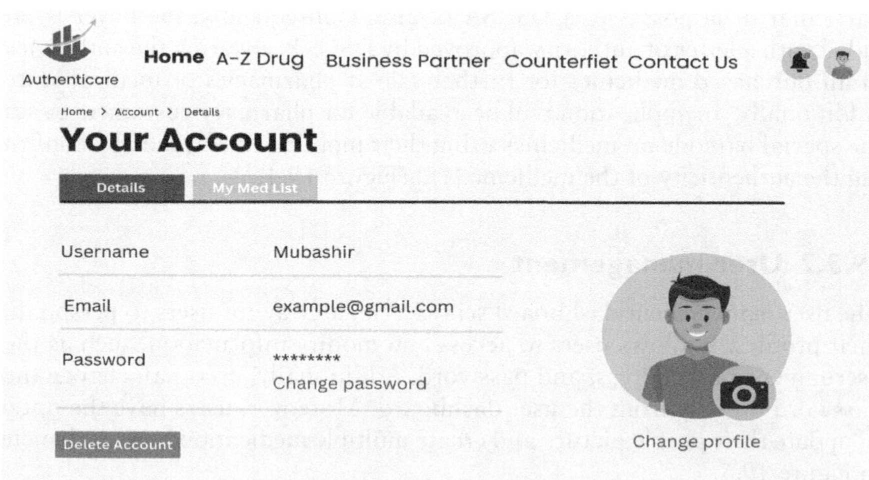

Figure 19.5 User management dashboard overview.

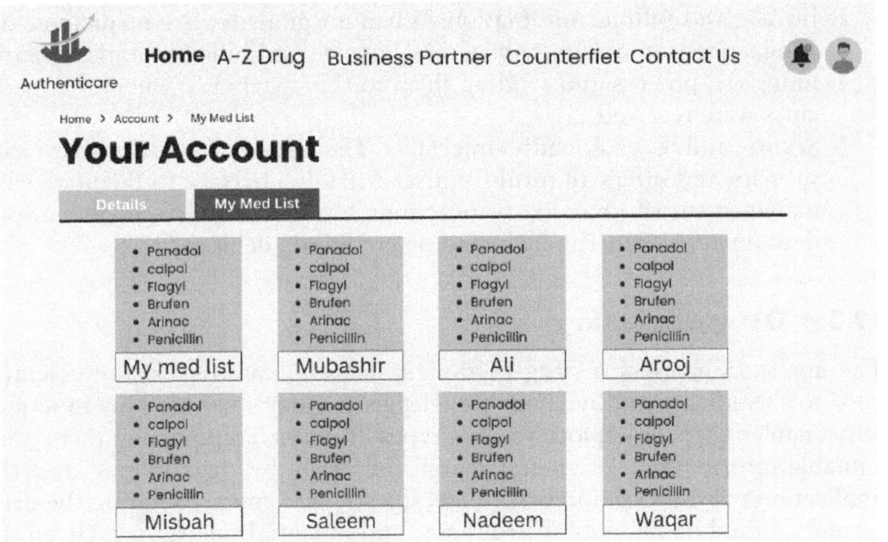

Figure 19.6 Creating multiple medication lists in user dashboard.

allowing them to register their account to a new email if desired. Furthermore, users can change or recover their password from the dashboard, facilitating password management. In cases where a user forgets their password, they can initiate password recovery or change it directly from the dashboard. Users also have the option to delete their account from their user dashboard. Moreover, the dashboard enables users to create multiple medication lists, as illustrated in Figure 19.6. Should a user wish to create separate lists for their family's medication, they can easily do so from the user dashboard.

19.3.3 Payment management

In this digital era, payment is paramount, and digital payment methods have become accessible worldwide. In our application, digital payment methods will be integrated specifically for business partners, encompassing manufacturers, importers, hospitals, and pharmacies. While the application itself is free for end-users, business partners will be required to pay for services such as data entry into the application. They can utilize APIs or administration access to input data into the application.

The need for a payment method arises from various factors:

1. Collaboration with Manufacturers and Vendors: The application collaborates with suppliers, manufacturers, importers, hospitals, and pharmacies, simplifying payment processes and fostering transparency between business partners and "Authenticare."

2. Invoice and Billing Automation: When hospitals receive medicines, the application automatically generates invoice and billing details, streamlining the process and sending them to the warehouse where the medicines were received.
3. Secure and User-Friendly Interface: The application prioritizes user security and offers an intuitive, user-friendly interface to facilitate easy management of financial transactions for partners. It accepts various digital payment methods, including credit and debit cards.

19.3.4 Drug dictionary

The application offers a drug dictionary feature, catering to the essential need for healthcare and medical knowledge in today's world. This functionality enables users to explore various types of drugs, empowering them with valuable information. Users can search for drugs by their names, and the application provides comprehensive details for each entry, including the drug formula, brand name, dose instructions, and potential side effects. Given the vast array of manufactured drugs, the application facilitates an A-to-Z drug search (Figure 19.7), ensuring users can easily find information on any drug they seek.

This feature serves a wide range of users, including health professionals, who can access up-to-date information about medicines, enhancing their knowledge base. Additionally, users/patients can utilize this functionality to educate themselves about medications and the associated side effects. This

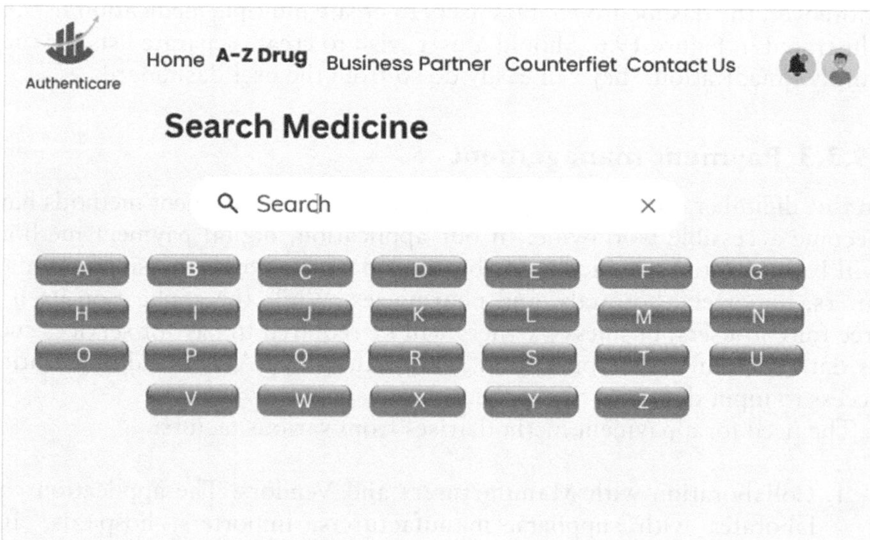

Figure 19.7 Drug search functionality.

knowledge equips individuals to make informed decisions and communicate effectively with their health professionals. By providing insights into medications, this functionality enhances the quality of healthcare professionals and empowers individuals to take charge of their health, enabling them to make informed decisions about their medications.

19.3.5 Alerts and notification

The application offers alerts and notifications (Figure 19.8) to users and stakeholders, ensuring they remain updated in real-time. This feature plays a crucial role in keeping business partners informed about various aspects of their partnerships, including deliveries, payments, and other services. The necessity for alerts and notifications in this application is evident in several aspects:

1. Delivery Alerts: This feature enables business partners to receive immediate notifications regarding the status of their shipments and deliveries. Alerts provide information on dispatch times and estimated delivery times, facilitating real-time visibility for enhanced supply chain management. This helps mitigate the risks of miscommunications and delays.
2. Payment Reminders: Business partners receive notifications related to financial transactions, such as invoices or subscription fees. These reminders ensure that partners never miss deadlines and can conveniently track records of their financial transactions.
3. Updates: Business partners are notified of any changes or updates within the application. These features enable partners to stay informed and maintain seamless collaboration.

Figure 19.8 Alerts and notifications feature.

19.3.6 Business partner management

In this system, we focus on the various business partners involved in the pharmaceutical industry, encompassing individuals contributing to the journey from the creation of medicine to its distribution (Figure 19.9). Among these partners, the manufacturer plays a pivotal role in the creation of medicine. Their primary responsibility is to produce medicines as per the contracts they have been awarded. The pharmaceutical company specifies the ingredients to be utilized in the medicine, and the manufacturer blends these ingredients according to predetermined stages to create the final product. Additionally, the manufacturer maintains records of the manufacturing process, including the machinery employed.

Furthermore, the manufacturer holds a digital permit approved by the DSCSA, which authorizes them to manufacture and sell the medicine. Once manufactured, the medicine is transferred to a distributor responsible for delivering it to the public. Distributors ensure that medicines reach pharmacies where they are accessible to the general population. Additionally, warehouses and hospitals may directly purchase medicines from the manufacturer, expanding the distribution network. Through collaboration among these stakeholders, the pharmaceutical supply chain operates efficiently, ensuring the availability of medicines to the public.

19.3.7 Distributer management

In the pharmaceutical industry, a distributor can either be an individual or an agency tasked with purchasing medicine from manufacturers and supplying it to pharmacies or medical stores. To engage in these activities, distributors

Figure 19.9 Pharmaceutical business partners.

must obtain digital authorization certification approved by the DSCSA. This certification confirms the distributor's permission to procure medicine directly from manufacturers and distribute it within a specific area. The system stores information about both manufacturers and distributors for future reference.

As elucidated in the authentication section, the supply chain comprises three key components: TH, TI, and TS. These components maintain records that trace the origin of the medicine back to the original manufacturer and document the distributor's procurement from the manufacturer, as well as the subsequent sale of medicine to various pharmacies and medical stores within a designated area. This approach fosters transparency and accountability throughout the supply chain, ensuring the integrity and reliability of pharmaceutical distribution processes.

19.3.8 Pharmacy management

Pharmacy management is a critical component of the pharmaceutical industry [16]. Pharmacies serve as the point of contact for end-users seeking medications. Pharmacists, who operate within pharmacies, are entrusted with the responsibility of dispensing the correct medications to patients. It is imperative for pharmacies to procure authentic medicines from various warehouses while ensuring the safety and quality of all drugs.

The application will aid pharmacists in maintaining comprehensive records of the origin and manufacturing details of medicines. Effective pharmacy management entails ensuring that the drug supply meets all authentic standards, thereby guaranteeing the safety and efficacy of medications provided to patients.

19.4 FUTURE WORK

Authentication of medicine represents a significant and challenging process, and implementing an effective solution is a crucial step for future endeavors. Following a thorough examination of this specific problem, it became apparent that identifying counterfeit medicine is a complex task, beyond the capabilities of an individual or a team. The need arises to develop robust software that can ascertain the authenticity of medicine, differentiating between genuine and counterfeit products. Taking into account all the necessary steps to identify authentic medicines, the next phase involves developing software that utilizes this method.

This software will utilize mobile phone capabilities, specifically designed for mobile devices, and can be installed on any mobile device. It aims to scan the unique code on the medicine and compare it against the corresponding information stored in a database. If all information aligns with the database, the software will display a message confirming the authenticity of the medicine; otherwise, it will indicate an error, signaling that the medicine is counterfeit.

The application will be accessible to every business partner involved in the entire process, from manufacturing to delivering the medication to end-users. Additionally, ordinary people can use this application to ensure the authenticity of the medication they are consuming.

To enhance these systems, the utilization of the eAntHocNet routing protocol in a mobile device network setup is proposed to detect and prevent the distribution of fake medications. The dynamic structure of the floating ad hoc network facilitates reliable communication between devices, enabling real-time data sharing and collaboration in the detection and tracking of counterfeit medications [17]. The Counterfeit Medicine Detection System can further improve its capabilities by integrating an intrusion detection system (IDS) to identify and filter out malicious activity associated with counterfeit medicines. The IDS can optimize its detection capabilities by reducing false alarms while maintaining an appropriate level of missed detections, thus enhancing the system's ability to recognize and prevent the spread of fake medications, ensuring public health and safety [18].

Implementing a dynamic clustering approach can enhance the accuracy and speed of identifying counterfeit medications by analyzing and optimizing information flow within the system. Optimization algorithms can be employed for resource allocation and decision-making, ensuring the system operates at peak performance. In summary, incorporating ideas from the articles can contribute to the evolution and improved functionality of the Counterfeit Medicine Detection System in the future [19].

19.5 CONCLUSION

Counterfeit medicine poses a significant global challenge, escalating with each passing day. Finding a solution to this problem is paramount. Our application's primary objective is to halt the flow of counterfeit medicine. Each of the methods integrated into this application—TH, TS, and TI—plays a pivotal role in thwarting the distribution of counterfeit medicine. These terminologies converge within a robust digital system crucial for combating counterfeit medicine. Through this application, users can scan the barcodes and unique numbers on medicines. Subsequently, the application will analyze and compare the data with that of business partners, providing results indicating whether the medicines are counterfeit or genuine.

REFERENCES

1. Asmami, M., & Wald, L. (1992). Interband calibration of the POLDER sensor. In Remote sensing for monitoring the changing environment of Europe (pp. 253–259). Balkema.
2. Sirrs, C.J., 2023. Fluid Fakes, Contested Counterfeits: The World Health Organization's Engagement with Fake Drugs, 1948–2017. Medicine Anthropology Theory, 10(3), pp. 1–29.

3. Yang, J. and Mishra, A., 2023. Diversion in Prescription Opioid Supply Chains: Evidence from the Drug Supply Chain Security Act. Available at SSRN 4377801.
4. Adeyeye, M., Kayode, J., Adeniran, A., Osho, F. and Udokwelu, W., 2023. Enabling Pharmaceutical Traceability in the Nigerian Supply Chain Using GS1 Global Standards: Lean Traceability Including In-Country Serialization of COVID-19 Vaccines. Journal of Regulatory Science, 11(1), pp. 1–14.
5. Ziavrou, K.S., Noguera, S. and Boumba, V.A., 2022. Trends in Counterfeit Drugs and Pharmaceuticals before and during COVID-19 Pandemic. Forensic Science International, 338, p. 111382.
6. World Health Organization, 2017. WHO Global Surveillance and Monitoring System for substandard and falsified medical products. In WHO global surveillance and monitoring system for substandard and falsified medical products.
7. Sarkar, S., 2022. Pharmaceutical Serialization: Impact On Drug Packaging. International Journal, 10(3).
8. Ofori-Parku, S.S., 2022. Fighting the Global Counterfeit Medicines Challenge: A Consumer-Facing Communication Strategy in the US Is An Imperative. Journal of Global Health, 12.
9. Rajora, N., 2022. Counterfeit and Illicit Drugs Trade: A Quantitative Data On How Counterfeit Drugs Impact Globally. International Journal, 10(2).
10. Sarkar, S., 2022. Supply Chain Security Act 2023: Interoperable Data Exchange for Drug Traceability. International Journal of Scientific Research in Computer Science Engineering and Information Technology, 8, pp. 471–476.
11. Lima, M.B.A. and Yonamine, M., 2023. Counterfeit Medicines: Relevance, Consequences and Strategies to Combat the Global Crisis. Brazilian Journal of Pharmaceutical Sciences, 59, p. e20402.
12. Botta, G.B., 2023. Pharmaceutical Medicine Traceability: An Overview of Global Compliance. World Journal of Biology Pharmacy and Health Sciences, 15(2), pp. 245–252.
13. Hillary, S.C. "Attitude and Challenges of Consumers and Pharmacists towards Reporting Counterfeit Medicines in Lagos State, Nigeria." PhD diss., 2023.
14. Kristensen, S.B., Clausen, A., Skjødt, M.K., Søndergaard, J., Abrahamsen, B., Möller, S. and Rubin, K.H., 2023. An Enhanced Version of FREM (Fracture Risk Evaluation Model) Using National Administrative Health Data: Analysis Protocol for Development and Validation of a Multivariable Prediction Model. Diagnostic and Prognostic Research, 7(1), p. 19.
15. Celeste, B. and Kessler, D., 2023. Closing a Key Gap in DSCSA Compliance with Credentialing. Blockchain in Healthcare Today, 6(1).
16. Astuti, E.K.A., Sriatmi, A. and Agushybana, F., 2023. Implementation of Patient Drug Prescription Services in Hospital Pharmacy Installations: Has It Been Managed Properly to Reduce Waiting Time?" Contagion: Scientific Periodical Journal of Public Health and Coastal Health, 5(1), pp. 189–203.
17. Khan, I.U., I.M. Qureshi, M.A. Aziz, T.A. Cheema and S.B.H. Shah., 2020. Smart IoT Control-Based Nature Inspired Energy Efficient Routing Protocol for Flying Ad Hoc Network (FANET). IEEE Access, 8, pp. 56371–56378, doi: 10.1109/ACCESS.2020.2981531.
18. Abdollahi, A. and Fathi, M., 2020. An Intrusion Detection System On Ping of Death Attacks in IoT Networks. Wireless Personal Communications, 112, pp. 2057–2070.
19. Hosseini, A.M. and Mohammadi, A., 2023. Dynamic Clustering and RRH Selection in Non-Coherent Ultra-Dense CRAN with Limited Fronthaul Capacity. Wireless Personal Communications, pp. 1–18.

Chapter 20

SDN-enabled intrusion detection system using machine learning and neural network schemes

Abida Tahsin Tawfik, Sadoon Hussein Abdullah, Ahmed Sami Nori, and Muhammad Allah Rakha

20.1 INTRODUCTION

Software-defined network (SDN) nodes are interconnected with the help of wireless communication technologies. IEEE 802.11 technology is basically used for the interconnectivity of SDN nodes. SDN is having applications in civil and military domains [1, 2]. SDN nodes used to have limited energy and other resources. Also, SDN is quite vulnerable to cyberattacks [3]. However, SDN can be deployed in smart hospitals, agriculture, transportation, smart homes, and smart cities [4–6]. An intruder tries to either unbalance or steal information through continuous spoofing. Due to various attacks from intruders, SDN does not work properly. Overall, the decision-making process will be disrupted. This is a critical concern because SDN is susceptible to cyberattacks. Sybil, Domain Name System (DNS), Denial of Service (DoS), Proof of Delivery (PoD), distributed denial of service (DDoS), and other attacks might compromise SDN resources [7–9]. Consequently, intrusion detection systems, or IDSs, are thought to be a potential means of identifying cyberattacks. What's more intriguing is that cyberattacks cannot be detected by conventional IDS methods. Fake data packets are easily detected by IDS techniques based on machine learning and neural networks. Advanced machine learning and neural network methods make it easy to detect unauthorized data packets. The three primary techniques for IDSs are hybrid, signature, and anomaly [10, 11]. However, IDS based on machine learning and neural networks falls under the anomalous category.

The chapter follows this structure: in Section 20.2, a comprehensive review of relevant literature is presented, Section 20.3 explains the approach, Section 20.4 covers datasets, and Section 20.5 provides further details regarding the tests and results. Section 20.6 deals with closing statements.

20.2 RELATED WORKS

Traditional SDN and datasets have many limitations which are as discussed as follows:

 DOI: 10.1201/9781032664293-23

The topological structure of SDN makes it vulnerable. Detection of DoS/DDoS attacks in the context of SDN is a challenging task. Smart automation in every field of study has raised some security issues and increased vulnerabilities in SDN. Previous researchers simulated many traditional algorithms to predict cyberattacks on SDN. However, the prediction/identification of intruders within the network is also considered a major threat. SDN has different physical structures in design. Moreover, due to many design issues, SDN is on security risk. SDN to some extent tries to safeguard networks from cyberattacks. Man-in-the-middle attack is known as DoS, which is used to disrupt overall communication. DoS/DDoS is usually applied either on bandwidth or SDN network resources. SDN topological design is disturbed due to DoS/DDoS attacks. These attacks create buffer zones in the SDN network. Also, illegal data packet flooding causes congestion [12–14].

Previously, researchers have used datasets like UNSW-NB15, NSL-KDD, KDD'99, KDDTest+, and KDDTest–21. Mainly artificial intelligence experts were focused on detecting possible cyberattacks from the mentioned datasets. Accordingly, machine learning, deep learning, hybrid models, neural networks, and optimization algorithms were utilized to balance a high accuracy rate. Also, metrics like false alarm, precision, recall, and F1-score are used during experimentation [15–24].

20.3 METHODOLOGY

This section provides comprehensive details regarding the technique. Overall simulation is performed on Python. Binary classification is used on both datasets. During simulation, null and missing values from datasets are removed. Division of training and testing datasets is made possible. Cyberattacks are identified from two sophisticated datasets using machine learning and neural network techniques. When compared to other lagged classifiers, random forest fared better. Performance is evaluated using criteria such as accuracy, recall, F1-score, and precision. Figure 20.1 also shows the methodology section's detailed flow chart.

20.3.1 Stochastic gradient descent

Stochastic gradient descent (SGD) is a neural network optimization technique that operates based on the objective function. Fundamentally, this function aims to attain both minimum and maximum values. SGD is having issues with data streaming [25, 26].

20.3.2 Naïve Bayes multinomial

The Naïve Bayes multinomial is a classification method for supervised learning. This algorithm can be implemented easily. Also, NBM can be applied to

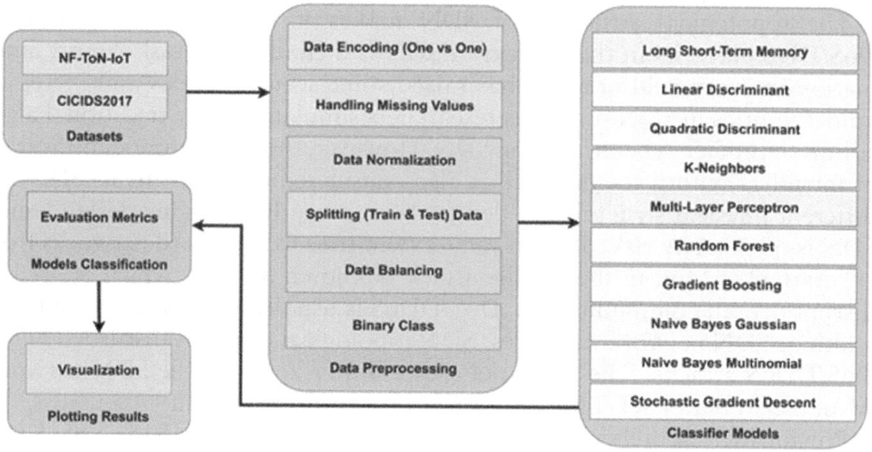

Figure 20.1 Methodology flow chart.

continuous and discrete data. NBM is scalable and can handle large datasets. Apart from that NBM has real-time applications in detection [27].

20.3.3 Naïve Bayes Gaussian

A generative probabilistic machine learning method is Naïve Bayes Gaussian. NBG has applications in filtering and detection. However, NBG uses the probability density function [28].

20.3.4 Gradient boosting

A machine learning method called gradient boosting is used to turn poor learners into strong learners.

Boosting can easily improve accuracy. Also, GB can be applied for detection in many civil and military applications [29].

20.3.5 Long short-term memory

LSTM is a neural network scheme that has input, output, and forget gates. LSTM predicts patterns and information in a time series manner. LSTM can be used for sequential data. It has issues in long sequences. Also, LSTM can be utilized for DDoS attack detection [30].

20.3.6 Random forest

Decision trees form the basis of the random forest machine learning technique. RF is the sub-extension of the bagging method. However, RF is used to split features easily. Also, RF can be used for cyber defense [31].

20.3.7 Linear discriminant analysis

An approach to supervised learning is called linear discriminant analysis. LDA can be usually applied to large datasets and used to separate different classes. LDA can be applied for detecting cyberattacks. Also, this method is used to reduce computational cost [32].

20.3.8 Quadratic discriminant analysis

Quadratic discriminant analysis is a similar technique to linear discriminant analysis. QDA is used to estimate the individual covariance matrix of every class. It can also be used to detect DDoS flooding [33].

20.3.9 K-neighbors

K-neighbor is a popular machine learning technique. Classification, detection, and regression problems can all be solved with this method. K-neighbor is used to balance data for detecting cyber trolling on social media platforms. Also, the detection system of cyberattacks in self-driving cars uses the k-neighbor mechanism [34, 35].

20.3.10 Multi-layer perceptron

Multi-layer perceptron is a neural network scheme. This technique is used to learn relationships between linear and non-linear. MLP can be used for both binary and multi-classification. MLP uses a backpropagation mechanism. Also, it can be utilized for cyberattack detection [36].

20.4 DATASETS

During experimentation, two datasets are used which include CICIDS2017 and NF-ToN-IoT. Data preprocessing, labeling, and removal of null values process are performed. Data is generally separated into testing and training categories. Cyberattacks are identified using machine learning and neural network techniques. There are 2830743 instances in the CICIDS2017 dataset. On the other hand, man-in-the-middle, DoS/DDoS, brute force, heartbleed, bot, and port scan threats are listed in CICIDS2017 [37]. The NF-ToN-IoT dataset comprises a total of 1,379,274 data points. This comprehensive dataset encompasses various types of cyberattacks, including DoS/DDoS, injection, MITM, password attacks, ransomware, scanning activities, XSS, backdoor intrusions, and injection attempts [38].

20.5 EXPERIMENTS AND RESULTS

Overall, dataset imbalance is a very major issue. However, most of the time minority and majority classes have unbalanced behavioral problems. Therefore, the CICIDS2017 and NF-ToN-IoT datasets are balanced using SOMTE-library. Data about the CICIDS2017 dataset is shown in Figures 20.2 and 20.3. Figures 20.4 and 20.5 illustrate imbalance behavior and solution using SOMTE-library using the NF-ToN-IoT dataset.

Figure 20.6 shows the CICIDS2017 dataset's various feature representations. A massive amount of data is displayed with the use of the CICIDS2017 dataset correlation matrix.

Figure 20.7, on the other hand, shows the correlation matrix for the NF-ToN-IoT dataset.

The confusion matrix of the random forest model for the CICIDS2017 and NF-ToN-IoT datasets is described in detail in Figures 20.8 and 20.9. Identification and classification of data can be clearly explained with the use of a confusion matrix.

Random forest employs the NF-ToN-IoT and CICIDS2017 datasets to achieve good accuracy and performance. Figure 20.10 illustrates that the accuracy of random forests is approximately 99.6%. Random forest has a higher accuracy level than other conventional methods. Conversely, Figure 20.11 shows the classifiers' accuracy level using the CICIDS2017 dataset.

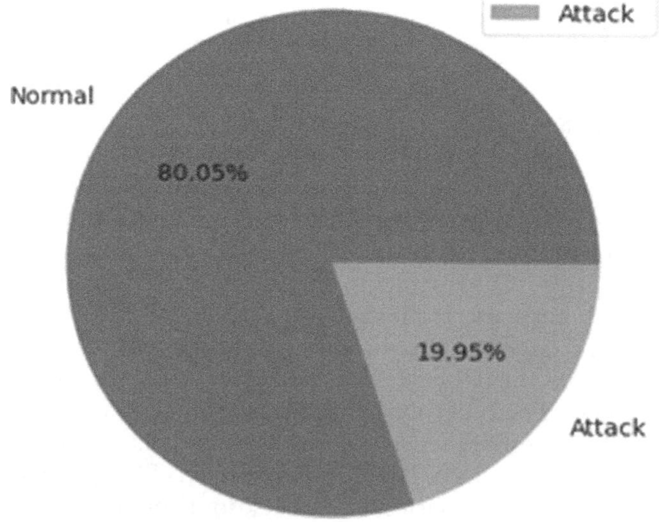

Figure 20.2 CICIDS2017 (imbalance).

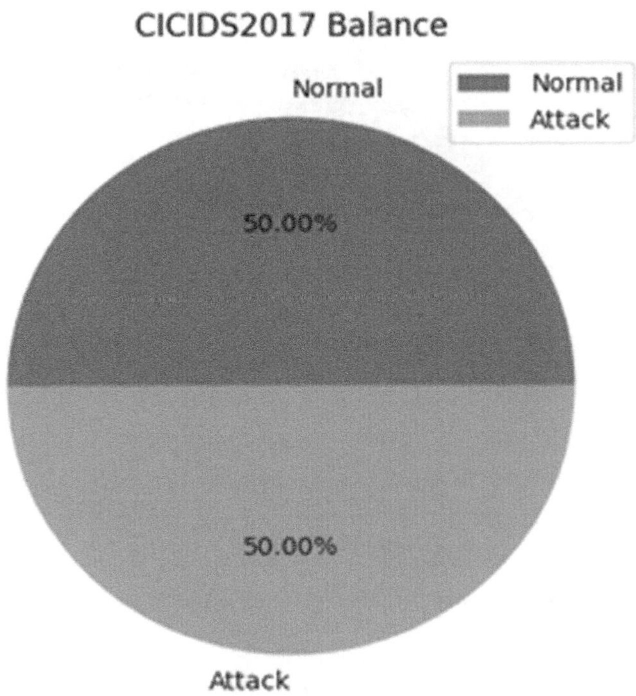

Figure 20.3 CICIDS2017 (balance).

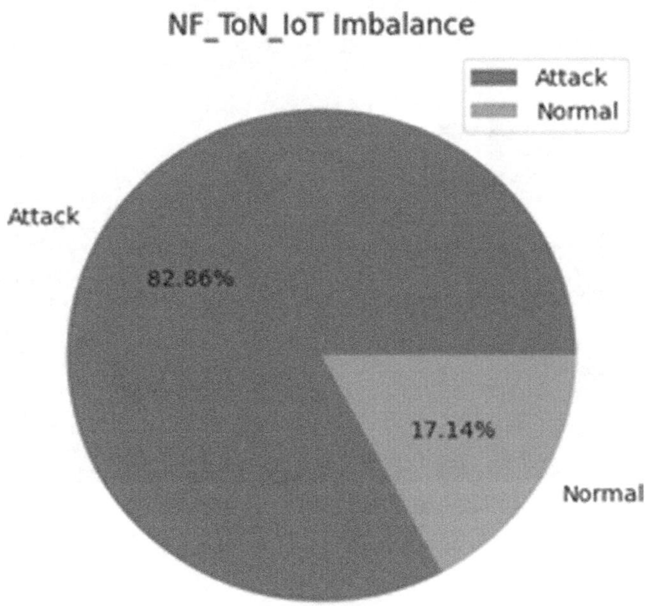

Figure 20.4 NF-ToN-IoT (imbalance).

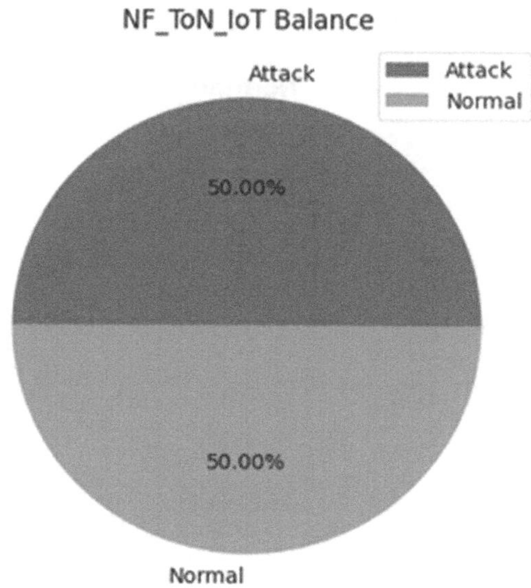

Figure 20.5 NF-ToN-IoT (balance).

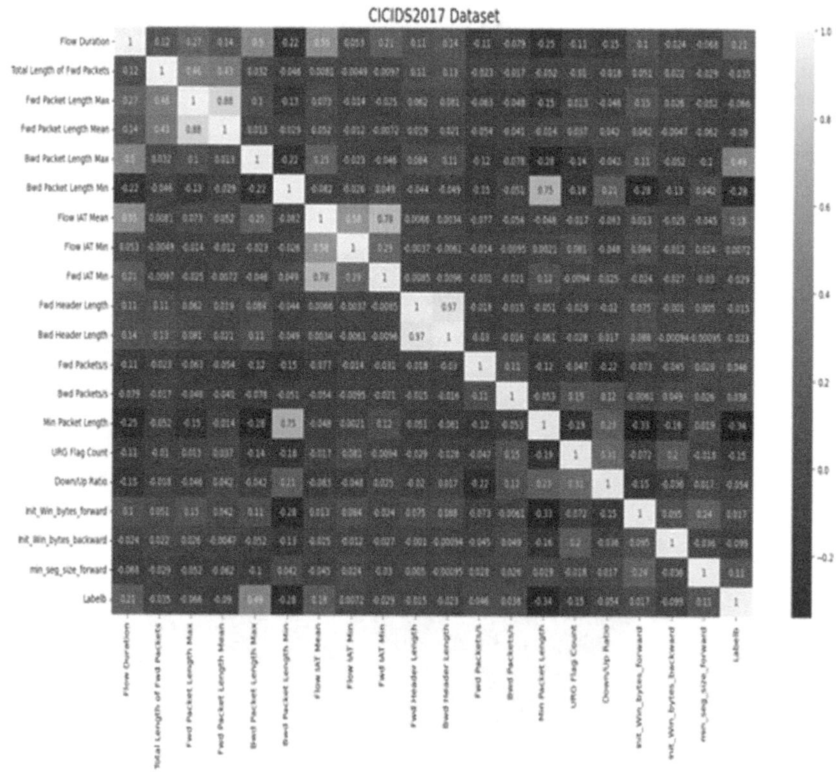

Figure 20.6 Correlation matrix of CICIDS2017 dataset.

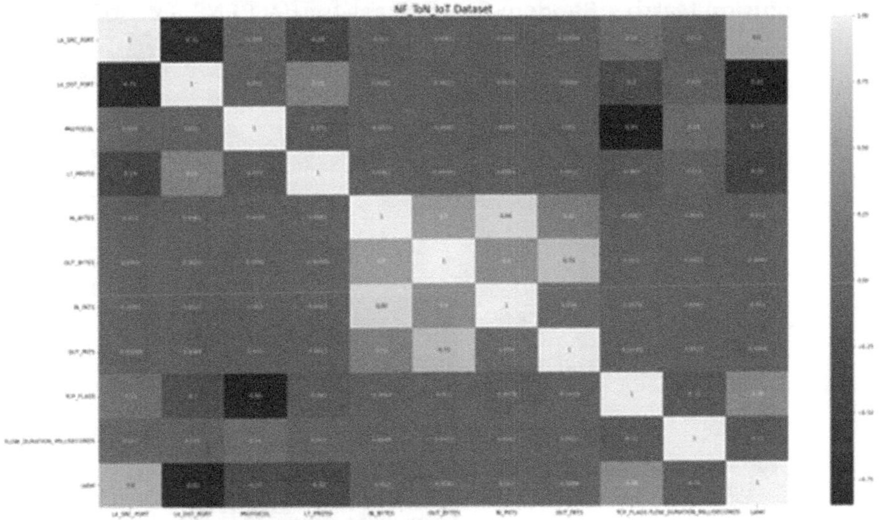

Figure 20.7 Correlation matrix of NF-ToN-IoT dataset.

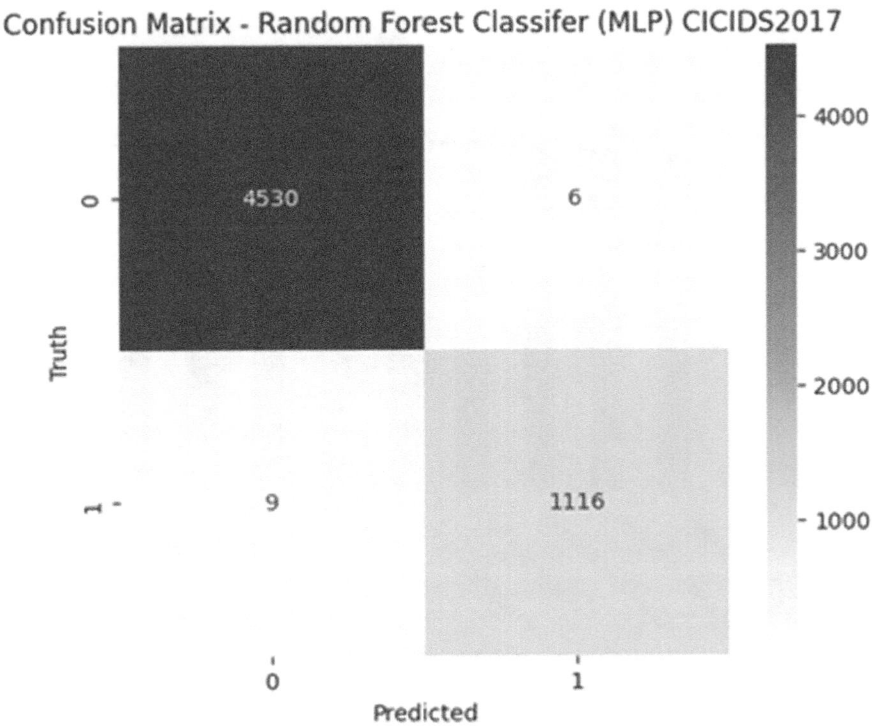

Figure 20.8 Confusion matrix of random forest using CICIDS2017.

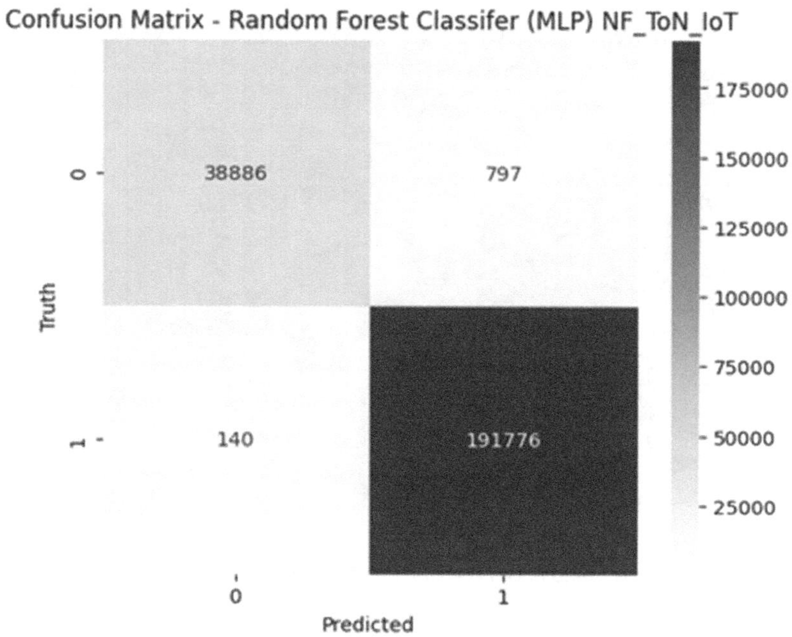

Figure 20.9 Confusion matrix of random forest using NF-ToN-IoT.

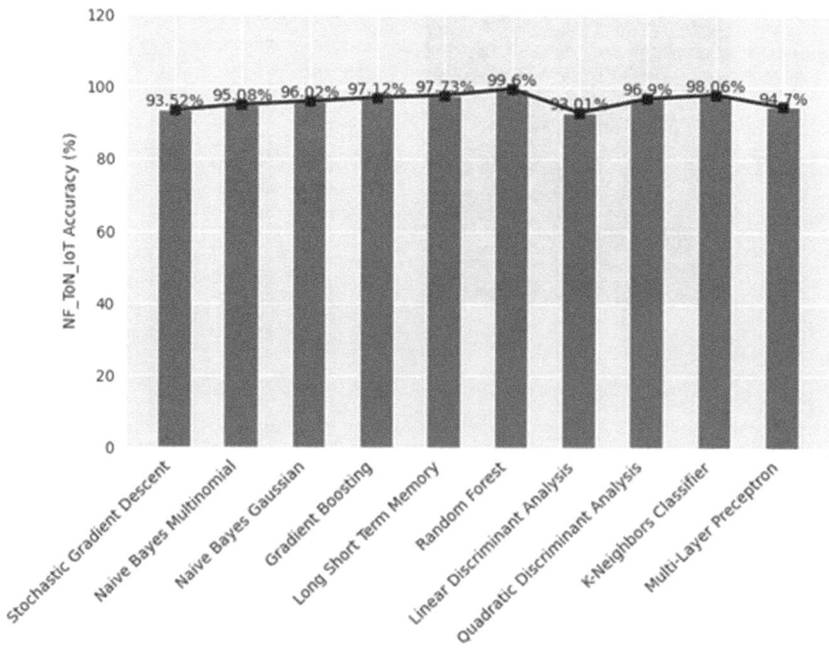

Figure 20.10 Performance comparison of machine learning models on NF-ToN-IoT dataset.

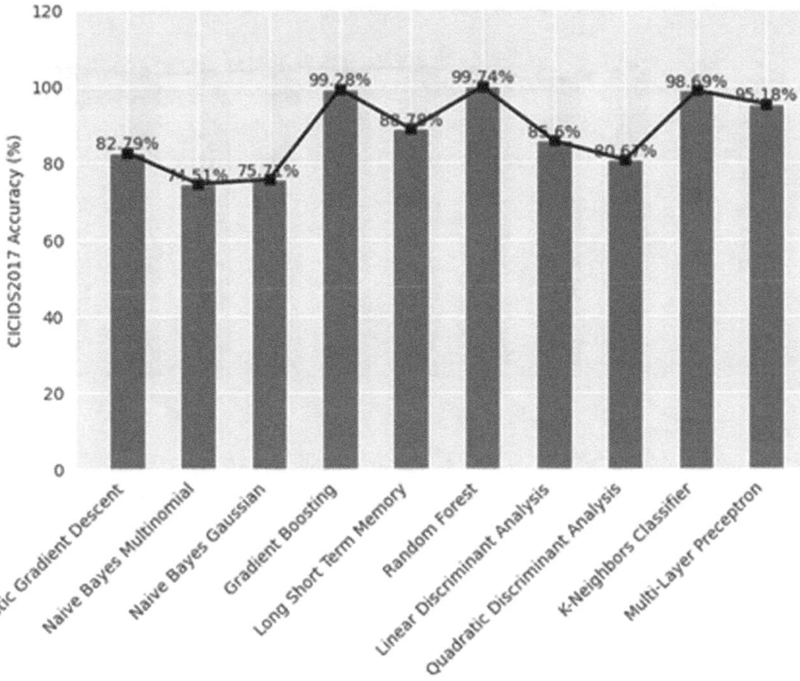

Figure 20.11 Accuracy comparison across models on CICIDS2017 dataset.

Using the CICIDS2017 dataset, the random forest classifier gets a superior accuracy of 99.74% compared to other approaches. Equation (20.1) presents the formula of accuracy which is used as performance metrics in both datasets. However, Equation (20.2) is about precision. The formulation of recall is illustrated in Equation (20.3). In Equation (20.4), F1-score is well explained.

$$Accuracy = (TP + TN)/(TP + TN + FP + FN) \quad (20.1)$$

$$Precision = TP/(TP + FP) \quad (20.2)$$

$$Recall = TP/(TP + FN) \quad (20.3)$$

$$F1 - score = 2 \times Precision \times Recall/(Precision + Recall) \quad (20.4)$$

Figures 20.12–20.14 illustrate the simulation outcomes for F1-score, recall, and precision utilizing the NF-ToN-IoT dataset. Random forest demonstrates superior performance compared to other classifiers in terms of F1-score and recall. However, MLP stands out notably in accuracy performance.

Additionally, findings for precision, recollection, and F1-score using the CICIDS2017 dataset are shown in Figures 20.15–20.17. Ninety-nine percent

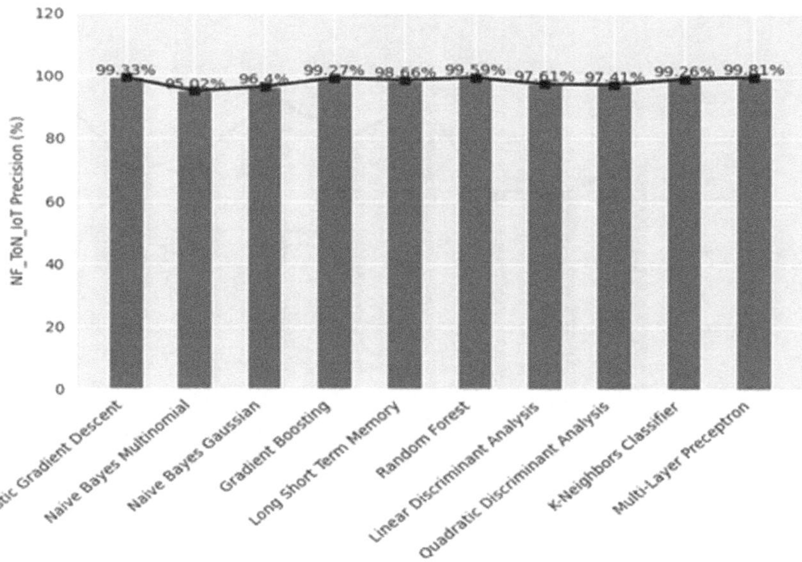

Figure 20.12 Precision analysis of classifiers on NF-ToN-IoT dataset.

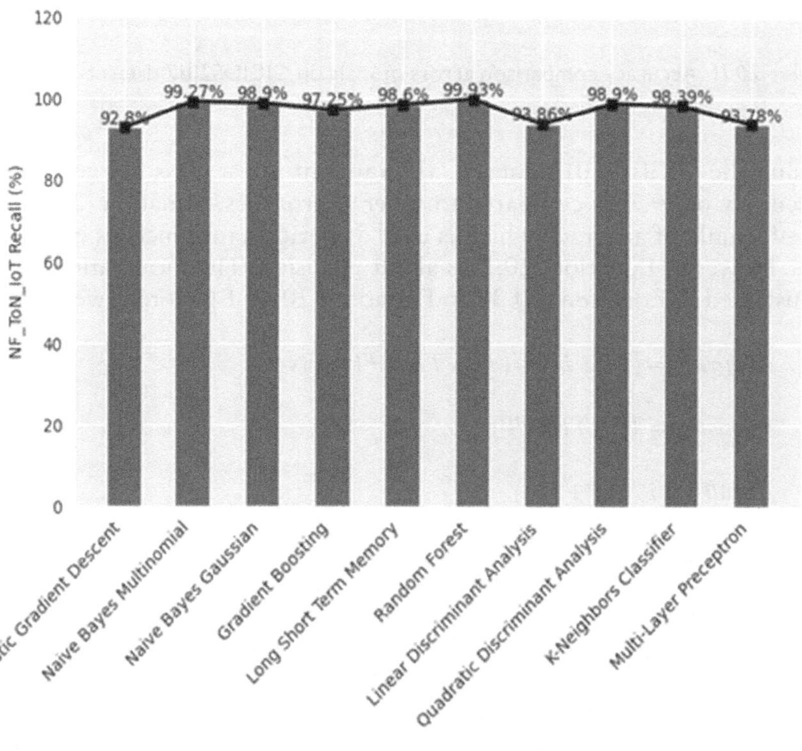

Figure 20.13 Recall assessment of classifiers on NF-ToN-IoT dataset.

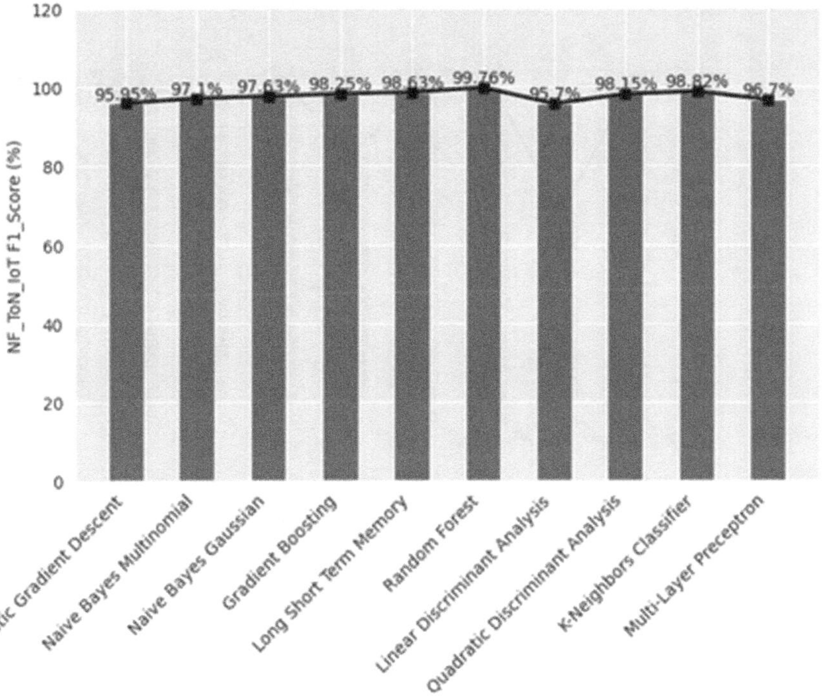

Figure 20.14 F1-score analysis of classifiers on NF-ToN-IoT dataset.

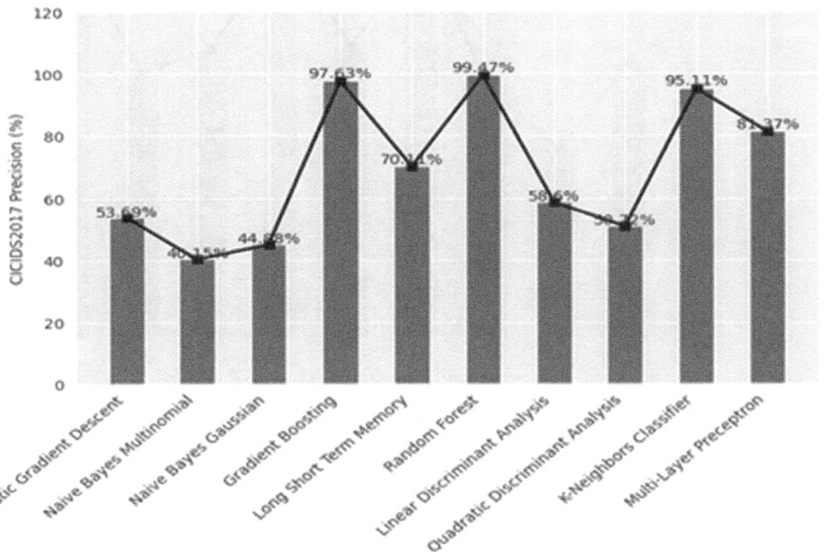

Figure 20.15 Precision assessment of classifiers on CICIDS2017 dataset.

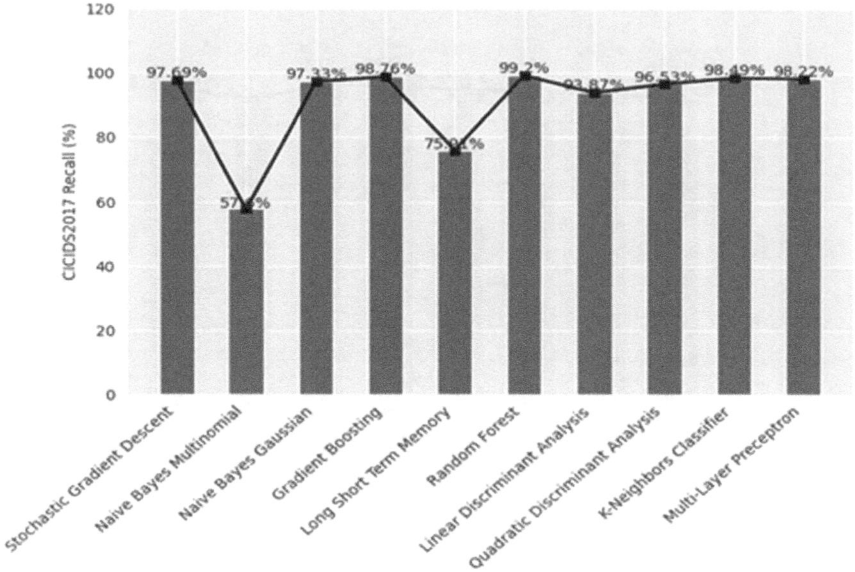

Figure 20.16 Recall analysis of classifiers on CICIDS2017 dataset.

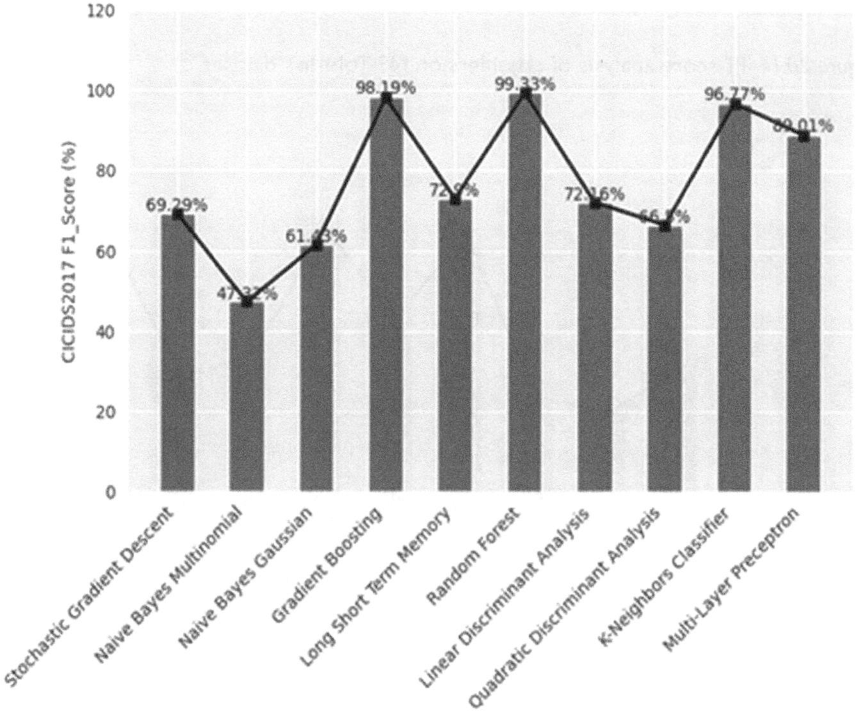

Figure 20.17 F1-Score evaluation of classifiers on CICIDS2017 dataset.

recall, a precision of around 99.47%, and an F1-score of 99.33% are attained by the random forest classifier in simulations.

20.6 CONCLUSION

This chapter presents ideas about software-defined-network-enabled IDSs. SDN is used to control huge amounts of data in a better way. Machine learning and neural-network-based IDS approaches are considered optimal approaches for detecting cyberattacks. During simulation, two advanced datasets CICIDS2017 and NF-ToN-IoT are used. A comprehensive comparative study is performed utilizing around ten classifiers. Random forest outperforms conventional methods significantly in terms of performance metrics such as recall, accuracy, precision, and F1-score. Deep learning systems may soon be replicated to identify cyberattacks. Also, engineers are required to design new advanced datasets.

REFERENCES

1. Kumar, Rajender, Alankrita Aggarwal, Karun Handa, Punit Soni, and Mukesh Kumar. "Software-defined networks and its applications." Software Defined Networks: Architecture and Applications 1, (2022): 63–96.
2. Modieginyane, Kgotlaetsile Mathews, Babedi Betty Letswamotse, Reza Malekian, and Adnan M. Abu-Mahfouz. "Software-defined wireless sensor networks application opportunities for efficient network management: a survey." Computers & Electrical Engineering 66 (2018): 274–287.
3. Sokappadu, Bhargava, Avishek Hardin, Avinash Mungur, and Sheeba Armoogum. "Software-defined networks: issues and challenges." In 2019 Conference on Next Generation Computing Applications (NextComp), pp. 1–5. IEEE, 2019.
4. Wang, An, Zili Zha, Yang Guo, and Songqing Chen. "Software-defined networking enhanced edge computing: a network-centric survey." Proceedings of the IEEE 107, no. 8 (2019): 1500–1519.
5. Usman, Muhammad, Anteneh A. Gebremariam, Usman Raza, and Fabrizio Granelli. "A software-defined device-to-device communication architecture for public safety applications in 5G networks." IEEE Access 3 (2015): 1649–1654.
6. Kurungadan, Basima, and Atef Abdrabou. "Using software-defined networking for data traffic control in smart cities with WiFi coverage." Symmetry 14, no. 10 (2022): 2053.
7. Gao, Shang, Zhe Peng, Bin Xiao, Aiqun Hu, Yubo Song, and Kui Ren. "Detection and mitigation of DoS attacks in software-defined networks." IEEE/ACM Transactions on Networking 28, no. 3 (2020): 1419–1433.
8. Xue, Peilei, and Zhongyuan Jiang. "Enhancing path reliability against Sybil Attack by improved multi-path-trees in SDN." In GLOBECOM 2020–2020 IEEE Global Communications Conference, pp. 1–6. IEEE, 2020.
9. Abdollahi, Asrin, and Mohammad Fathi. "An intrusion detection system on ping of death attacks in IoT networks." Wireless Personal Communications 112 (2020): 2057–2070.
10. Cahyo, Aditya Nur, Anny Kartika Sari, and Mardhani Riasetiawan. "Comparison of hybrid intrusion detection system." In 2020 12th International Conference on Information Technology and Electrical Engineering (ICITEE), pp. 92–97. IEEE, 2020.

11. Khan, Inam Ullah, Asrin Abdollahi, Ryan Alturki, Mohammad Dahman Alshehri, Mohammed Abdulaziz Ikram, Hasan J. Alyamani, and Shahzad Khan. "Intelligent detection system enabled attack probability using Markov chain in aerial networks." Wireless Communications and Mobile Computing 2021 (2021): 1–9.
12. Jmal, Rihab, Walid Ghabri, Ramzi Guesmi, Badr M. Alshammari, Ahmed S. Alshammari, and Haitham Alsaif. "Distributed blockchain-SDN secure IoT system based on ANN to mitigate DDoS attacks." Applied Sciences 13, no. 8 (2023): 4953.
13. Kaur, Sukhveer, Krishan Kumar, and Naveen Aggarwal. "Analysis of DDoS attacks in software defined networking." In 2022 IEEE Delhi Section Conference (DELCON), pp. 1–6. IEEE, 2022.
14. Kareem, Mohammed Ibrahim, and Mahdi Nsaif Jasim. "The current trends of DDoS detection in SDN environment." In 2021 2nd Information Technology to Enhance e-learning and Other Application (IT-ELA), pp. 29–34. IEEE, 2021.
15. Roy, Bipraneel, and Hon Cheung. "A deep learning approach for intrusion detection in internet of things using bi-directional long short-term memory recurrent neural network." In 2018 28th International Telecommunication Networks and Applications Conference (ITNAC), pp. 1–6. IEEE, 2018.
16. Le, Hai-Viet, Quoc-Dung Ngo, and Van-Hoang Le. "IoT Botnet detection using system call graphs and one-class CNN classification." International Journal of Innovative Technology and Exploring Engineering 8, no. 10 (2019): 937–942.
17. Miao, Qiguang, Jiachen Liu, Ying Cao, and Jianfeng Song. "Malware detection using bilayer behavior abstraction and improved one-class support vector machines." International Journal of Information Security 15 (2016): 361–379.
18. Burnaev, Evgeny, and Dmitry Smolyakov. "One-class SVM with privileged information and its application to malware detection." In 2016 IEEE 16th International Conference on Data Mining Workshops (ICDMW), pp. 273–280. IEEE, 2016.
19. Almiani, Muder, Alia AbuGhazleh, Amer Al-Rahayfeh, Saleh Atiewi, and Abdul Razaque. "Deep recurrent neural network for IoT intrusion detection system." Simulation Modelling Practice and Theory 101 (2020): 102031.
20. Pajouh, Hamed Haddad, Reza Javidan, Raouf Khayami, Ali Dehghantanha, and Kim-Kwang Raymond Choo. "A two-layer dimension reduction and two-tier classification model for anomaly-based intrusion detection in IoT backbone networks." IEEE Transactions on Emerging Topics in Computing 7, no. 2 (2016): 314–323.
21. Xu, Congyuan, Jizhong Shen, Xin Du, and Fan Zhang. "An intrusion detection system using a deep neural network with gated recurrent units." IEEE Access 6 (2018): 48697–48707.
22. Li, Zhida, Prerna Batta, and Ljiljana Trajkovic. "Comparison of machine learning algorithms for detection of network intrusions." In 2018 IEEE International Conference on Systems, Man, and Cybernetics (SMC), pp. 4248–4253. IEEE, 2018.
23. Elmasry, Wisam, Akhan Akbulut, and Abdul Halim Zaim. "Empirical study on multiclass classification-based network intrusion detection." Computational Intelligence 35, no. 4 (2019): 919–954.
24. Qureshi, Ayyaz-ul-Haq, Hadi Larijani, Jawad Ahmad, and Nhamoinesu Mtetwa. "A heuristic intrusion detection system for Internet-of-Things (IoT)." In Intelligent Computing: Proceedings of the 2019 Computing Conference, volume 1, 2019, pp. 86–98. Springer International Publishing.
25. Huh, G. "Enhanced stochastic gradient descent with backward queried data for online learning." In 2020 IEEE International Conference on Machine Learning and Applied Network Technologies (ICMLANT), Hyderabad, India, pp. 1–5, 2020, doi: 10.1109/ICMLANT50963.2020.9355978.

26. Yazan, Ersan, and M. Fatih Talu. "Comparison of the stochastic gradient descent based optimization techniques." In 2017 International Artificial Intelligence and Data Processing Symposium (IDAP), Malatya, Turkey, 2017, pp. 1–5, doi: 10.1109/IDAP.2017.8090299.
27. Akhter, Arnisha, UZZAL, K. Acharjee, et POLASH, Md Masbaul A. Cyber bullying detection and classification using multinomial Naïve Bayes and fuzzy logic. Int. J. Math. Sci. Comput, 2019, vol. 5, no 4, p. 1–12.
28. Mumtaz, Gohar, Sheeraz Akram, Waseem Iqbal, M. Usman Ashraf, Khalid Ali Almarhabi, Ahmed Mohammed Alghamdi, and Adel A. Bahaddad. "Classification and prediction of significant cyber incidents (SCI) using data mining and machine learning (DM-ML)." IEEE Access, (2023), vol. 11, p. 94486–94496.
29. Upadhyay, Darshana, Jaume Manero, Marzia Zaman, and Srinivas Sampalli. "Gradient boosting feature selection with machine learning classifiers for intrusion detection on power grids." IEEE Transactions on Network and Service Management 18, no. 1 (2020): 1104–1116.
30. Liang, Xiaoyu et ZNATI, Taieb. A long short-term memory enabled framework for DDoS detection. In : 2019 IEEE global communications conference (GLOBECOM). IEEE, 2019. p. 1–6. doi: 10.1109/GLOBECOM38437.2019.9013450.
31. Faruq, Zaidan Fadhlurohman, MANTORO, Teddy, BHAKTI, Muhammad Agni Catur, et al. Random forest classifier evaluation in ddos detection system for cyber defence preparation. In : 2022 IEEE 8th International Conference on Computing, Engineering and Design (ICCED). IEEE, 2022. p. 1–5. 2022, doi: 10.1109/ICCED56140.2022.10010341.
32. TAN, Zhiyuan, JAMDAGNI, Aruna, HE, Xiangjian, et al. Network intrusion detection based on LDA for payload feature selection. In : 2010 IEEE Globecom Workshops. IEEE, 2010. p. 1545-1549. doi: 10.1109/GLOCOMW.2010.5700198.
33. Sangodoyin, Abimbola O., AKINSOLU, Mobayode O., PILLAI, Prashant, et al. Detection and classification of DDoS flooding attacks on software-defined networks: A case study for the application of machine learning. IEEE Access, 2021, vol. 9, p. 122495–122508. doi: 10.1109/ACCESS.2021.3109490.
34. Luqyana, Wanda Athira, AHMADIE, Beryl Labique, et SUPIANTO, Ahmad Afif. K-nearest neighbors undersampling as balancing data for cyber troll detection. In: 2019 International Conference on Sustainable Information Engineering and Technology (SIET). IEEE, 2019. p. 322–325. doi: 10.1109/SIET48054.2019.8986079.
35. PAWAR, Yashaswini S., HONNAVALLI, Prasad, et ESWARAN, Sivaraman. Cyber Attack Detection On Self-Driving Cars Using Machine Learning Techniques. In: 2022 IEEE 3rd Global Conference for Advancement in Technology (GCAT). IEEE, 2022. p. 1–5.
36. Palenzuela, F. et al. "Multilayer perceptron algorithms for cyberattack detection." In 2016 IEEE National Aerospace and Electronics Conference (NAECON) and Ohio Innovation Summit (OIS), Dayton, OH, USA, pp. 248–252, 2016, doi: 10.1109/NAECON.2016.7856806.
37. Sharafaldin, Iman, Arash Habibi Lashkari, and Ali A. Ghorbani. "Toward generating a new intrusion detection dataset and intrusion traffic characterization." In 4th International Conference on Information Systems Security and Privacy (ICISSP), Portugal, January 2018.
38. Sarhan, Mohanad, Siamak Layeghy, Nour Moustafa, and Marius Portmann. "Netflow datasets for machine learning-based network intrusion detection systems." In Big Data Technologies and Applications: 10th EAI International Conference, BDTA 2020, and 13th EAI International Conference on Wireless Internet, WiCON 2020, Virtual Event, December 11, 2020, Proceedings 10, pp. 117–135. Springer International Publishing, 2021.

Chapter 21

Information security awareness in higher education

The need for a tailor-made suit

Reismary Armas and Hamed Taherdoost

21.1 INTRODUCTION

Today's speed at which technology develops across industries is extraordinarily significant, and although each industry develops at its own pace, these changes are still forceful and fast. The race to cover emerging needs for digitization, innovation, and process automation also opens the doors for more crimes related to Information Security to occur worldwide every day. A significant proof of registered losses due to cybercrime is the one reported by IC3 and FBI (2023), which indicates that only in the United States, in 2022, did the registered losses represent 10,300 million US dollars. Additionally, they indicate that cyberattacks have increased drastically in recent years.

Additionally, the annual report of IBM Security (2023) indicates that the main attack vector is related to phishing and stolen or compromised credentials and where the main reason for the attack is to be able to steal personal information, personal identification information of clients, and that of the employees.

In the case of the higher education industry, although it is not the industry at the top of registered attacks, this sector finds itself at the crossroads of innovation and vulnerability, becoming an interesting new target for cyber attackers, as reported by IBM Security (2023). The convergence of education and technology has guided in a new era of opportunities and challenges, currently allowing this sector to carry out online learning, remote work, and interconnect systems to make this possible while simultaneously challenging its operation due to the new landscape of cybersecurity threats. This paradigm shift in the higher education sector's digital scenery necessitates a comprehensive understanding of the cybersecurity threat landscape and the urgent need for tailored awareness and training programs.

By the need described, this chapter analyzes global trends, reports, and studies, aiming to shed light on the vulnerabilities faced by higher educational institutions, the nature of the attacks they encounter, and the implications of these threats for students, faculty, administrative staff, and the institutions. Additionally, it will explore the current state of cybersecurity awareness and training within higher education, examining the existing gaps and the potential

DOI: 10.1201/9781032664293-24

consequences of inadequate preparation. They are followed by an explanation of the need for this industry to tailor their cybersecurity awareness and training efforts, demonstrating the need for a framework that focuses on building user awareness and training programs designed specifically for the HE industry.

21.2 CYBERSECURITY THREAT LANDSCAPE AND THE POSITION OF HIGHER EDUCATION

As has been happening in different industries, the education area has undergone an important revolution in terms of digital transformation. Although before the 2020 pandemic there was already a trend toward online classes in certain programs, most administrative staff continued to work face-to-face in offices. However, the latest progress has been driven by the Covid-19 pandemic, where most universities and colleges, among other educational institutions, found themselves in need of not only adapting to changes to survive during this period but also adapting to the new needs of the market, flexibility, and mobility of students, professors, and staff.

The need to maintain the operation of organizations and rapidly adapt to the new environment caused these to become more exposed, adding new security breaches and threats to existing ones worldwide. As reported by Apricorn (2022) and Taherdoost (2022), IT teams had to provide, in minimal time, the tools and adapt mechanisms, processes of access to information, and connection with the business to enable staff, professors, and students to continue their activities remotely, a condition which is an established practice today in many industries or the case of education in many programs. The Capricorn report also shows that one in four employees acknowledges that there is (Baker Hostetler, 2022) an information security policy but needs to adhere to it, fully complying with the best practices for information security, privacy, and confidentiality. Also, 72% of employees do not consider themselves the target of an attacker (Apricorn, 2022).

Moreover, Baker Hostetler (2022) published the top five most common attacks registered in the United States, with network intrusion being the first one with 56% followed by phishing attacks with 24% of the share, in addition, where 37% of these attacks are the first step to implant ransomware, 27% to the theft of data, 21 to get unauthorized access to accounts, and 7% to install malware. All these actions are directly related to the lack of training and user awareness, which demonstrates a need to reinforce knowledge on information security and cybersecurity issues, where users are the focal point, specifically adapted to the specific category and business.

Although the education industry is not the most attacked, it ranks sixth among the most, with a share of 7.3% worldwide from 2018 to 2022 (IBM Security, 2023).

Another relevant insight reveals a trend in dramatically increasing attacks on the education industry. As mentioned above, most cyberattacks do not occur in the education industry. However, even so, this industry has increased its malware attacks by 175% year-over-year in the world and occupied the top with the higher percentage of customers targeted by malware in 2021 (SonicWall, 2023), having an average of 2,314 malware attacks weekly worldwide in 2022 (Check Point Software Technologies, 2023). It is the most affected area by business email compromise cyberattack attempts worldwide registered in the same period (SonicWall, 2023). Also, Zandt (2021) argues that education was the second industry most affected by ransomware in the United States.

On the other hand, IBM Security (2023) reports that the cost of data breaches affecting public and private colleges and universities was $3.86 million in 2022 and $3.65 million so far in 2023. Moreover, only in the United States, education ranked fifth in the quantity of data violations from 2020 to 2022 (Petrosyan, 2023; Taherdoost, 2022).

These data and many others published periodically by organizations dedicated to tracking and monitoring information security and the cybersecurity market reflect that the education industry is increasingly desirable as the object of security attacks. In contrast, although all industries indicate that one of the biggest concerns is related to the awareness and training of people on these issues (Statista, 2023), the fact is that the users themselves are not being adequately trained in compliance with the security policies of the organizations, and therefore, by not feeling like a target of a cyberattack or not knowing in depth the consequences of a cyberattack, they do not adhere to the organization policies and good practices accordingly.

21.3 PERCEPTION OF CYBERSECURITY POLICIES AND AWARENESS PROGRAMS IN HIGHER EDUCATION

A review of previous studies in different countries was conducted to analyze the perception of students and staff in higher education. The first one is in the United States, where Amorosa and Yankson (2023) recently investigated human errors related to the increase in data breaches in higher education, this study highlights the need for investment in the creation and adjustment of security and privacy policies, awareness and training of users across the organization, and most of the concretized attacks start with some factor of lack of knowledge and bad practices of the end users.

In addition, this study carried out its research over three institutions as a source of analysis, which were called A, B, and C, where individuals in institution A were surveyed; this sample involved both students and faculties from one of the departments of the institution, of which the sample size was the 10% of the individuals of which 50% completed the surveys, having a total of 25 valid interviewees. From this sample, it was possible to obtain that by asking the individuals if they had received any training related to cybersecurity during their onboarding or classes at the university, which resulted in only 8% of the students answering yes, 56% answering no, and 16% did not answer.

In the case of the employees, 44% claim to have received some induction in information security from the university, while 36% answered no, and 20% did not answer. In general, the total participant results are divided into 8% of participants indicated they had received induction, 76% said they had not, and 16% did not answer. As numbers reveal, in this case, many students need to be trained in these matters, while some faculties receive information or training in information security university policies (Amorosa & Yankson, 2023).

Besides Amorosa and Yankson (2023) include a research review of public policies and training processes in the three studies institutions, finding that they have structured cybersecurity policies based on NIST's (National Institute of Standards and Technology's) Cybersecurity Framework V1. However, these policies are not linked to a training program and continuous awareness of the individuals who live in the study centers; this inference is related to the survey results of institution A.

In another case study in Poland surveying 119 students (almost 70% of them are students at the University of Zielona Góra), the researchers found that 25.2% of the students surveyed have been victims of fraud, and 76.5% have known of someone who has been the victim of cybercriminals. However, in general, they feel confident with their devices and freely expose themselves to risks associated with cybersecurity without relevant concern about it. This indicates the perception that the individuals surveyed in this study are not aware of information security issues and the consequences that this can bring for them as a person and for the institution (Zarębska et al., n.d.).

On the same train of ideas, Xiaoyu et al. (2023) examined the knowledge, attitudes, and behavior of undergraduates in China regarding the disclosure of personal data online. The study analyzed five universities, where they got 156 valid responses to questionnaires. The research gathered 110 individuals in liberal arts, 10 in engineering, and 36 in natural science. The authors found that although there is some awareness in the institutions about security issues, more is needed to protect the critical data of the students. Furthermore, students need to improve in understanding and self-concern regarding personal data privacy; this effectively demonstrates the need to reinforce security policies and establish a training process for students and staff.

In addition, Matyokurehwa et al. (2023) found a direct relationship between social engineering, malware, the Internet of Things (IoT), and cybersecurity awareness through an analysis conducted in three universities in Zimbabwe (A, B, and C), where the sample constituted 322 valid questionnaires. Indicating that there is a need for a process of awareness and training in university students on topics of cyberattacks related to these three edges (social engineering, malware attacks, and IoT) and additionally that the perspective that students have on Information Security can be valuable inputs for the formulation of information security policies adapted to the specific industry.

Adding to what was previously analyzed, López et al. (2023) focus on the analysis of cybersecurity perception among Generation Z in university students by researching two universities in Mexico on a sample of 563 individuals to whom questions were applied to identify their self-perception in terms of their levels of knowledge about computer security, computer virus, and how likely it is that their computers become infected with some malware. Individuals could score their level on a scale between 0 and 10 according to their feelings about their awareness of the subject.

In this case, the average rating regarding the level of knowledge about information security and computer viruses did not reach a 5/10 score. Additionally, individuals rated a value greater than 5/10 as an average probability of being a victim of malware. On the other hand, students showed concern about exposition to vulnerabilities and information stolen since the average rating for this section was 6.7/10. Moreover, it was found that there is awareness regarding how long a password should be since, on average, respondents choose passwords that are approximately 10 characters long (López et al., 2023), and according to (Kaspersky, 2019), a strong password should contain at least eight characters. However, the study does not analyze this issue deeply, where other factors such as complexity, frequency of password changes, and non-use of personal information are also relevant when taking care of the information and identity behind a password. Another relevant finding is that most students perceive themselves as respectful of information security regulations by complying with the security policies of said universities rating this category mostly a 7/10 (López et al., 2023).

In general, analyzing all the information collected in the different studies done in various parts of the world, it agrees that the community behind higher education institutions is aware, and these institutions have information security policies. However, even feeling that they comply with the requirements established in said policies, there is a concern about the possibility of being victims of cyber attackers and a lack of training and leveling knowledge on these issues.

On the other hand, it is also evident that the institutions comply with documentation and security policies, some of which are aligned with world-recognized standards such as NIST. However, these need to be specifically adapted to the field, their needs, and how the information assets of these institutions are managed. They are not linked to a continuous training process, which is updated at least periodically according to market trends. However, they are being forced organically according to the circumstances generated in a corrective and non-proactive way.

21.4 WHY DOES THE HIGHER EDUCATION INDUSTRY NEED A TAILOR-MADE SUIT FOR INFORMATION SECURITY AWARENESS AND TRAINING PROCESSES?

As demonstrated in the first section of this chapter, the higher education industry has become a target for cybercrime, registering a greater number of attacks year after year. This is because this area handles a large number of valuable information assets, with live data that changes or increases at least every six months, storing useful information that attackers see as a source of resources to carry out fraud (Alexei, 2021), most of them related to financial purposes, data recorded by (Verizon, 2022), indicating that 70% of all attacks are for said reasons.

Secondly, there is notable dissatisfaction regarding the effectiveness of security policies applied in institutions, lack of training, and constant updating that strengthens the user layer with training to be truly aware of the risks associated with information security and it's time to harden your skills to recognize threats.

Cross-higher education industry knowledge, which adapts to the type of user who lives in higher education institutions (students, professors, business, and administrative), would allow the language of information security to be adapted to the target audience, contributing a more user-friendly way and would provide a tool to train cognitive skills that enable and continuously enhance correct threat detection (Ben-Asher & Gonzalez, 2015).

In addition, as (Nipon, 2019) argues, the user training process should not be based solely on theoretical knowledge but also on practical experiences that truly make it possible to identify the level of maturity of the individuals in the subject matter and, from there, draw up a continuous improvement roadmap that allows the processes, policies, and awareness and training in information security remain current, and this is achieved with the customization or adaptation of the current frameworks to the reality, scope, and limitations of the industry.

Finally, the preceding leads to what many authors and statistics explain, technologies, trends, threats, and needs change constantly and at a faster pace as time goes by; therefore, it is necessary to build an awareness and training process that, in turn, allows be updated and adapted to the environment in which it lives.

21.5 CONCLUSION

As this chapter demonstrates, the evolving cybersecurity threat landscape has deeply impacted the higher education sector due to the shift toward online learning, remote work, and increased connectivity that has exposed educational institutions to increased risk of cyberattacks and security breaches; despite being a less desirable sector than others, the industry has recently become one of the main targets of cybercriminals due to the large amount of valuable information that flows through it.

Although some of the institutions have implemented security policies, aligned with world-renowned security frameworks, more is needed to control or track the dynamism of threats. Furthermore, many students and teachers still need to receive adequate training since they are unaware of the risks to which they are exposed. To counter these challenges, the higher education sector must develop tailored cybersecurity training and awareness programs capable of adapting to changes in the market, resources, and industry-specific evolution.

Awareness and training programs go beyond disseminating theoretical knowledge; it is necessary to incorporate practical experiences that help people identify vulnerabilities and threats in real-world scenarios.

REFERENCES

Alexei, A. (2021). Cyber Security Strategies for Higher Education Institutions. Journal of Engineering Science, XXVIII, 74–92. https://doi.org/10.52326/jes.utm.2021.28(4).07

Amorosa, K., & Yankson, B. (2023). Human Error – A Critical Contributing Factor to the Rise in Data Breaches: A Case Study of Higher Education. 14(1). https://doi.org/10.2478/hjbpa-2023-0007

Apricorn. (2022). 2022-IT-security-survey-whitepaper. Apricorn. https://apricorn.com/content/infographic/2022-IT-security-survey-whitepaper.pdf

Baker Hostetler. (2022). 2022 data security incident response report. Digital Assets and Data Management. https://f.datasrvr.com/fr1/722/85356/2022_DSIR_Report.pdf

Ben-Asher, N., & Gonzalez, C. (2015). Effects of Cyber Security Knowledge on Attack Detection, 51–61. https://doi.org//10.1016/j.chb.2015.01.039

Check Point Software Technologies. (2023). Average weekly number of malware attacks in organizations worldwide in 2022, by industry. In. Statista: Statista.

IBM Security. (2023). X-Force threat intelligence index 2023. I. Security. https://www.ibm.com/downloads/cas/DB4GL8YM

IC3 & FBI. (2023). Annual amount of monetary damage caused by reported cybercrime in the United States from 2001 to 2022. In. Statista.

Kaspersky. (2019). How to create a strong password. https://support.kaspersky.com/common/windows/3730

López, A., Roque, R., Prieto, M., & Salazar, R. (2023). Cybersecurity among University Students from Generation Z: A Comparative Study of the Undergraduate Programs in Administration and Public Accounting in two Mexican Universities. 12(1), 503–511. https://doi.org/10.18421/TEM121-60

Matyokurehwa, K., Rudhumbu, N., Gombiro, C., & Mlambo, C. (2023). Cybersecurity Awareness in Zimbabwean Universities: Perspectives from the Students. 4(2), 1–12. https://doi.org/10.1002/spy2.141

Nipon, N. (2019). How to Increase Cybersecurity Awareness. ISACA. https://www.isaca.org/resources/isaca-journal/issues/2019/volume-2/how-to-increase-cybersecurity-awareness

Petrosyan, A. (2023). Number of cases of data violation due to cyber-attacks in the United States from 2020 to 2022, by industry. In Statista.

SonicWall. (2023). 2023 SonicWall cyber threat report. by área. https://www.sonicwall.com/medialibrary/en/white-paper/2023-cyber-threat-report.pdf

Statista. (2022). Cybersecurity investments planned by companies in the United States in 2022, by area. https://www.statista.com/statistics/1338388/planned-cybersecurity-investments-us-tech-companies-by-area/

Taherdoost, H. (2022). Understanding Cybersecurity Frameworks and Information Security Standards – A Review and Comprehensive Overview. Electronics, 11(14), 2181.

Verizon. (2022). DBIR report 2022 – Results and analysis – Basic web application attacks. https://www.verizon.com/business/en-nl/resources/reports/dbir/2022/results-and-analysis-basic-web-application-attacks/

Xiaoyu, L., Qin, A., Wilson, H., Yunfeng, Z., Kolletar, Z., & Xiaoshu, X. (2023). Undergraduates Knowledge Attitude and Behavior (KAB) towards the Disclosure of Personal Data Online in China. 370, 46–63. https://doi.org/10.3233/FAIA230168

Zandt, F. (2021). The industries most affected by Ransomware. Retrieved from Statista. In. Statista.

Zarębska, J., Howis, N., & Barska, M. (n.d.). Research on students perception of information technology security – A new era of threats. Scientific papers of Silesian University of Technology, 170, Organization & Management/Zeszyty Naukowe Politechniki Slaskiej.

For Product Safety Concerns and Information please contact our
EU representative GPSR@taylorandfrancis.com Taylor & Francis
Verlag GmbH, Kaufingerstraße 24, 80331 München, Germany